USMG   NATO ASI

LES HOUCHES

Session XLV

1985

TRAITEMENT DU SIGNAL

SIGNAL PROCESSING

VOLUME II

CONFÉRENCIERS

Johann Friedrich BÖHME
Vito CAPPELLINI
G. Clifford CARTER
Tariq Salim DURRANI
Yves GRENIER
Claude GUEGUEN
Mos KAVEH
Steven M. KAY
Gary E. KOPEC
Sun-Yuan KUNG
Wolfgang MECKLENBRÄUKER
José M.F. MOURA
Bernard PICINBONO
Stewart F. REDDAWAY
Louis L. SCHARF
Anastasios N. VENETSANOPOULOS

USMG NATO ASI

LES HOUCHES

SESSION XLV

12 Août–6 Septembre 1985

TRAITEMENT DU SIGNAL

SIGNAL PROCESSING

édité par

JEAN-LOUIS LACOUME, TARIQ SALIM DURRANI
et RAYMOND STORA

VOLUME II

1987

NORTH-HOLLAND
AMSTERDAM · OXFORD · NEW YORK · TOKYO

ISBN 0 444 87027 X (Set)
  0 444 87058 X (Volume I)
  0 444 87059 8 (Volume II)

Published by:
North-Holland Physics Publishing
a division of
Elsevier Science Publishers B.V.
P.O. Box 103, 1000 AC Amsterdam
The Netherlands

Sole distributors for the USA and Canada:
Elsevier Science Publishing Company, Inc.
52 Vanderbilt Avenue
New York, NY 10017
USA

Library of Congress Cataloging in Publication Data
Ecole d'été de physique théorique (Les Houches, Haute-
  Savoie, France) (45th : 1985)
  Traitement du signal – Signal processing.

  English and French.
  At head of title: USMG, NATO ASI. Les Houches,
session XLV, 12 août-6 septembre 1985.
  Lectures presented at the Ecole d'été de physique
théorique.
  Includes bibliographies.
  1. Signal processing. I. Lacoume, Jean-Louis.
II. Durrani, Tariq S. III. Stora, Raymond, 1930–
IV. NATO Advanced Study Institute. V. Title. VI. Title:
Signal processing.
TK5102.5.E26 1985   621.38′043   87-13993

ISBN 0 444 87027 X (Set)
ISBN 0 444 87058 X (Volume I)
ISBN 0 444 87059 8 (Volume II)

Printed in The Netherlands

LES HOUCHES
ÉCOLE D'ÉTÉ DE PHYSIQUE THÉORIQUE

ORGANISME D'INTÉRÊT COMMUN DE L'UNIVERSITÉ
SCIENTIFIQUE ET MÉDICALE DE GRENOBLE ET DE
L'INSTITUT NATIONAL POLYTECHNIQUE DE
GRENOBLE
AIDÉ PAR LE COMMISSARIAT À L'ÉNERGIE
ATOMIQUE

SESSION XLV
INSTITUT D'ÉTUDES AVANCÉES DE L'OTAN
NATO ADVANCED STUDY INSTITUTE
12 Août–6 Septembre 1985

Directeurs scientifiques de la session: Jean-Louis Lacoume, CEPHAG, BP
46, F38402 Saint Martin d'Hères Cedex, France; Tariq Salim Durrani,
Department of Electronic and Electrical Engineering, University of
Strathclyde, 204 George Street, Glasgow G1 1XW, UK

SESSIONS PRÉCÉDENTES

---

*Sessions ayant reçu l'appui du Comité Scientifique de l'OTAN.

LECTURERS

BÖHME, Johann Friedrich, Ruhr Universität Bochum, Lehrstuhl für Signaltheorie, Postfach 102148, D-4630 Bochum 1, Fed. Rep. Germany.

CAPPELLINI, Vito, Istituto di Ricerca, Sulle Onde Elettromagnetiche, Via Panciatichi 64, 50127 Florence, Italy.

CARTER, G. Clifford, Code 331, Naval Underwater Systems Center, New London, CT 06320, USA.

DURRANI, Tariq Salim, University of Strathclyde, Department of Electronic and Electrical Engineering, 204 George Street, Glasgow G1 1XW, UK.

GRENIER, Yves, Ecole Nationale Supérieure des Télécommunications, 46 Rue Barrault, 75634 Paris Cedex 13, France.

GUEGUEN, Claude, Ecole National Supérieure des Télécommunications, 46 Rue Barrault, 75634 Paris Cedex 13, France.

KAVEH, Mos, University of Minnesota, Department of Electrical Engineering, 123 Church Street S.E., Minneapolis, MN 55455, USA.

KAY, Steven M. Rhode Island University, Department of Electrical Engineering, Kelley Hall, Kingstone, Rhode Island 02881, USA.

KOPEC, Gary E., Schlumberger Palo Alto Research, 3340 Hillview Avenue, Palo Alto, CA 94304, USA.

KUNG, Sun Yuan, University of Southern California, Department of Electrical Engineering Systems, University Park, Los Angeles, CA 90089-0272, USA.

MECKLENBRÄUKER, Wolfgang, Technical University of Vienna, Gusshausstrasse 25/389, A-1040 Vienna, Austria.

MOURA, José M.F., Laboratory for Information and Decision Systems, MIT, Room 35-213, 77 Massachusetts Avenue, Cambridge, MA 02139, USA.

PICINBONO, Bernard, Laboratoire des Signaux et Systèmes, Université de Paris-Sud, E.S.E., Plateau du Moulon, 91190 Gif-sur-Yvette, France.

REDDAWAY, *Stewart*, Defence Technology Centre, ICL Limited, Eksdale Road, Wokingham, Berkshire RG11 5TT, UK.

SCHARF, *Louis L.*, University of Colorado, Department of Electrical and Computer Engineering, Campus Box 425, Boulder, CO 80309, USA.

VENETSANOPOULOS, *Anastasios*, Department of Electrical Engineering, University of Toronto, 35 St. George Street, Toronto, Ontario, Canada, M5S 1A4.

PARTICIPANTS

Amengual Rigo, *Mateo*, E.T.S. Ingen. Telec., c/o Jorge Girona Salgado s/n, Barcelona, Spain.

Anarin, *Emin*, Hava Harp Okulu (Air Force Academy), Yesilyurt, Istanbul, Turkey.

Barral, *Henri*, Ecole Nationale Supérieure des Telécommunications, Dept. SYC, 46 Rue Barrault, 75634 Paris Cedex 13, France.

Baronti, *Stefano*, Nat. Res. Council, Istituto Ricerca, Sulle Onde Elettromagnetiche, Via Panciatichi 64, Florence 50127, Italy.

Benidir, *Messaoud*, Laboratorie des Signaux et Systèmes, Université de Paris-Sud, E.S.E., Plateau du Moulon, 91190 Gif-sur-Yvette, France.

Bergmans, *Jan*, Philips Research Laboratories PO Box 80 000, 5600 JA Eindhoven, The Netherlands.

Certenais, *Joël*, Thomson Sintra ASM, Route du Conquêt, 29000 Brest, France.

Chollet, *Gérard*, Ecole Nationale Supérieure des Télecommunications, Dept. SYC, 46 Rue Barrault, 75634 Paris Cedex 13, France.

Clarke, *Ira*, Royal Signals and Radar Establishment, St. Andrews Road, Great Malvern, Worcestershire WR14 3PS, UK.

Dechambre, *Monique*, CRPE, 4 Av. de Neptune, 94107 Saint-Maur Cedex, France.

De Villiers, *Geoffrey*, Royal Signals and Radar Establishment, St. Andrews Road, Great Malvern, Worcestershire WR14 3PS, UK.

Dewey, *Richard*, Philips Research Laboratoires, Cross Oak Lane, Redhill, UK.

Djuric, *Petar*, Bors Kidric Inst., Dept. 270, PO Box 522, 11001 Belgrad, Yugoslavia.

Doroslavacki, *Milos*, Boris Kidric Inst., Dept. 270, PO Box 522, 11001 Belgrad, Yugoslavia.

Evangelista, *Gianpaolo*, Dipartimento di Fisica, Università di Napoli, Mostra d'Oltremare, pad. 20, 80125 Naples, Italy.

Feig, *Ephraim*, IBM, PO Box 218, Yorktown Heights, NY 10598, USA.

Flandrin, *Patrick*, Laboratoire Traitement du Signal, ICPI, 25 Rue du Plat, 69288 Lyon Cedex 02, France.

Gimenez, Gérard, Laboratoire Traitement du Signal et Ultrasons, INSA, Bâtiment 502, 69621 Villeurbanne Cedex, France.

Gingras, Denis, Lehrstuhl für Signaltheorie, Elektrotechnik, Ruhr Universität, D-4630 Bochum 1, Fed. Rep. Germany.

Glangeaud, François, CEPHAG, BP 46, 38402 Saint-Martin-d'Hères Cedex, France.

Haerle, Norbert, Lehrstuhl für Signaltheorie, Ruhr Universität, Norbert Härle, Postfach 102148, D-4630 Bochum 1, Fed. Rep. Germany.

Joudon, Alain, CEN SACLAY, SEIPE/Dph.Pe, Bâtiment 141, 91191 Gif-sur-Yvette Cedex, France.

Keroub, Israel Herbert, Haifa Radio Observatory, POB 911, Haifa, Israel.

Klippenberg, Nils, Norwegian Defense Research Establishment, P.O. Box 25, 2007 Kjeller, Norway.

Kummert, Anton, Lehrstuhl für Nachrichtentechnik, Postfach 102148, D-4630 Bochum 1, Fed. Rep. Germany.

Lefaudeux, François, GERDSM, Le Brusc, 83140 Six Fours Les Plages, France.

Lerch, Reneé, Fraunhofer Institute for Microelectronic Circuits and Systems, c/o B.J. Hosticka, Bismarck Str. 69, D-4100 Duisburg, 1, Fed. Rep. Germany.

Li, Xing, Dept. Génie Informatique, U.T.C., 60206 Compiègne Cedex, France.

Lourtie, Isabel, CAPS, Complexo I, Inst. Sup. Técnico, Av. Rovisco Pais 1, P-1096 Lisboa Codex, Portugal.

Ludeman, Lonnie, Electrical and Computer Engineering Department, New Mexico State University, Las Cruces, NM 88003, USA.

Marcos, Sylvie, Laboratoire des Signaux et Systèmes, Université de Paris-Sud, E.S.E., Plateau du Moulon, 91190 Gif-sur-Yvette, France.

Mather, John, Royal Signals and Radar Establishment, St. Andrews Road, Great Malvern, Worcestershire WR14 3PS, UK.

Morin, Anne, IRISA, Université de Rennes I, 35042 Rennes Cedex, France.

Munoz Gracia, M. Pilar, Univ. Politec. Catal., Facultat d'Informatica, Jordi Girona Salgado no. 31, 08034 Barcelona, Spain.

Pallas, Marie-Agnès, CEPHAG, BP 46, 38402 Saint-Martin-d'Hères Cedex, France.

Prieels, René, Université Catholique de Louvain, 2 Chemin du Cyclotron, FYNU, B-1348 Louvain-la-Neuve, Belgium.

Ribeiro, Maria Isabel, CAPS, Complexo I, Inst. Sup. Técnico, Av. Rovisco Pais 1, P-1096 Lisboa Codex, Portugal.

Roberto, Vito, Dipartimento di Fisica Teorica, Strada Costiera 11, 34014 Trieste, Italy.

Sacco, Vincenzo Maria, IROE-CNR, Via Panciatichi 64, Florence, Italy.

Sankur, Bülent, Bogaziçi Univ., Dept. Elec. Eng., Bebek, PK 2, Istanbul, Turkey.

Siohan, Pierre, CCETT, 2 R. du Clos Courtel, BP 59, 35510 Cesson-Sevigne, France.

Sommen, Piet, Philips Research Laboratories, PO Box 80 000, 5600 JA Eindhoven, The Netherlands.

Stewart, Kenneth Anderson, Signal Processing Group, Royal College, Strathclyde University, 208 George Street, Glasgow, Scotland.

Thiel, Jacques, GREMI, Université d'Orléans, UER SFA, R. de Chartres, 45046 Orleans Cedex, France.

Toppano, Elio, Istituto di Fisica, Univ. Degli Studi di Udine, Via Larga 36, 33100 Udine, Italy.

Vallet, Robert, Ecole Nationale Supérieure des Télécommunications, 46 rue Barrault, 75634 Paris Cedex 13, France.

Vazquez Grau, Gregori, ETS de Ingeniera de Telecomunicacion, Dto Tratamiento y Transmission de la Informacion, Barcelona 08071, Spain.

Ward, Jeremy, Royal Signals and Radar Establishment, St. Andrews Road, Great Malvern, Worcestershire WR14 3PS, UK.

Yaminysharif, Mohammad, Digital Signal Processing Group, Dept. Electronic and Electrical Engineering, Strathclyde University, Glasgow, Scotland.

PRÉFACE

L'Ecole d'Eté des Houches, organisme d'intérêt commun de l'Institut National Polytechnique de Grenoble et de l'Université Scientifique et Médicale de Grenoble, organise chaque année, deux sessions d'une durée de quatre à huit semaines, qui font le point sur un secteur de la Physique. L'enseignement permet à de jeunes chercheurs et à des non spécialistes d'accéder au niveau de la recherche la plus vivante et la longueur des sessions facilite un travail en profondeur.

Au cours des années 80, le Conseil d'Administration de l'Ecole d'Eté a souhaité élargir le domaine d'activité de l'Ecole en s'ouvrant sur des domaines de recherche relevant des Sciences Physiques de l'Ingénieur. La première session illustrant cette orientation a pris pour thème le Traitement du Signal et son organisation a été confiée à T.S. Durrani (Université de Strathclyde) et J.L. Lacoume (Institut National Polytechnique de Grenoble). Cette session a été animée par 16 professeurs issus d'universités ou de centres de recherche industriels des Etats-Unis, de France, d'Angleterre, de République Fédérale d'Allemagne, d'Italie, d'Autriche et du Portugal. Les 50 auditeurs de l'Ecole étaient des jeunes chercheurs, universitaires ou industriels, ou des utilisateurs du traitement du signal. Ils provenaient essentiellement des pays d'Europe Occidentale avec, pour cette première session, une faible participation américaine.

L'enseignement de l'Ecole a été organisé en cours et groupes de travail. On peut classer les cours en:
– fondements du traitement du signal,
– principaux thèmes du traitement du signal,
– techniques du traitement du signal.
Les cours fondamentaux ont porté sur:
– la théorie du signal,
– la détection et l'estimation,
– le filtrage non linéaire,
– la représentation temps–fréquence.
Dans son cours sur les signaux déterministes et aléatoires, B. Picinbono a tout d'abord défini, d'une maînière très synthétique et unificatrice, les

notions de base: classification des signaux, stationnarité, modélisation AR/ARMA, grandeurs spectrales. Il a ensuite présenté les idées fondamentales de l'estimation linéaire en moyenne quadratique dans une vision nouvelle du problème du filtrage linéaire statistique permettant d'englober dans une même approche, tous les problèmes classiques (Wiener causal et non causal) et en donnant une élégante généralisation, basée sur l'idée de contrainte linéaire.

La théorie de la détection et de l'estimation est apparue dans le cours de L.L. Scharf profonde puissante et vivante. Dans un domaine riche, parfois impénétrable, L.L. Scharf a dégagé des lignes de force comme les notions de suffisance (résumé exhaustif) et d'invariance, permettant à la fois de guider l'intuition et de contribuer à la preuve. L.L. Scharf nous a montré l'efficacité des outils principaux (maximum de vraisemblance, estimation en moyenne quadratique) dans la résolution des problèmes posés par les modèles paramétriques. Enfin, il a dégagé des concepts fondamentaux liés à l'antagonisme complexité–précision, en présentant une approche originale de la détermination optimale du rang d'un modèle.

Le filtrage linéaire et non linéaire est un thème central en traitement du signal et dans des disciplines connexes comme l'automatique. Allant à l'essentiel, J.M.F. Moura a couvert l'ensemble des techniques et résultats fondamentaux: calcul stochastique, intégrales de Itô et Stratanovitch, équations de Kolmogorov, Fokker–Planck et Chandrasekhar.... L'approche de J.M.F. Moura éclaire d'un jour nouveau la notion de processus stochastique vue sous l'angle de l'estimation linéaire en moyenne quadratique. Ce paradigme de la théorie de fonctions aléatoires se retrouvant au coeur des approches de B. Picinbono, L.L. Scharf et J.M.F. Moura montre à la fois son rôle central et ses multiples facettes.

Les représentations temps–fréquence ont depuis longtemps retenu et captivé l'attention des spécialistes, théoriciens ou praticiens, du traitement du signal. Ce thème d'étude déborde largement le domaine du traitement du signal dans ses multiples applications à l'analyse et à la synthèse de la parole, aux recherches en géophysique, dans les sonars biologiques et de manière générale dans l'étude de tous les systèmes non stationnaires. On retrouve également les mêmes concepts au coeur de la mécanique quantique dans les notions fondamentales de variables conjuguées, de commutation ou non commutation, d'opérateurs associés aux relations d'incertitude de Heisenberg.

W. Mecklenbräuker et Y. Grenier ont contribué à éclairer la subtile dialectique opposant les principes généraux guidant le choix des méthodes (positivité, localisation, causalité, invariance) à la résistance de la nature intransigeante sur l'irréductibilité de l'incertitude temps–fréquence. L'approche classique de W. Mecklenbräuker décrit les principales techniques dans le cadre unifié des formes quadratiques conduisant naturellement à la représentation de Wigner–Ville. Y. Grenier propose une définition nouvelle des processus non stationnaires basée sur l'idée de modèle stationnaire étendu. Sur cette base, Y. Grenier nous présente une large gamme de résultats nouveaux, prenant leur source dans la définition précise et rigoureuse de la densité d'énergie dans le plan temps–fréquence (dénommé relief) et aboutissant à une élégante formulation paramétrique permettant de "restationnariser" une large classe de processus non stationnaires. A partir de cette généralisation fondamentale, Y. Grenier développe de nouvelles techniques de mesure et de représentation, illustrées par des applications en traitement des signaux sismiques et dans le domaine de la parole.

Avant de passer aux principaux thèmes du traitement du signal, arrêtons nous sur un cours destiné à réaliser une ouverture sur une nouvelle technique, dont les applications se développent dans de nombreux domaines: l'intelligence artificielle. G. Kopec a été un pionnier dans l'utilisation des méthodes et techniques de l'intelligence artificielle en traitement du signal en développant des outils de base: système intégré de traitement du signal ISP, langage de représentation des signaux SRL, base de données SDB. Le cours de G. Kopec, en définissant les objectifs et les domaines d'application de l'intelligence artificielle, en décrivant les techniques de recherche, de représentation des connaissances et en donnant enfin des exemples d'application des outils qu'il a développés, permettra, soyons-en sûr, aux spécialistes du traitement du signal de se familiariser avec l'intelligence artificielle et contribuera à faire entrer cette technique dans le domaine du traitement du signal (ce cours ne figure pas dans la présente édition, on pourra se reporter aux références*).

La deuxième partie de l'Ecole portait sur les principaux thèmes du traitement du signal. Tout en cherchant à couvrir au mieux l'ensemble des domaines sur lesquels se focalise l'activité actuellement productrice

---

*G.E. Kopec, The Integrated Signal Processing System ISP, IEEE Trans. Acoust. Speech & Signal Process., ASSP 32, no 4 (August 1984).
G.E. Kopec, The Signal Representation Language SRL, IEEE Trans. Acoust. Speech & Signal Process., ASSP 33, no 4 (August 1985).

en traitement du signal, nous n'avons vraisemblablement pas pu être totalement exhaustifs. Nous avons retenu:
- les traitements numériques,
- l'analyse spectrale,
- le traitement d'antenne,
- les systèmes adaptatifs,
- la cohérence et les mesures de retard,
- le traitement d'image.

Une des caractéristiques principales de l'évolution du traitement du signal dans les vingt dernières années, induite par la mise au point de nouveaux composants, est le développement – sinon la généralisation – des méthodes numériques de traitement du signal. Le cours de V. Cappellini nous donne les bases de la description des systèmes numériques de traitement, les méthodes de conception de ces systèmes, avant de présenter un panorama de leurs principaux domaines d'applications. Les similitudes et les différences entre les systèmes analogiques et numériques sont recensées et utilisées. La connexion intime avec la technologie au niveau du matériel et du logiciel est mise en évidence par V. Cappellini qui décrit les possibilités envisageables dans un proche avenir au niveau des processeurs élémentaires et des moyens de stockage de l'information. Un tour d'horizon des multiples applications des traitements numériques ouvre le vaste champ couvert par ce domaine que nous aurons l'occasion de retrouver et de développer dans plusieurs autres cours.

S. Kay présente un panorama complet et critique des nombreuses méthodes d'analyse spectrale des signaux aléatoires qui sont venues, depuis une vingtaine d'années, concurrencer les méthodes plus anciennes mais toujours compétitives, basées sur la transformation de Fourier et le théorème de Wiener–Kintchine: périodogramme moyenné ou lissé, corrélogramme. En décortiquant les principes des différentes méthodes, en analysant leurs différentes mises en oeuvre, en donnant leurs propriétés, le cours de S. Kay fait la somme des tendances actuelles et ouvre de nouvelles voies. L'analyse que donne S. Kay de l'extraction de sinusoïdes d'un bruit établit le lien entre l'analyse spectrale et le traitement d'antenne. Le texte de S. Kay n'étant pas disponible, nous renvoyons les lecteurs à l'article de synthèse (donné en référence*) qui recouvre très largement l'ensemble des sujets traités par S. Kay.

Le traitement d'antenne est, comme l'analyse spectrale, un domaine dans lequel se développent actuellement de multiples recherches. J.F.

---

*S.M. Kay and J.L. Marple, Spectrum Analysis. A Modern Perspective, Proc. IEEE 69, no. 1 (1981).

Böhme en présente une vue synthétique d'une remarquable clarté et ouvre la voie des recherches les plus modernes en nous présentant les fondements et les applications des méthodes paramétriques qu'il a développées. Partant de la modélisation de la propagation en ondes planes, J.F. Böhme aboutit à la formulation paramétrique la plus générale et la plus réaliste, prenant en compte les propriétés des sources, du milieu de propagation et du réseau de capteurs. J.F. Böhme parcours ainsi tout l'éventail des méthodes: formation de voies, méthodes haute résolution, méthodes paramétriques. Dans son approche paramétrique, il donne un cadre théorique rigoureux à cette méthode et décrit des moyens pratiques de résolution. Dans cette partie de son cours, J.F. Böhme nous donne de nombreux résultats originaux et ouvre de nouvelles voies de recherche.

Dans son cours sur les traitements adaptatifs, T.S. Durrani part de la définition de la notion de système adaptatif. T.S. Durrani situe les systèmes adaptatifs dans le cadre général de la théorie des systèmes optimaux et il nous décrit les principales méthodes utilisées pour implanter un sytème adaptatif. L'étude complète des propriétés des systèmes optimaux en moyenne quadratique (LMS) permet à T.S. Durrani d'introduire de manière simple l'analyse de la vitesse de convergence et du "désajustement" ainsi que la relation entre ces deux paramètres. Des résultats nouveaux sont ensuite donnés sur les systèmes adaptatifs en treillis qui constituent une classe particulièrement importante de systèmes adaptatifs. Les perspectives de développement de la technique des systèmes adaptatifs associés à des systèmes experts dans une structure "intelligente" montrent une des tendances nouvelles dans le développement de cette technique.

Un problème essentiel par la simplicité de sa formulation et par son importance pratique dans de nombreux domaines d'application est celui de la mesure du retard entre deux signaux. Ce sujet recouvre des applications fondamentales dans les domaines du sonar ou du radar et il a ouvert la voie à de nombreuses applications industrielles, particulièrement en vélocimétrie. La grandeur fondamentale conduisant à la solution de ce problème et permettant d'en caractériser les performances est l'intercorrélation et plus particulièrement la cohérence (densité spectrale de puissance d'interaction normée).

G.C. Carter a articulé son cours sur la définition et la mesure de la cohérence conduisant aux estimateurs de retard et s'ouvrant sur la mesure conjointe (à l'aide d'un réseau de capteurs) de l'azimuth et de la distance d'une source. Dans la partie consacrée à la cohérence, G.C. Carter nous donne un panorama complet des résultats théoriques et empiriques sur les algorithmes d'estimation et les propriétés moyennes (biais et variance) de l'estimateur. L'estimateur optimal d'un retard (au sens maximum de vraisemblance) est ensuite établi et ses propriétés développées (borne de Cramer–Rao, borne de Ziv–Zakai, etc.) au plan théorique et pratique. Enfin G.C. Carter nous présente une ouverture vers la mesure conjointe de l'azimut et de la distance d'une source avec un réseau de capteurs. Cela lui permet de donner une approche originale appuyée sur des résultats de simulation des propriétés de cette estimateur et de présenter une étude critique de la géométrie optimale de l'antenne.

Les images sont actuellement un sujet très important. Ce thème est au carrefour de la recherche théorique sur les méthodes et les algorithmes et de l'étude d'architectures et de techniques rapides. Il s'ouvre également sur une grande variétés d'applications dans la vie économique, la connaissance de l'environnement et de l'univers, la protection de la santé et la diffusion de la culture. Comme l'indique A.N. Venetsanopoulos en introduction à son cours sur le traitement digital des images, si, au cours de sa vie, un individu utilise 2×10^{12} bits sous forme de signaux de parole, il échange environ 10^{16} bits d'informations sous forme d'images avec le monde extérieur....

Deux cours donnent les fondements de ce secteur: le traitement digital des images et la reconstruction des images.

La contribution de A.N. Venetsanopoulos est un panorama des traitements digitaux. Le champ du traitement d'image est tout d'abord cerné et décrit dans sa variété d'objectifs (amélioration, inversion, analyse, interprétation) et de méthodes (linéaires, non linéaires, locales ou non, adaptatives,...). A.N. Venetsanopoulos passe ensuite en revue les composantes essentielles de l'activité de recherche et développement en traitement d'images. L'amélioration des images est présentée dans sa généralité en insistant particulièrement sur le filtrage linéaire multidimensionnel. La restauration des images pose le problème des critères perceptuels, les méthodes sont présentées et une application particulière en radiologie est développée. En couvrant largement le domaine de l'analyse et de l'interprétation des images, A.N. Venetsanopoulos nous conduit aux confins du domaine actuellement exploré et nous donne enfin les clés de l'avenir centré sur l'utilisation d'outils conceptuels et matériels nouveaux: intelligence artificielle, morphologie mathématique, composants, architectures, pour surmonter les défis posés par la formidable quantité d'information à traiter et la complexité des procédures d'évaluations.

La reconstruction des images qui a connu un grand développement dans les problèmes de tomographie médicale s'applique également de

plus en plus dans le traitement des sondages sismiques, en radio-astrono-
mie, en microscopie électronique et dans le contrôle non destructif.
M. Kaveh nous donne dans son cours une vision concise et complète de
ce problème et des principales méthodes mises en oeuvre. La première
tâche à accomplir en vue de reconstruire un objet à partir de ses
projections est de se donner un modèle de l'interaction entre l'objet
étudié et les ondes (rayons X, ondes acoustiques, élastiques, électrons)
utilisés pour réaliser les "projections". M. Kaveh nous présente le
modèle de rayons (optique géométrique) et les extensions (Born, Rytov)
prenant en compte la diffraction. Dans ces deux situations le problème
de reconstruction peut être résolu (théorème de Radon), mais les con-
traintes pratiques de l'échantillonnage, du temps de calcul et de la
capacité limité de stockage de données limitent la précision de la
reconstruction. M. Kaveh nous présente les méthodes de reconstruction
par rétropropagation, les méthodes algébriques (par inversion de
matrices) en insistant sur les méthodes les plus utilisées qui se placent
dans le domaine de Fourier (Transformée de Fourier bidimensionnelle).
Il conclut sur la supériorité pratique de la méthode de Fourier tout en
insistant sur la plus grande généralité des méthodes algébriques. Enfin, il
insiste sur les voies de recherche essentielles qui concernent l'affinage des
modèles physiques d'interaction ondes-milieu et la prise en compte de
ces modèles plus élaborés dans les méthodes de reconstruction.

Le traitement du signal a connu une croissance considérable sous
l'impulsion du développement des moyens de traitement. Les moyens de
traitement du signal sont conditionné par les algorithmes et les systèmes.
Les algorithmes bénéficient de plusieurs dizaines (ou centaines) d'années
de recherche en mathématiques appliquées. On retrouve dans leur
dénomination la trace des plus grands mathématiciens (Gauss, Choleski,
Schur, Gram-Smith,...). Malgré les progrès extraordinaires des moyens,
la vitesse de calcul reste un objectif prioritaire qui sous-tend les recherches toujours très actives sur les algorithmes rapides.

L'introduction à la notion de rang de déplacement et les algorithmes
rapides associés, présentés par C. Gueguen, nous situe au coeur des
progrès actuels dans ce domaine. La multiplicité des méthodes
développées rendent une synthèse à la fois difficile et nécessaire. C.
Gueguen suscite notre intérêt en nous montrant la relation directe entre
les problèmes d'algorithmiques et les grands sujets d'intérêt pratique en
traitement du signal. Le concept central de matrice de Toeplitz sert de
pivot au cours de C. Gueguen. Après nous avoir montré leur rôle
essentiel (en prédiction, analyse spectrale, traitement d'antenne,...) il
nous présente un panorama de leurs propriétés. Une ouverture est alors

faite sur les matrices "proches de Toeplitz". La mesure de proximité, difficile à définir, est quantifiée par la notion récente et fondamentale de rang de déplacement. A partir de ces idées de base, C. Gueguen développe et ordonne les principaux algorithmes rapides. Si, dans ce domaine et pour les T.S., l'outil fondamental demeure l'algorithme de Levinson–Durbin factorisant l'inverse de la matrice de corrélation, C. Gueguen nous montre les potentialités importantes des extensions de cet algorithme à des matrices proches de Toeplitz et, dans une analyse intéressante des axes de recherche actuels, il nous convie à étendre largement les principes qu'il contient à des situation très différentes. Ce domaine si riche et si exploré a-t-il révélé tous ses secrets? En nous posant en conclusion la question fondamentale – encore ouverte – de la recherche des structures de matrices se prêtant à des algorithmes d'inversion rapides, C. Gueguen nous montre le chemin des recherches futures dans ce domaine.

Quand on dispose d'algorithmes rapides, il est nécessaire, pour les mettre en oeuvre, de les implanter sur des systèmes performants. Ce sujet d'étude conclut la série des cours avec deux contributions sur l'architecture des systèmes de traitement du signal.

Le cours de S.Y. Kung sur les "Processeurs Vectoriels VLSI en Traitement du Signal" nous introduit directement au coeur des problèmes, des succès et des enjeux de ce domaine essentiel de la recherche et de la technologie. S.Y. Kung nous fait pénétrer dans ce domaine, en nous donnant les lignes de force d'un monde souvent dominé par l'empirisme. Nous ne pouvons pas ne pas être frappés par l'ouverture d'esprit extraordinaire que nécessite les progrès dans cette voie de recherche où ressurgissent les plus grandes découvertes des Mathématiques Appliquées qui viennent s'incarner dans le monde réel et géométrique des VLSI. Le processeur s'intègre alors naturellement dans la physique des ondes en y devenant un transformateur de données régi par la loi de la propagation des ondes où il atteint ses limites de capacité et de vitesse. C'est ainsi qu'apparaissent naturellement les structures fondamentales des systèmes systoliques et des processeurs à "front d'onde". Le grand mérite du cours de S.Y. Kung est de nous aider à appréhender la philosophie fondamentale de ces systèmes, de les situer l'un par rapport à l'autre, et de nous donner à travers les diagrammes de flux et leur utilisation, les méthodologies d'étude et de conception de ces systèmes. Cette vision d'ensemble est illustrée et éclairée par le développement d'exemples concrets. Enfin, les divers niveaux du système global, processeur vectoriel, "puce", sont présentés et replacés dans le cadre de leurs applications dans le traitement numérique des signaux. Nous voyons apparaître, comme nous y invite la conclusion de S.Y. Kung, la nécessaire et fructueuse harmonie entre les recherches sur l'architecture et le développement des VLSI. La base conceptuelle des futurs progrès dans le développement de super-calculateurs, basée sur les processeurs vectoriels, est ainsi posée en nous donnant une vision plus claire, des tendances futures de ce domaine primordial de recherche et développement.

Dans le même domaine, le cours de S.F. Reddaway sur "Le Traitement du Signal par Processeurs Vectoriels" nous donne un panorama de la famille des processeurs vectoriels distribués (DAP). Le principe qui préside à la conception de ces systèmes est l'optimisation des transferts de données, tâche qui constitue le goulot d'étranglement des systèmes numériques de traitement du signal. Les DAP sont définis autour d'une mémoire "en tranches" optimisant les transferts de données. La puissance et la flexibilité des capacités de mémoire des DAP, les placent, selon S.F. Reddaway, en meilleure position que les systèmes systoliques (SAP) et les processeurs à front d'onde (WAP) quant à leur polyvalence et à leur capacité à traiter de grandes quantités de données. S.F. Reddaway nous montre également l'importante de la mise en oeuvre de langages de programmation adaptés à ce type de processeurs. Ainsi la manipulation directe de tableau de données (signal, image) vues comme des entités élémentaires introduit une économie important dans la formulation et une grande souplesse d'utilisation que vient encore étendre l'usage de mots de longueur variable. La description des principes caractérisant ces processeurs est concrétisée par l'étude de cas portant sur la transformation de Fourier (FFT), à 1, 2 ou 3 dimensions, opération fondamentale en traitement du signal, et l'analyse d'un système global de traitement de signaux radar. Les tableaux chiffrés qui illustrent ces cas nous donnent une idée précise des possibilités de ces systèmes replacés dans le contexte de la technologie actuelle.

Cet ensemble de cours sur les tendances et les possibilités actuelles de la technologie fait apparaître l'interaction profonde entre le matériel (VLSI, architecture), le logiciel (langage) qui prennent leur source dans l'algorithmique dont les nouvelles frontières ont été explorées par C. Gueguen. Nous voyons ainsi s'estomper ou disparaître la distinction traditionnelle entre recherche fondamentale et appliquée et nous prenons conscience de la très large intégration entre la connaissance et l'action réalisée par les techniques de traitement numérique des signaux.

Après une laborieuse et fructueuse journée de travail, éventuellement agrémentée par une excursion vespérale au sein d'un des plus beaux massif des Alpes, les participants à la session de l'Ecole des Houches sur le Traitement du Signal ont eu plusieurs fois le plaisir d'admirer la magie toujours renouvelée du coucher de soleil sur la chaîne du Mont-Blanc. Quand l'ombre du soir gagnait les prairies de la vallée et redonnait aux forêts des premières pentes leur profondeur et leur mystère, les sommets, couronnés de neiges éternelles, apparaissaient comme autant de repères marquant les axes d'une chaîne une et diverse, mystérieuse et désirable. Qui n'a alors associé l'appel de ces cimes à l'exigence de connaissance et de plénitude qu'irradiait dans son être la découverte d'une discipline aussi une et diverse, mystérieuse et désirable que le Traitement du Signal?

Remerciements

Notre gratitude va aux professeurs pour le soin qu'ils ont mis à la préparation et à la rédaction de leurs cours, ainsi qu'aux participants qui ont su maintenir une ambiance scientifique et humaine à la fois stimulante et souriante au cours des quatre semaines qu'a duré cette session XLV de l'Ecole de Physique Théorique des Houches.

La réalisation de cette session et la publication de ce volume sont le résultat de nombreuses contributions:

– le soutien financier de l'Université Scientifique et Médicale de Grenoble, et les subventions de la Division des Affaires Scientifiques de l'OTAN, qui a inclus cette session dans son programme d'Instituts d'Etudes Avancées, du Commissariat à l'Energie Atomique et du Ministère des Relations Extérieures;

– l'orientation et le soutien effectif du Conseil de l'Ecole;

– le soin apporté par Michèle Carret dans la préparation et la frappe des manuscrits;

– l'aide d'Henri Coiffier, de Nicole Leblanc, d'Anny Glomot, et de toute l'équipe qui a rendu la vie de tous les jours aussi comfortable que possible.

Jean-Louis Lacoume
Tariq Salim Durrani
Raymond Stora

PREFACE

The Les Houches Summer School of theoretical physics, organised jointly by the "Institut National Polytechnique" and the "Université Scientifique Technique et Médicale", both in Grenoble, holds during the summer, each year, a couple of sessions lasting between four and eight weeks each which cover in depth and update a couple of hot spots in physics. Tutorial courses and seminars allow young researchers and non-specialists to reach the level of the most advanced research in the chosen field.

In the eighties, the board of trustees of the Summer School expressed the wish to broaden the domain of activities of the School and open it in particular towards physical sciences applied to engineering. The first theme chosen to materialize this orientation was Signal Processing and the organization was trusted to T.S. Durrani, from the University of Strathclyde and J.L. Lacoume from the "Institut National Polytechnique", in Grenoble. The lectures were given by some sixteen professors originating from universities and industrial research centers by the USA, France, the United Kingdom, the Federal Republic of Germany, Italy, Austria and Portugal. Some fifty participants, young researchers from universities or industrial establishments, and users of signal processing in other fields did benefit from the course. They mostly originated from Western Europe, with a rather small participation from the United States.

The teaching was organized into courses and study groups.

The courses fall into essentially three classes:
- the foundations of signal processing,
- the main themes in signal processing,
- techniques in signal processing.

The main courses covered:
- signal theory,
- detection and estimation,
- nonlinear filtering,
- the time–frequency representation.

In his course on deterministic and random signals, B. Picinbono first defined in a synthetic and unifying fashion such basic notions as classifying signals, stationarity, AR/ARMA modelization, spectral characteristics. He then presented the main ideas concerning linear estimates in the quadratic mean, shedding a new light on the problem of statistical linear filtering and allowing to encompass within a single approach all the classical problems (causal and noncausal Wiener) as well as to give an elegant generalization based on the idea of a linear constraint.

Detection and estimation theory has appeared as deep, powerful and lively, in the course given by L.L. Scharf. From such a rich and often impenetrable field, L.L. Scharf has extracted several lines of forces such as the notions of sufficiency and invariance, which allow both to guide intuition and contribute to proofs. L.L. Scharf has well shown the efficiency of the main tools (maximum likelihood, estimate in the quadratic mean) in solving the problems posed by parametric models. Finally he has sorted out the fundamental concepts connected with the complexity–accuracy antagonism, presenting an original approach to the optimal model order determination.

Linear and nonlinear filtering are central in signal processing as well as in connected fields such as automatic control. Getting to the point right away, J. Moura covered the bulk of techniques and main results: stochastic calculus, Itô and Stratanovitch integrals, Kolmogorov, Fokker–Planck and Chandrasekhar equations, etc. J.M.F. Moura's approach sheds a new light on the notion of stochastic processes seen from the point of view of linear estimation in the quadratic mean. This paradigm in the theory of random functions, which lies at the core of the expositions of B. Picinbono, L.L. Scharf and J.M.F. Moura, thus plays a central rôle and exhibits a remarkable diversity.

The time–frequency representations have for a long time captured the attention of specialists in signal processing, both theorists and practitioners. This topic reaches far beyond signal processing, with applications to speech analysis and synthesis, geophysical research, biological sonars, and, more generally, to the study of all nonstationary phenomena. Similar concepts are found at the core of quantum mechanics where the fundamental notions of conjugate variables, of commutativity versus noncommutativity of the associated operators lie at the roots of the Heisenberg uncertainty principle.

W. Mecklenbräuker and Y. Grenier have focused on the subtle dialectics opposing the general principles which condition the choice of a method (positivity, localization, causality, invariance) to the resistance of an unforgiving nature on the irreducible time–frequency uncertainty.

The classical approach of W. Mecklenbräuker describes the main techniques within the unified framework of quadratic forms, which leads in a natural way to the Wigner–Ville representation. Y. Grenier proposes a new definition of nonstationary processes based on the idea of an extended stationary model. On these grounds, Y. Grenier presents a broad spectrum of new results originating from an accurate and rigorous definition of the energy density in the time–frequency plane – the so-called "relief plot" – and leading to an elegant parametric formulation which allows to "re-stationarize" a large class of non stationary processes. Starting from this generalization, Y. Grenier develops new techniques of measurements and representations and illustrates them by applications to the treatment of seismic signals and speech.

Before we start on the main topics in signal processing proper, a new technique deserves our attention in view of its applications in numerous domains; we have named Artificial Intelligence. G. Kopec has been a pioneer in the use of the methods and techniques pertaining to artificial intelligence in signal processing. He has developed several basic tools such as ISP (integrated systems for signal processing), SRL (signal representation languages), SDB (signal data basis). G. Kopec's course defines the goals and domains of application of artificial intelligence, describes techniques in research and representation of knowledge and gives several examples of applications of the tools he has developed. It will surely allow specialists in signal processing to become familiar with artificial intelligence and will contribute to incorporate these techniques into the main body of signal processing. Although the lectures given by G. Kopec are not included in this volume, their content can be traced back to the bibliography*.

The second part of the course involved signal processing proper. Although we tried our best to cover the main areas on which most of the productive research is focused, we had to choose and decided to retain the following:

- numerical treatment,
- spectral analysis,
- antenna array processing,
- adaptive systems,
- coherence and delay measurements,
- image processing.

---

*G.E. Kopec, The Integrated Signal Processing System ISP, IEEE Trans. Acoust. Speech & Signal Process. ASSP 32, no. 4 (1984).
G.E. Kopec, The Signal Representation Language SRL, IEEE Trans. Acoust. Speech & Signal Process. ASSP 33, no. 4 (1985).

One of the main characteristics in the evolution of signal processing over the last twenty years, induced by the production of new components, is the development – if not the generalization – of numerical methods. The course given by V. Cappellini gives the basis for the description of numerical systems for signal processing, the methods for conceiving these systems and an overview of the main domains of application. Similarities and differences between analogue and numerical systems are sorted out and used. The intimate connection with technology at the level of both software and hardware is set out by V. Cappellini who describes the potentialities one may envisage in a near future at the level of elementary processors and information storage systems. A quick review of the many applications of numerical treatments opens up the broad perspectives of this domain we shall have the opportunity to meet again and see developed in several other courses.

S. Kay presents an exhaustive and critical overview of many methods in the spectral analysis of random signals which have, during the last twenty years, seriously competed with the older but still operative methods based on Fourier transform and the Wiener–Khintchine theorem: averaged out or smoothened periodograms, correlograms. In analyzing the principles of these different methods, as well as their applications, and in describing their properties, S. Kay's course integrates the present trends and opens up new pathways. The analysis given by S. Kay of the process of extracting sine waves from a noise establishes the link between spectral analysis and antenna array processing. Since the text of these lectures is not available we refer the reader to the review article* which broadly overlaps with the bulk of the topics treated by S. Kay.

Antenna array processing, just as spectral analysis, is a field in which active research is currently pursued. J.F. Böhme presents a synoptic, remarkably clear view of this activity, and opens the way to the most recent research through a description of the foundations and applications of the parametric methods he has developed. Starting from the modeling of plane wave propagation, J.F. Böhme reaches the most general and realistic parametric formulation, taking into account the properties of the sources, of the medium and of the array of receivers. J.F. Böhme thus scans the whole spectrum of methods: channel formation, high resolution methods, parametric methods. In his parametric approach he gives that method a rigorous theoretical framework and describes practical techniques for achieving high resolution. In this part

*S.M. Kay and J.L. Marple, Spectrum Analysis. A modern perspective, Proc. IEEE 69, no. 1 (1981).

of his course, J.F. Böhme gives a number of original results and opens new research alleys.

In his course on adaptive treatments, T.S. Durrani starts from the definition of an adaptive system. He then embeds adaptive systems within the general framework of the theory of optimal systems and describes the main methods used to devise adaptive systems. A complete study of the properties of optimal systems in the quadratic mean allows T.S. Durrani to introduce in a simple manner an analysis for the speed of convergence and the misadjustment as well as of the correlation between these two parameters. New results are then given on adaptive lattices which constitute a particularly important class of adaptive systems. Perspectives on the development of the technique associating adaptive systems with expert systems within an intelligent structure show one of the new trends in the development of this field.

One of the most important problems, both simple to formulate and of practical importance in many domains of application, is that of measuring the time delay between two signals. This topic covers fundamental applications in sonar as well as radar techniques and has opened the way to many industrial applications, in particular in velocity measurements. The fundamental quantity which leads to a solution of this problem and allows to characterize its performances is the measure of cross-correlation, and more specifically of coherence (i.e., the normalized spectral density of interaction power).

G.C. Carter centered his course on the definition and measurement of coherence, leading to delay estimators and opening on the joint measurement (with the help of an array of sensors) of the azimuth and distance from a source. In the section devoted to coherence, G.C. Carter gives a complete overview of both theoretical and empirical results on estimation algorithms and the average properties (bias and variance) of the estimators. The optimal estimator of a delay (in the sense of maximum likelihood) is then established and its properties are exhibited (Cramer–Rao bound, Ziv–Zakai bound, etc.,...) on both theoretical and practical grounds. Finally, G.C. Carter presents an opening towards the joint estimation of the azimuth and distance from a source with the help of an array of sensors. This allows him to give an original approach relying on results from simulations of the properties of this estimator, and to present a critical study of the optimal geometry of an antenna array.

Image processing is at the moment a most important topic. This theme is at the crossroads of theoretical research on methods and algorithms and of the study of architectures and fast algorithms. It also opens up a variety of applications in economical life, in the knowledge of environment and universe, in health protection and in the spreading of culture. As indicated by A.N. Venetsanopoulos in the introduction to his course on digital image processing, if an individual uses throughout his lifetime 2×10^{12} bits in the form of speech signals, he does exchange with the outside world around 10^{16} bits of information in the form of images . . . .

Two courses are devoted to the foundations of this area: image digital processing and image reconstruction.

A.N. Venetsanopoulos' contribution consists of an overview of digital image processing. The field of image processing is first defined and described in his variety of goals (improvement, inversion, analysis, interpretation) and methods (linear, nonlinear, local or nonlocal, adaptive, . . .). A.N. Venetsanopoulos then reviews the essential components of research activities and development in image processing. The improvement of images is presented in all its generality with particular emphasis on multidimensional linear filtering. The restoration of images poses the problem of perception criteria, the methods are presented and a particular application to radiology is developed. Covering broadly the domain of image analysis and interpretation, A.N. Venetsanopoulos leads us to the boundaries of the domain now being explored and gives us the keys to a future centered around the use of conceptual tools and new equipment: artificial intelligence, mathematical morphology, components, architectures, likely to overcome the challenge proposed by the enormous amount of information to be dealt with and the complexity of evaluation processes.

Image reconstruction which has undergone a spectacular development in medical tomography also applies more and more in seismic exploration, radioastronomy, electronic microscopy and non-destructive testing. M. Kaveh has conveyed in his course a concise and complete vision of this problem and of the main methods involved. The first task to be performed in order to reconstruct an object from its projections is to give a model of the interaction between the object under study and the waves used to realize the projections (X-rays, acoustic, elastic or electromagnetic waves). M. Kaveh presents a model of rays (geometrical optics) and its extensions (Born, Rytov) which takes diffraction into account. In both situations the reconstruction problem can be solved (Radon's theorem) but the practical constraints due to sampling, computing time and limited data storage capacity put limits on the accuracy of the reconstruction. M. Kaveh presents reconstruction methods based on backward propagation, algebraic methods (by matrix inversion), with emphasis on the most frequently used methods which rely on two-dimensional Fourier analysis. He concludes by recognizing the practical superiority of the Fourier method, although he insists on the greater generality of the algebraic methods. Finally, he insists on the most promising research alleys concerned with refining the physical models for the wave–medium interaction and taking these more refined models into account throughout the application of the reconstruction methods.

Signal processing has witnessed a considerable growth under the impulse of the expansion of the tools available for processing. These tools are conditioned by algorithms and systems. Algorithms have benefited from several tens (or hundreds) of years of research in applied mathematics. One can find in their labelling the trace of the greatest known mathematicians (Gauss, Choleski, Schur, Gram-Schmidt). In spite of the outstanding progress in the tools, computational speed remains an objective of first priority which underlies the always very active research on fast algorithms.

An introduction to the notion of displacement rank, and the associated fast algorithms, presented by C. Gueguen, throws us right at the heart of the present progress in this domain. The multiplicity of the methods which have been and are still being developed, makes a synthesis both difficult and necessary. C. Gueguen arouses our interest in showing the direct relationship between algorithm complexity and important topics of practical interest in signal processing. The central concept of a Toeplitz matrix serves as an axis around which C. Gueguen's course is built. After showing its central role (in prediction, spectral analysis, antenna treatment, . . .) he presents an overview of its properties. Then an opening on matrices "close to Toeplitz" takes place. A measure of proximity, which is hard to define, becomes quantitative thanks to the recent and fundamental notion of displacement rank. Starting from these basic ideas, C. Gueguen develops and orders the main fast algorithms. If, in this domain and for the T.S. the fundamental tool remains the Levinson–Durbin algorithm which factorizes the inverse of the correlation matrix, C. Gueguen shows the important potentialities of extensions of this algorithm to matrices close to Toeplitz, and, in an interesting analysis of the present research directions, he invites us to broadly extend the principles in question to quite different situations. Has such a rich and well explored domain revealed all its secrets? Asking, in conclusion, the fundamental question, – still open – of the search for the structure of matrices which allow for fast inversion algorithms, C. Gueguen points to the directions of future research in this domain.

When one has fast algorithms at one's disposal, it is necessary, in order to be able to use them, to embed them into high performance systems. This question concludes the series of courses with two contributions to the architecture of signal processing systems.

S.Y. Kung's course on VLSI parallel processors in signal processing takes us right to the heart of the problems, successes and issues of this domain, which is essential in research and technology. S.Y. Kung, as he takes us into this domain, gives the lines of force of a world often under the reign of empiricism. We cannot but be struck by the extraordinary openmindedness required by the progress witnessed in this research where the most prominent discoveries in Applied Mathematics reappear and are embodied in the real and geometrical world of VLSI. The processor becomes an integrated agent in wave physics, as it becomes a transformer of data ruled by the laws of wave propagation and reaches the limits of capacity and velocity. This is the way by which the fundamental structures of systolic systems and wave front processors make their appearance. The great merit of S.Y. Kung's course is to help us apprehend the fundamental philosophy of these systems, to compare them, and to give us, through flow diagrams and their use, methodologies proper to the study and conception of these systems. This global vision is illustrated and enlightened by developments of concrete examples. Finally, the various levels of the global system, vectorial processor, "cells", are presented and set back within the framework of their applications in digital signal processing. We can see, as S.Y. Kung's conclusion invites us to do, the necessary and fruitful harmony between research in architecture and the development of VLSI. The conceptual basis for future progress in the development of supercomputers, based on vectorial processors, is thus founded and gives us a clearer vision of the future trends in this domain so crucial for research and development.

In the same vein, the course given by S.F. Reddaway on "Signal processing on a processor array" gives an overview of the family of Distributed Array Processors (DAP). The principle which rules the conception of such systems is the optimization of data transfer, a task which is the bottleneck of the computing systems of signal processing. DAP's are defined around memory blocks which optimize data transfer. The power and flexibility of DAP memories place them, according to S.F. Reddaway, in a better position than systolic systems (SAP), wave front processors (WAP), in so far as versatility and the ability to deal with large amounts of data are concerned. S.F. Reddaway also shows the importance to use programming languages adapted to these types of processors. Thus, the direct manipulation of data (signal, image) seen as elementary operations, introduces a substantial saving in the formulation and a remarkable flexibility in the use, still extended by the use of variable word lengths. The description of the principles which characterize these processors is concretized by the study of cases bearing on the Fast Fourier Transform (FFT), one-, two-, or three-dimensional, an operation which is fundamental in signal processing, and by the analysis of a system for radar signal processing. The numerical tables which illustrate these cases give us a precise idea of the potentialities of these systems within the context of present technology.

This collection of courses on the present trends in and possibilities of technology shows the strong interaction between the hardware (VLSI, architecture), and the software (language) which have their sources in the algorithms whose new boundaries have been explored by C. Gueguen. Thus we see fade away or disappear the traditional distinction between fundamental and applied research and become conscious of the substantial integration between knowledge and achievements made possible by the techniques of numerical signal processing.

After a laborious and fruitful working day, eventually embellished by an afternoon excursion in one of the most beautiful massifs of the Alpes, the participants of this session of the Les Houches School on Signal Processing, had more than once the pleasure to admire the ever renewed magic of a sunset on the Mont-Blanc ridge. When the evening shadows would invade the meadows in the valley and give to the woods on the first slopes their depth and mystery, the summits, covered by the eternal snows, would appear as many landmarks marking the axes of a one and diverse ridge, mysterious and desirable. Who then could refrain from associating the call of the summits with the demand for knowledge and plenitude irradiated within his being by the discovery of such a rich, diverse, mysterious and desirable discipline as Signal Processing?

Acknowledgements

We wish to express our gratitude to the professors for the care with which they have prepared and written up their lectures, as well as to the participants who maintained throughout the four weeks of this XLV session of the School of Theoretical Physics in Les Houches a human and scientific atmosphere which proved both stimulating and relaxed.

This session and the publication of this volume of lecture notes would not have been possible without:
 – the financial support from the Université Scientifique et Médicale de Grenoble, and NATO Scientific Affairs Division (which included this

session in its Advanced Studies Institute Programme), the Commissariat à l'Energie Atomique, and the Ministère des Relations Extérieures;
 – the guidance of the Board of Trustees of the school;
 – the careful preparation and typing of the manuscripts by Michèle Carret;
 – the help of Henri Coiffier, Nicole Leblanc, Anny Glomot and the whole team at Les Houches who made everyday life as comfortable as possible.

Jean-Louis Lacoume
Tariq Salim Durrani
Raymond Stora

CONTENTS

Course 3. Linear and nonlinear stochastic filtering, by J.M.F. Moura 205

Course 4. *A tutorial on non-parametric bilinear time–frequency signal representations, by W. Mecklenbräuker* 277

Course 5. Parametric time–frequency representations, by Y. Grenier   *339*

PART II – MAIN TOPICS IN SIGNAL PROCESSING

Course 10. Digital image processing and analysis, by A.N. Venetsanopoulos *573*

PART III – TECHNIQUES IN SIGNAL PROCESSING

PART II

MAIN TOPICS IN SIGNAL PROCESSING

COURSE 6

DIGITAL SIGNAL–IMAGE PROCESSING

Vito CAPPELLINI

Dipartimento di Ingegneria Elettronica
University of Florence, Florence, Italy
and Istituto di Ricerca sulle Onde Elettromagnetiche – C.N.R.
Via Panciatichi 64, 50127 Florence, Italy

J.L. Lacoume, T.S. Durrani and R. Stora, eds.
Les Houches, Session XLV, 1985
Traitement du signal / Signal processing
© *Elsevier Science Publishers B.V., 1987*

Contents

1. Introduction

The area of digital signal–image processing is of increasing importance and interest due to the impressive advances in digital methods and technology. There are in particular to be outlined the following aspects: definition of efficient algorithms performing digital filtering of linear and non-linear type; wide software or hardware implementation capabilities due to the large expansion and evolution of standard computers, mini-computers, array-processors, microprocessors and very large scale integration circuits (VLSI) at a decreasing cost; great application flexibility and adaptivity in several important fields such as speech processing, communications, radar–sonar systems, remote sensing, biomedicine, moving object recognition and robotics.

In this course the representation of discrete time signals and systems [one-dimensional (1-D) case] as well as of sampled images and discrete space systems [two-dimensional (2-D) case] is described. Digital transformations [as fast Fourier transform (FFT)] are outlined. The design methods of digital filters (mainly 1-D case) are also presented. Finally, implementation aspects are considered together with some application examples.

2. Discrete functions and fundamental properties of 1-D and 2-D digital systems

Discrete functions are defined only for discrete values of their variables and, in general, are sequences of numbers. These numbers can be obtained as quantized samples of an analogue signal $f(t)$ or an image $f(x, y)$, according to the sampling theorem [1]: for a signal, a sampling frequency equal to or greater than two times the maximum signal frequency is to be used.

The notation, which can be used for 1-D sequences is the following: $\{f(n)\}, n = N, \ldots, N'$, where $N = -\infty$ and $N' = \infty$ for infinite sequences. For 2-D sequences, the notation is $\{f(n_1, n_2)\}, n_1 = N_1, \ldots, N_1'$

402

and $n_2 = N_2, \ldots, N_2'$, where $N_1 = N_2 = -\infty$ and $N_1' = N_2' = \infty$ for double infinite sequences.

We can observe that within the terms sequences of numbers and numerical transformations of sequences the *quantization* aspect is conceptually enclosed due to the implicit finite precision needed for any numerical representation. In the development of the theory of discrete systems, sequences of numbers with infinite precision are in general considered as a first approach.

Further in the above defined sequences the time sampling interval T or the space sampling interval X does not appear, because, to simplify the notation, a normalized frequency scale νT for a signal and normalized frequencies $\nu_1 X$ and $\nu_2 X$ for an image are used. In practice it is assumed $T = X = 1$: when an absolute scale is required, it is easy to transform the normalized frequencies to the actual frequencies.

In the following the fundamental properties of 1-D and 2-D digital systems are presented.

2.1. Fundamental properties of 1-D digital systems

A 1-D digital system can be defined as an *operator*, which transforms an input sequence $\{f(n)\}$ to an output sequence $\{g(n)\}$ [1]:

$$\{g(n)\} = T_0[\{f(n)\}] \tag{1}$$

where T_0 is the operator symbol. A system of this type is said to be *linear*, denoted by the symbol T_L, if and only if the principle of *superposition* holds. This is equivalent to saying that if $\{g_1(n)\}$ and $\{g_2(n)\}$ are the outputs of the system with inputs $\{f_1(n)\}$ and $\{f_2(n)\}$, respectively, then

$$T_L[\{af_1(n) + bf_2(n)\}] = aT_L[\{f_1(n)\}] + bT_L[\{f_2(n)\}]$$
$$= a\{g_1(n)\} + b\{g_2(n)\}, \tag{2}$$

where a and b are arbitrary constants. This is a combination of the *additivity* and *homogeneity* properties of linear systems.

Since linearity is valid (for absolutely convergent systems) for the sum of an infinite number of terms, the linear system results to be uniquely determined by its response to the *unit sample sequence*, defined as

$$\delta(n) = 1, \quad n = 0$$
$$\delta(n) = 0, \quad n \neq 0 \tag{3}$$

Thus, by observing that the generic sample $f(n)$ of the input signal can be written in the form

$$f(n) = \sum_{k=-\infty}^{\infty} f(k)\delta(n-k) \tag{4}$$

another input–output relation can be obtained as

$$g(n) = T_L\left[\left\{\sum_{k=-\infty}^{\infty} f(k)\delta(n-k)\right\}\right]$$

$$= \sum_{k=-\infty}^{\infty} f(k)T_L[\{\delta(n-k)\}]$$

$$= \sum_{k=-\infty}^{\infty} f(k)h(k,n), \tag{5}$$

where $\{h(k,n)\}$ is the sequence, representing the response to the unit sample sequence.

One useful relation is obtainable for a subclass of linear systems, the shift-invariant linear systems, which are characterized by the property that if $\{g(n)\}$ is the response to $\{f(n)\}$, then $\{g(n-k)\}$ is the response to $\{f(n-k)\}$. In other words, this implies that the output of the system to an input signal is *independent* of the time instant of its application to the system. In this case the *response of the system to the impulse* $\{\delta(n-k)\}$ is $\{h(n-k)\}$ and the input–output relation, *convolution*, is [1]

$$g(n) = \sum_{k=-\infty}^{\infty} f(k)h(n-k) = \sum_{k=-\infty}^{\infty} h(k)f(n-k). \tag{6}$$

It is interesting to discuss the notion of *causal systems*, which are defined as systems whose output depends only on the present and past values of the input. This means that in relation (6) $g(n)$ should not depend on $f(k)$ for $k > n$; therefore for the impulse response $\{h(n)\}$,

$$h(n) = 0, \quad n < 0, \tag{7}$$

and relation (6) becomes (limiting also the processing to N data)

$$g(n) = \sum_{k=0}^{N-1} h(k)f(n-k). \tag{8}$$

It is interesting to observe that the notion of causality is less fundamental in the digital domain than in the analogous one. In this second case, causality is equivalent to the *physical realizability* of the system and it corresponds to the intuitive notion that the output at a time instant t' can depend only on the past inputs. In the digital domain, the data to be processed are often stored on some support (memory of the processor, magnetic tape, disk, etc.) and in this case the notion of past, present and future becomes purely conventional. For instance, sequences of numbers (representing signals) can be processed going in the negative indices direction (that is from the future to the past).

A very important class of linear shift-invariant systems is the one described by the following difference equation,

$$g(n) = \sum_{k=0}^{N-1} a(k)f(n-k) - \sum_{k=1}^{M-1} b(k)g(n-k), \tag{9}$$

where $\{f(n)\}$ are the samples of the input signal, $\{g(n)\}$ are the output data (representing the samples of the output signal), and $\{a(k)\}$ and $\{b(k)\}$ are the *coefficients*, which define the system. It can be shown [1], in the hypothesis of initial rest conditions, that the output of the system described by relation (9) can be computed *recursively* from the knowledge of the previously computed samples and/or from the initial conditions. It can also be observed that, formally, when $\{b(k)\} = 0$ and $\{a(k)\} = \{h(k)\}$ relation (9) reduces to relation (8).

According to the properties of the z-transform [1], in the z-domain, the convolution relation (6) reduces to the form,

$$G(z) = H(z)F(z), \tag{10}$$

where

$$H(z) = \sum_{k=-\infty}^{\infty} h(k)z^{-k} \tag{11}$$

is the z-transform of the unit impulse response. $H(z)$ is called the *z-transfer function* of the digital system, completely defining its performance. In analogous way, for the digital system defined by relation (8), the transfer function $H(z)$ results to be

$$H(z) = \sum_{k=0}^{N-1} h(k)z^{-k}. \tag{12}$$

Further, if we take the z-transform of the two sides of relation (9), a relation of type (10) can easily be obtained, where $H(z)$ is now expressed as [1]

$$H(z) = \frac{\sum_{k=0}^{N-1} a(k)z^{-k}}{1 + \sum_{k=1}^{M-1} b(k)z^{-k}}, \tag{13}$$

which represents the z-transfer function of the digital system defined by the difference equation (9).

By considering the expression of the variable z as a function of the Laplace variable $p = \sigma + j\omega$ (where σ is the real part and ω the angular frequency, $\omega = 2\pi\nu$):

$$z = e^{(\sigma + j\omega)T}, \tag{14}$$

where T is the sampling interval, and by setting $\sigma = 0$ and $T = 1$ (normalized frequency scale), the *frequency response* of the digital system can be obtained. For the systems defined by relations (12) and (13) the following frequency responses result, [1]:

$$H(e^{j\omega}) = \sum_{k=0}^{N-1} h(k) e^{-jk\omega}, \tag{15}$$

and

$$H(e^{j\omega}) = \frac{\sum_{k=0}^{N-1} a(k) e^{-jk\omega}}{1 + \sum_{k=1}^{M-1} b(k) e^{-jk\omega}}. \tag{16}$$

Function $H(e^{j\omega})$ describes the change in *amplitude* and *phase*. In general, $H(e^{j\omega})$ is a complex number and two representations are possible, that is, the representation in terms of its *real part* and its *imaginary part*:

$$H(e^{j\omega}) = H_R(e^{j\omega}) + H_I(e^{j\omega}), \tag{17}$$

or in terms of its *magnitude* and *phase*:

$$H(e^{j\omega}) = |H(e^{j\omega})| e^{-j \, \mathrm{Arg}[H(e^{j\omega})]}. \tag{18}$$

$H(e^{j\omega})$ is a continuous function of ω and further it is a *periodic function* of ω with period 2π, that is, $H(e^{j\omega}) = H(e^{j(\omega + 2k\pi)})$, with $k = -\infty, \infty$. This is the reason why the correct symbol for the frequency response of the digital system would be indeed $H(e^{j\omega})$, even if, for notational convenience, we use in the following in general the simplified symbol $H(\omega)$. The periodicity of $H(\omega)$ implies that the system behaves in the same way for all angular frequencies with multiples of 2π added to the basic angular frequency.

Finally, a very important property of a digital system is represented by its *stability*. Several definitions can be used for defining the stability. The most useful one is the BIBO (bounded input, bounded output) *stability criterion*. This can be expressed by saying that a digital system is stable if its response to a limited input is also limited. For linear shift-invariant digital systems, the BIBO criterion corresponds to [1]

$$\sum_{n=-\infty}^{\infty} |h(n)| < \infty, \tag{19}$$

where $\{h(n)\}$ is the impulse response of the system.

It is important to observe that obviously this stability criterion is always satisfied if $\{h(n)\}$ has a finite number of terms, as it is for systems defined by relations (8) or (12).

The situation is in general more involved for digital systems represented by relations (9) or (13), that is, when the z-transfer function has the form of a rational function of the following type

$$H(z) = \frac{A(z)}{B(z)}, \tag{20}$$

where $A(z)$ and $B(z)$ are polynomials in z. In this case the system can be considered as the cascade of two systems: $H(z) = H_1(z)H_2(z)$, where $H_1(z) = A(z)$ and $H_2(z) = 1/B(z)$. According to the previous considerations, the stability of the system $H(z)$ is only related to the stability of $H_2(z)$, since $H_1(z)$ is always stable. If we indicate with $\{h_2(n)\}$ the impulse response of the $H_2(z)$ system, the following relation can be written for the definition of impulse response and z-transform,

$$\sum_{n=0}^{\infty} h_2(n)z^{-n} = \frac{1}{B(z)}, \tag{21}$$

and $h_2(n)$ can be obtained by inverting $B(z)$ (see also relation 13)

$$B(z) = \sum_{n=0}^{M-1} b(n)z^{-n}. \tag{22}$$

By assuming $b(0) = 1$, let us consider first the case $B(z) = 1 + kz^{-1}$. We can write [1]

$$\frac{1}{B(z)} = \frac{1}{1 + kz^{-1}} = 1 - kz^{-1} + k^2z^{-2} - k^3z^{-3} + \cdots$$

$$= \sum_{n=0}^{\infty} (-1)^n (kz^{-1})^n, \tag{23}$$

and, by equating the coefficients of equal powers in relations (21) and

(23), $\{h_2(n)\}$ can be obtained in the form

$$h_2(n) = (-k)^n. \tag{24}$$

The series (23) converges if $|k| < 1$, but it does not converge for $|k| > 1$. The first-order system $1/(1 + kz^{-1})$ is stable for $|k| < 1$ (minimum phase), while it is unstable for $|k| > 1$ (maximum phase) [1].

It is very simple to relate the previous considerations to the position of poles of $H(z)$, that is of zeros of $B(z)$, in the z-plane. For the first-order factor of the form $1/(1 + kz^{-1})$, its denominator has a zero for $z = -k$. If $|k| < 1$, then it is stable and the pole of this factor lies inside the unit circle in the z-plane, while, if $|k| > 1$ it is unstable and the pole lies outside the unit circle.

Since any $H(z)$ can be expressed in a *cascade of first-order factors*, this means that a digital system, in general of the causal type, having a transfer function of the form (20), is *stable* if and only if all the poles are *inside the unit circle* in the z-plane, while it is *unstable* if its poles are *outside the unit circle*. The *stability test* is therefore relatively simple and efficient: it is in fact a direct evaluation of pole positions in the z-plane and very useful computational methods exist for finding the roots of polynomials [1].

2.2. Fundamental properties of 2-D digital systems

Most of the properties defined for 1-D digital systems can be easily extended to 2-D digital systems, as it is briefly outlined in the following.

A 2-D digital system can be defined as an *operator*, which transforms an input 2-D sequence $\{f(n_1, n_2)\}$ to an output 2-D sequence $\{g(n_1, n_2)\}$:

$$\{g(n_1, n_2)\} = T_0[\{f(n_1, n_2)\}], \tag{25}$$

where T_0 is the operator symbol. A system of this type is said to be *linear*, denoted by the symbol T_L, if and only if the principle of *superposition* holds.

A *convolution* relation can be written if the 2-D digital system is *shift invariant*, that is, if to the input $\{f(n_1 - k_1, n_2 - k_2)\}$ there exists a corresponding output $\{g(n_1 - k_1, n_2 - k_2)\}$, whose form is independent of the position of the input. In this case, if $\{h(n_1 - k_1, n_2 - k_2)\}$ is the output of the system to the unit pulse $\{\delta(n_1 - k_1, n_2 - k_2)\}$, defined by the relation

$$\delta(n_1, n_2) = 1, \quad n_1 + n_2 = 0,$$

$$\delta(n_1, n_2) = 0, \quad n_1 + n_2 \neq 0, \tag{26}$$

then the convolution sum can be obtained (in truncated form) as

$$g(n_1, n_2) = \sum_{k_1=0}^{N_1-1} \sum_{k_2=0}^{N_2-1} f(k_1, k_2) h(n_1 - k_1, n_2 - k_2)$$

$$= \sum_{k_1=0}^{N_1-1} \sum_{k_2=0}^{N_2-1} h(k_1, k_2) f(n_1 - k_1, n_2 - k_2). \tag{27}$$

A very important class of linear shift-invariant 2-D systems is described by the following difference equation [1]:

$$g(n_1, n_2) = \sum_{k_1=0}^{N_1-1} \sum_{k_2=0}^{N_2-1} a(k_1, k_2) f(n_1 - k_1, n_2 - k_2)$$

$$- \sum_{\substack{k_1=0 \\ k_1+k_2 \neq 0}}^{M_1-1} \sum_{k_2=0}^{M_2-1} b(k_1, k_2) g(n_1 - k_1, n_2 - k_2), \tag{28}$$

where $\{a(k_1, k_2)\}$ and $\{b(k_1, k_2)\}$ are the coefficients, which define the system. The system described by relation (28) is *recursive*, in the hypothesis of suitable initial rest conditions (zero initial conditions are defined outside the support of the impulse response, which corresponds to the *first quadrant*).

According to the properties of the 2-D z-transform [1], the following relation can be obtained, connecting the input and output z-transforms:

$$G(z_1, z_2) = H(z_1, z_2) E(z_1, z_2), \tag{29}$$

where $H(z_1, z_2)$ is the 2-D z-transfer function of the digital system, completely defining its performance. The expressions of $H(z_1, z_2)$ for the 2-D systems, defined by relations (27) and (28), result to be respectively [1]

$$H(z_1, z_2) = \sum_{k_1=0}^{N_1-1} \sum_{k_2=0}^{N_2-1} h(k_1, k_2) z_1^{-k_1} z_2^{-k_2}, \tag{30}$$

and

$$H(z_1, z_2) = \frac{\sum_{k_1=0}^{N_1-1} \sum_{k_2=0}^{N_2-1} a(k_1, k_2) z_1^{-k_1} z_2^{-k_2}}{1 + \sum_{k_1=0}^{M_1-1} \sum_{k_2=0}^{M_2-1} b(k_1, k_2) z_1^{-k_1} z_2^{-k_2}}, \qquad (31)$$
$$k_1 + k_2 \neq 0$$

By considering the expression of the variables z_1 and z_2 as a function of the Laplace variables $p_1 = \sigma_1 + j\omega_1$ and $p_2 = \sigma_2 + j\omega_2$ (where σ_1 and σ_2 are the real parts, and $\omega_1 = 2\pi\nu_1$ and $\omega_2 = 2\pi\nu_2$ are the angular frequencies),

$$z_1 = e^{(\sigma_1 + j\omega_1)X},$$

$$z_2 = e^{(\sigma_1 + j\omega_2)X}, \qquad (32)$$

where X is the space sampling interval, and by setting $\sigma_1 = \sigma_2 = 0$ and $X = 1$ (normalized frequencies), the frequency response of the digital system can easily be obtained. For instance for 2-D digital systems, defined by relations (27) or (30), the frequency response is resulting to be

$$H(e^{j\omega_1}, e^{j\omega_2}) = \sum_{k_1=0}^{N_1-1} \sum_{k_2=0}^{N_2-1} h(k_1, k_2) e^{-j(k_1\omega_1 + k_2\omega_2)}, \qquad (33)$$

which is a continuous function of ω_1 and ω_2, *periodic* with period 2π.

A very important property of 2-D digital systems is represented by their *stability*. The BIBO (bounded input, bounded output) stability criterion can be expressed in the form

$$\sum_{n_1=-\infty}^{\infty} \sum_{n_2=-\infty}^{\infty} |h(n_1, n_2)| < \infty, \qquad (34)$$

where $\{h(n_1 n_2)\}$ is the impulse response of the 2-D system.

If the system is of the type (27) or (30), condition (34) is always satisfied.

For systems defined by relations (28) or (31) the stability is representing a problem quite more difficult with respect to the 1-D case. Indeed, in the 2-D case the formulation of the stability conditions, connected to relation (34), does not directly produce an efficient stability test, as in the 1-D case, due to the lack of an appropriate *factorization theorem of algebra* [1]. There are, however, different approaches for solving this problem: one useful method is based on the use of the properties of the *complex cepstrum* [1, 2].

3. Digital transformations

A *discrete linear transformation* can be defined through the following relations (1-D and 2-D) [3, 4]:

$$F(k) = \sum_{n=0}^{N-1} f(n)A(n, k),$$ (35a)

$$F(k_1, k_2) = \sum_{n_1=0}^{N_1-1} \sum_{n_2=0}^{N_2-1} f(n_1, n_2)A(n_1, n_2; k_1, k_2),$$ (35b)

where $f(n)$ or $f(n_1, n_2)$ are the input data (in general, the samples of the signal or the image to be processed) and $A(n, k)$ or $A(n_1, n_2; k_1, k_2)$ is the operator or *forward transform kernel* defining the transformation. If the transformation is *invertible*, as we suppose, the inverse transformation is expressed by

$$f(n) = \sum_{k=0}^{N-1} F(k)B(n, k),$$ (36a)

$$f(n_1, n_2) = \sum_{k_1=0}^{N_1-1} \sum_{k_2=0}^{N_2-1} F(k_1, k_2)B(n_1, n_2; k_1, k_2),$$ (36b)

where $B(n, k)$ or $B(n_1, n_2; k_1, k_2)$ represents the inverse operator or *inverse transform kernel*.

2-D transformations are called *separable* if $A(n_1, n_2; k_1, k_2)$ and $B(n_1, n_2; k_1, k_2)$ can be decomposed in the following way:

$$A(n_1, n_2; k_1, k_2) = A_c(n_1, k_1)A_r(n_2, k_2),$$ (37)

$$B(n_1, n_2; k_1, k_2) = B_c(n_1, k_1)B_r(n_2, k_2),$$ (38)

where the subscripts r and c indicate row and column 1-D transform operations.

Many discrete linear transformations have been defined in a precise way, along the general relations (35) and (36): the more important transformations such as Fourier, Hadamard–Walsh, Haar, Karhunen–Loéve transforms are briefly presented in the following.

The *discrete Fourier Transform* (DFT) is defined, in 1-D and 2-D forms, as [1, 4]

$$F(k) = \sum_{n=0}^{N-1} f(n) e^{-j(2\pi kn/N)}, \tag{39}$$

$$F(k_1, k_2) = \sum_{n_1=0}^{N_1-1} \sum_{n_2=0}^{N_2-1} f(n_1, n_2) e^{-j2\pi(k_1 n_1/N_1 + k_2 n_2/N_2)}. \tag{40}$$

It is interesting to observe that the above transforms are obtained directly from the z-transforms of the sequences $\{f(n)\}$ or $\{f(n_1, n_2)\}$, taking N equidistant points along the unit circle, or N_1 and N_2 equidistant points along the two unit circles. The *inverse discrete Fourier transform* (IDFT), in 1-D and 2-D forms, has the following expression:

$$f(n) = \frac{1}{N} \sum_{k=0}^{N-1} F(k) e^{j(2\pi kn/N)}, \tag{41}$$

$$f(n_1, n_2) = \frac{1}{N_1 N_2} \sum_{k_1=0}^{N_1-1} \sum_{k_2=0}^{N_2-1} F(k_1, k_2) e^{j2\pi(k_1 n_1/N_1 + k_2 n_2/N_2)}. \tag{42}$$

These inverse relations can be verified directly, by substituting for $F(k)$ in eq. (41) expression (39), or for $F(k_1, k_2)$ in eq. (42) expression (40), and using the *orthogonality relations* for the exponential functions.

In the practical application of the above DFT and IDFT transforms, the n, n_1 and n_2 indices correspond to discrete time or space values ($t = nT$, $x = n_1 X$ and $y = n_2 X$, T being the time sampling interval and X the space sampling interval), while the k, k_1 and k_2 indices correspond to discrete frequencies ($\nu = k\Delta\nu$, $\nu_1 = k_1\Delta\nu$ and $\nu_2 = k_2\Delta\nu$, $\Delta\nu$ being a constant frequency or space frequency interval) (see also section 2).

With suitable symmetry properties in the data sequences (samples of a signal or an image) the *discrete cosine transform* (DCT) can be defined as

$$F(k) = 2 \sum_{n=0}^{N-1} f(n) \cos\left[\frac{k\pi}{N}\left(n + \tfrac{1}{2}\right)\right], \tag{43}$$

$$F(k_1, k_2) = 2 \sum_{n_1=0}^{N_1-1} \sum_{n_2=0}^{N_2-1} f(n_1, n_2) \cos\left[\frac{k_1\pi}{N_1}\left(n_1 + \tfrac{1}{2}\right)\right]$$

$$\times \cos\left[\frac{k_2\pi}{N_2}\left(n_2 + \tfrac{1}{2}\right)\right]. \tag{44}$$

The *Hadamard transform* is based on the properties of the Hadamard matrix (square form with elements equal to ± 1, having orthogonality among the rows and the columns) [3]. A normalized Hadamard matrix H, of $N \times N$ size satisfies the relation

$$HH^T = 1. \tag{45}$$

The orthonormal Hadamard matrix of lowest order is the 2×2 Hadamard matrix

$$H_2 = \frac{1}{\sqrt{2}} \begin{bmatrix} 1 & 1 \\ 1 & -1 \end{bmatrix}. \tag{46}$$

This transform is also known in the literature as the *Walsh transform* [3]. Through a frequency interpretation of the preceding Hadamard matrix, the number of signs changing along any Hadamard matrix row, divided by 2, is called the *sequence* of the row. The rows of a Hadamard matrix of order N can also be considered to be obtained as samples of rectangular functions having a sub-period equal to $1/N$; these functions are called *Walsh functions*.

The *Haar transform* is based on the Haar matrix, which contains elements equal to ± 1 and 0 [3].

One of the most efficient transforms is represented by the *Karhunen–Loéve transform*, which can be defined in the following way (1-D and 2-D) [3]:

$$F(k) = \sum_{n=0}^{N-1} f(n)K(n, k), \tag{47}$$

$$F(k_1, k_2) = \sum_{n_1=0}^{N_1-1} \sum_{n_2=0}^{N_2-1} f(n_1, n_2)K(n_1, n_2; k_1, k_2), \tag{48}$$

where $K(n, k)$ and $K(n_1, n_2; k_1, k_2)$ kernels satisfy the relations

$$\lambda(k)K(n, k) = \sum_{n'=0}^{N-1} C(n, n')K(n', k), \tag{49}$$

$$\lambda(k_1, k_2)K(n_1, n_2; k_1, k_2)$$
$$= \sum_{n_1'=0}^{N_1-1} \sum_{n_2'=0}^{N_2-1} C(n_1, n_2; n_1', n_2')K(n_1', n_2'; k_1, k_2), \tag{50}$$

where $C(n, n')$ and $C(n_1, n_2; n_1', n_2')$ denote the *covariance functions* of 1-D or 2-D data to be processed, $\lambda(k)$ and $\lambda(k_1, k_2)$ are suitable constants (*eigenvalues* of covariance functions) for fixed values of k, k_1 and k_2.

In the practical application of the above 2-D discrete transformations, in general it is set $N_1 = N_2 = N$.

The previously considered discrete transformations can be, in general, evaluated in a *fast form* [1, 3]. The procedures for computing the transformations in a fast way are based on the division of the computing operations into a sequence of subsequent computing steps, in such a way that the results of the first computing steps (*partial results*) can be utilized repetitively in the subsequent steps. Efficient software packages are available for fast Fourier transform (FFT), fast cosine transform (FCT), fast Walsh transform (FWT) and fast Haar transform (FHT). For instance, the number of operations required to evaluate 1-D FFT results in $N \log_2 N$ instead of N^2 and the number of operations required to evaluate 2-D FWT results in $2N^2 \log_2 N$ instead of $2N^3$. No fast form is available for the Karhunen–Loéve transform.

The interest of applying FFT's (1-D and 2-D) to signal and image processing is represented by the fact that in this way a precise knowledge of the frequency behaviour of the analysed signal and image is achieved. Indeed, the *maximum frequency* of the signal and image *spectrum* can be obtained, which is very useful to verify if a correct sampling was performed, and for several digital processing operations. Further, if a part of the spectrum components is taken out, a suitable filtering operation is obtained: taking out high frequency components a kind of low-pass filtering is performed, while taking out low-frequency components high-pass filtering is achieved.

The above considered fast transforms (FFT, FWT, FHT) can also be very useful to perform a *data reduction* or *data compression* [4]. In fact, the most meaningful transformed data are, in general, concentrated in a relatively small region of the transformed domain with respect to the extension of the original input data: the number of significant transformed data is appreciably smaller than the number of original input data. A trivial example is represented by a signal constituted by a few sine waves: while in the time domain a large number of samples is required to represent the signal, in the frequency domain (use of FFT) few frequency values (amplitude and phase) specify the signal. Analogously, an image constituted by sudden variations in the gray levels (near rectangular form of gray level variation along the rows and/or the columns) will be represented in the Walsh domain (use of FWT) by a few transformed data. Furthermore, the transformed data can be compressed in a stronger way by applying simple algorithms such as *thresholding* (setting to zero the transformed data having a value under a

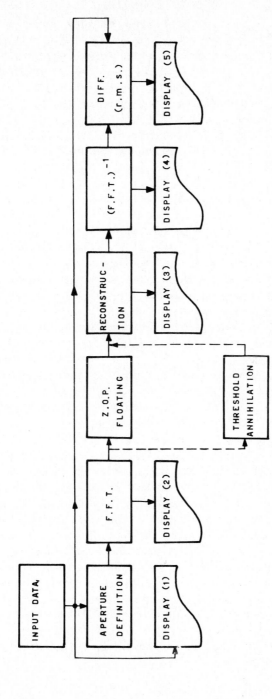

Fig. 1. Block diagram for the application of FFT to input data, using thresholding or ZOP algorithm to reduce the amount of transformed data.

small threshold, i.e. 1%) or *prediction-interpolation*. Figure 1 shows a block diagram for the application of FFT to input data, using threshold-ing or zero-order prediction (ZOP) to reduce the transformed data: the reconstruction of data is performed for comparison with the original ones and the eventual difference values are obtained (rms values). The data available at the different processing steps can be displayed for processing performance analysis [4].

The practical utility of performing data compression is constituted by the more efficient local *storage* of data or *data transmission*, which can be achieved.

4. Design methods of digital filters

Filtering is a very important operation for processing signals and images. By using suitably defined digital systems (see section 2), *digital filters* can be obtained for processing 1-D and 2-D data.

Digital filters can be classified in two main types, according to the properties of their impulse response $\{h(n)\}$ or $\{h(n_1, n_2)\}$: *finite impulse response* (FIR) digital filters and *infinite impulse response* (IIR) digital filters. FIR filters, as defined by relations (8), (12) or (27), (30), have a finite or limited impulse response: a *discrete convolution* is substantially performed between the input data and the impulse response data (coefficients) and no *feedback* on previous outputs is present. These filters are always stable (see relations 19, 34). IIR filters, as defined by relations (9), (13) or (28), (31), have an infinite or unlimited impulse response: this fact can be easily verified by applying at the input of the filter a unit pulse sequence as specified by relations (3) or (26), and observing that—due to the *feedback* action—data (representing the impulse response) are continuously coming from the output of the filter. The stability of IIR filters is to be carefully analysed and, while in the 1-D case stability tests and stabilization procedures are relatively easy to be performed (see subsection 2.1), stability aspects are quite more difficult in the 2-D case (see subsection 2.2).

In the following the more important design methods of digital filters are presented, mainly for the 1-D case with some considerations for the 2-D case.

4.1. Design methods of 1-D digital filters

Let us consider first the design methods of FIR digital filters and second of IIR digital filters.

Several design methods of FIR digital filters have been defined: filters using *windows*, filters with *frequency sampling, optimum filters, equiripple filters*.

FIR digital filters can be easily designed to have *linear phase*. This means that their frequency response (see relation 15) can be considered as the product of a pure delay term and a term which is either real or completely imaginary.

In the *window method* we start from the observation that, the frequency response (15) being periodic, it is possible to represent it as a Fourier series, whose coefficients, according to the sampling theorem, are proportional to the samples of the impulse response of the filter. Therefore, it is possible to obtain, analytically or by using an *approximation method* based on the inverse discrete Fourier transform (IDFT), the sampled impulse response, starting from the frequency-domain specifications. The problem is that the resulting impulse response is, in general, of infinite order and has to be *truncated* to obtain a digital filter that is usable in practice. If the truncation is performed using a *rectangular window* with an abrupt transition between the value equal to 1 in the zone where the impulse response has to be retained, and equal to zero in the truncation region, quite large errors in the frequency response are obtained for the practical applications (this is due to the fact that a convolution in the frequency domain corresponds to the multiplication in the time domain).

The procedure, therefore, should be able to truncate the impulse response by introducing minima errors in the frequency response. To this purpose, the obtained $\{h(n)\}$ sampled impulse-response values are multiplied by the samples $w(n)$ of a window function, whose transform presents a suitable trade-off between the width of the main lobe and the area under the sidelobes [1]. Many window functions have been defined to design FIR digital filters: Hanning window, Hamming window, Blackman window, Fejer window, Lanczos window, etc. Good efficiency is presented, in particular, by the Lanczos-extension window (Cappellini window 1), the Kaiser window and the Weber-type approximation windows (Cappellini windows 2 and 3).

The *Lanczos-extension window* has the following expression [1]:

$$w_1(n) = \left(\frac{\sin \dfrac{2n\pi}{N-1}}{\dfrac{2n\pi}{N-1}} \right)^m , \tag{51}$$

where m is a positive parameter that controls the correction performance

Table 1
Expression of the $w_3(t)$ window: $w_3(t) = at^3 + bt^2 + ct + d$.

Parameter	Time interval	
	$0 \leqslant t \leqslant 0.75$	$0.75 \leqslant t \leqslant 1.5$
a	1.783724	−0.041165
b	−3.604044	1.502131
c	0.076450	−4.591678
d	2.243434	3.651582

(trade-off between the obtained width of the transition band and the errors in the approximation).

The *Kaiser window* has the form [5]:

$$w_K(n) = \frac{I_0\left\{ \omega_a[(N-1)/2]T\sqrt{1 - [2n/(N-1)]^2} \right\}}{I_0\{ \omega_a[(N-1)/2]T \}}, \qquad (52)$$

where I_0 is the modified Bessel function of the first kind and zero order, and ω_a is a positive parameter controlling the window performance.

The *Weber-type approximations* $w_2(t)$ and $w_3(t)$, in analog form, are close representations of a window which gives a minimum value of the uncertainty product, suitably defined [6]. The expression of $w_3(t)$, defined in the time interval 0 to 1.5, is reported in table 1.

Higher efficiency is obtained by using $w_K(n)$, $w_2(n)$ and $w_3(n)$ windows, while good efficiency is achieved by using the $w_1(n)$ window, which has the advantage—for practical applications—to have a very simple expression. An example of a FIR digital filter designed by means of the $w_1(n)$ window is shown in fig. 2.

In the *frequency sampling method* we start from the consideration that DFT relations give a direct method for obtaining the impulse response of a filter whose transfer function is identical to N imposed values at N frequency points. In fact, if we chose N values of the frequency response $\{H(k)\}$, the finite impulse response of this filter is obtained from the IDFT relation (see relation 41) [1],

$$h(n) = \frac{1}{N} \sum_{k=0}^{N-1} H(k)\, e^{j(2\pi kn/N)}. \qquad (53)$$

This filter naturally has a frequency response which assumes the values $H(k)$ at N equidistant values of the frequency axis. Unfortunately, this

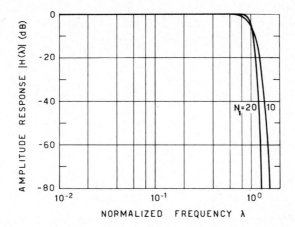

Fig. 2. Example of the frequency response (amplitude) of a FIR digital filter of low-pass type, designed by means of the $w_1(n)$ window ($m = 2$): $N_1 = (N - 1)/2$, normalized frequency $\lambda = \omega/\omega_t$ (where ω_t is the cutoff angular frequency).

direct procedure is not of practical interest because it is impossible to predict the frequency response among the chosen samples and, more-over, the behaviour within these intervals is in general not very satisfactory. The design procedures are based on the suitable selection of the frequency samples. By making reference to the low-pass filter design, the discontinuity in the frequency domain (implying a long impulse response and hence approximation errors) is avoided by allowing some frequency samples within a suitable width transition band to assume values differ-ent from 0 or 1, thereby smoothing the transition between pass-band and stop-band. The transition-band samples are selected to produce a mini-mum error in the approximation of the pass-band and stop-band: *steepest descent algorithm* and *linear programming methods* can be used. Figure 3 shows an example of frequency response of a FIR low-pass digital filter, designed by means of the frequency sampling method using linear programming with 3 variable coefficients in the transition band.

The frequency sampling filters considered above are *optimum filters* in the case they are designed by means of the linear programming methods. The meaning of the word optimum in this case is that they are filters for which the maximum error is minimum in the pass-band and/or stop-band, amongst all filters obtainable by varying frequency samples in the transition band. Indeed, several classes of optimum filters can be defined,

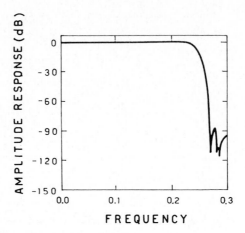

Fig. 3. Example of frequency response (amplitude) of a FIR low-pass digital filter, designed by means of the frequency sampling method (linear programming with 3 variable coefficients in the transition band): $N = 64$.

which are based on different choices of variable parameters and on the *optimality criterion* [1]. An interesting class of optimum filters is represented by the filters which are optimum in the minimum $\|L_p\|$ norm sense, where the definition of the $\|L_p\|$ norm is given by

$$\|L_p\| = \frac{1}{\omega_s} \int_0^{\omega_s} \left[D(e^{j\omega}) - H(e^{j\omega}) \right]^p d\omega, \tag{54}$$

where $D(e^{j\omega})$ is the *desired frequency response*, $H(e^{j\omega})$ is the *approximation function*, whose coefficients have to be obtained to yield the minimum $\|L_p\|$, and ω_s is the sampling angular frequency. The problem is relatively simple for $p = 2$, where the norm reduces to a problem of *minimum square error minimization* and consequently, after differentiation, to the solution of a set of linear equations. If p is different from 2, the problem is becoming non-linear and it is necessary to use a non-linear optimization approach.

In the area of optimum digital filter design, the Chebyshev or *equiripple design method* is of high interest. The method is based on linear programming: the variables in the optimization are now not some samples of the frequency response, as it happens in the frequency sampling method, but the entire set of the coefficients of the impulse response or equivalently all the frequency samples of its IDFT [1].

Several design methods of IIR digital filters have also been defined: methods using *transformations from continuous filters*, methods using *frequency transformations* and *analytic methods*.

The methods using transformations from continuous filters start from the existence of analogue continuous time filters and define suitable transformations to obtain the IIR digital filter. A very important transformation is the *bilinear z-transformation*: in the frequency response of the continuous or analog filter $H(p)$, the complex variable p is replaced by [1]

$$p \to k\frac{1 - z^{-1}}{1 + z^{-1}}, \tag{55}$$

where k is a real positive constant chosen as follows. The angular frequency axis Ω in the analogue case from relation (55) transforms to the unit circle $z^{-1} = \exp(-j\omega T)$, with T the sampling interval, so that

$$\Omega = k \tan\frac{\omega T}{2}, \tag{56}$$

and hence a critical frequency Ω_c in the analogue filter response will correspond to a critical frequency ω_c in the digital filter response, where

$$\Omega_c = k \tan\frac{\omega_c T}{2}. \tag{57}$$

Thus knowing the two critical frequencies, constant k can be calculated from relation (57). Analog filters of specific kind (low-pass, high-pass, etc.) can hence be transformed in digital filters of the same kind (low-pass, high-pass, etc.). Stability of the digital filter obtained in this way is guaranteed if the original continuous filter prototype is stable, because stable regions in one domain are *mapped* into stable regions in the other domain through the transformation (55).

The methods using *frequency transformations* permit to transform a given digital filter of some specific critical frequency to another filter of different critical frequency. This is in general achieved by mapping the complex variable z^{-1} through appropriate functions [7]: low-pass filters can in this way be transformed to high-pass and band-pass filters.

Analytic methods for designing IIR digital filters are based on the fact that the square of the amplitude response of an IIR digital filter can be determined analytically from the kind of approximation employed in the pass-band and in the stop-band [1]. For example *Butterworth filters* are easily defined by setting the square of the amplitude response as

$$|H(e^{j\omega})|^2 = \frac{1}{1 + \left[\dfrac{b(\omega)}{b(\omega_c)}\right]^{2n}}, \tag{58}$$

where $b(\omega)$ can be $\tan(\omega/2)$ or $\sin(\omega/2)$, and the angular frequency ω_c is the required cut-off frequency of the digital filter. From relation (58) it is straightforward to determine the location of the pole and zero positions on the z-plane by substituting the appropriate expression for $b(\omega)$ in terms of z^{-1} (as $\tan(\omega/2) = -j(1 - z^{-1})/(1 + z^{-1})$) and hence the poles in the stable region and the zeros can be chosen to construct the transfer function. In addition to the Butterworth filters, Chebyshev and elliptic filters have been designed by this method.

4.2. Design methods of 2-D digital filters

Let us consider first the design methods of FIR digital filters and second of IIR digital filters.

Regarding the design of FIR digital filters, the methods used for 1-D design can be easily extended to the 2-D case. In particular 2-D digital filters using *windows* can be designed, extending 1-D windows to the 2-D form. A 2-D window, having circular symmetry properties, can be defined starting from a 1-D $w(t)$ window through the relation (continuous form),

$$w(x, y) = w\left(\sqrt{x^2 + y^2}\right). \tag{59}$$

An example of a 2-D FIR digital filter, designed by means of the Lanczos-extension window, $w_1(n_1, n_2)$ ($m = 1.6$), using e.g. relations (27) and (30) with $N_1 = N_2 = 16$, is shown in fig. 4.

2-D FIR *frequency sampling* and *optimum digital filters* can be designed along the lines outlined for the 1-D case, with the greater complexity connected to the 2-D domain.

An interesting technique to obtain 2-D FIR digital filters is based on the *McClellan transformation* from 1-D filters to 2-D filters [1]:

$$\cos \omega = A \cos \omega_1 + B \cos \omega_2 + C \cos \omega_1 \cos \omega_2 + D, \tag{60}$$

where A, B, C and D are real constants. Nearly circular symmetry filters can be obtained in this way.

Regarding the design of 2-D IIR digital filters, the situation is quite more difficult, also for the stability problems outlined in subsection 2.2.

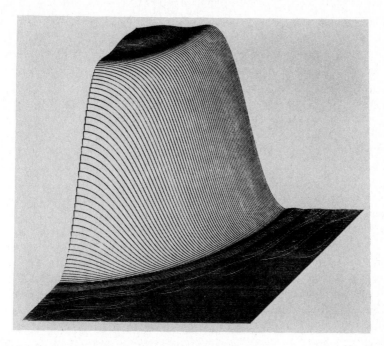

Fig. 4. Example of spatial frequency response (amplitude in one quadrant) of a circular 2-D FIR low-pass digital filter, using the $w_1(n_1, n_2)$ window ($m = 1.6$): $N_1 = N_2 = 16$.

Some design methods are: *differential correction methods*, *factorization methods*, *rotation methods* or *transformations* from 1-D to 2-D filters [1]. An interesting technique to design circularly symmetric 2-D IIR filters is based on the cascading of a number of elementary filters, which are called *rotated filters* (because they are designed by rotating 1-D continuous filters), and using the 2-D z-transform to obtain the corresponding digital filter: stable 2-D IIR digital filters are obtained [8]. Another useful method is based on transformations of the squared magnitude function of 1-D IIR digital filters to 2-D domain and on a suitable decomposition in four stable digital filters [1].

5. Implementation techniques

Implementation techniques are very important for the practical use of the above described digital signal–image processing methods. In the following some aspects are outlined for *software* and *hardware* implementation.

5.1. Software implementation

In the software implementation, general purpose computers or mini-computers are in general used to perform digital signal–image processing through suitable *programs*.

We can distinguish three types of programs regarding digital processing: *design programs, simulation programs* and *processing programs*. Design programs are used to design digital operations such as transformations, digital filtering, data reduction, etc., with the purpose to determine the digital operation parameters best suited for the desired application. Simulation programs are used to evaluate the performance of a given digital operation for processing some types of signals or images, in general by varying the operation parameters and obtaining the corresponding performance results. Processing programs are indeed used to actually perform the desired digital operation on the input signals or images, by means of the parameters defined through the design programs and tested through the simulation programs. As a typical example, fig. 5 shows the *flow-chart* of a program for designing 1-D IIR digital filters by Chebyshev polynomials (see subsection 4.1).

Some general criteria used to define good computer programs are the following ones: *modularity*, i.e. to decompose the program in suitable sub-routines to be used in several different programs, *flexibility*, i.e. to organize the program in such a way that it can be applied for several different processing purposes and *transportability*, i.e. to be able to use the same program on different computer systems.

A very important implementation aspect is represented by the *processing speed*, which is becoming an actual constraint in some practical situations such as *real-time signal processing* or processing of many large-size images. In this framework, the software definition can be very significant for solving speed problems. For instance in digital filtering applications, the use of IIR digital filters is, in general, giving a faster processing with respect to FIR digital filters (IIR filters require a lower number of coefficients) and, if FIR digital filters are to be used for their linear phase and stability advantages, the implementation in the transformed domain (that is the multiplication of the filter response by the signal or image FFT) can be more efficient than the direct convolution implementation, especially when large amounts of data (as images of great size) are to be processed.

For the above purposes, in general computers having fast CPU (central processing unit) and high data transfer-rates with disks and other periph-

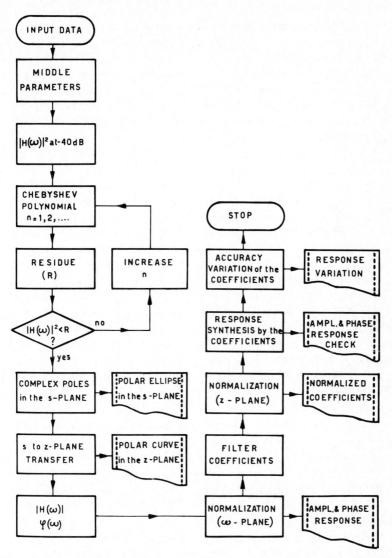

Fig. 5. Flow-chart of a computer program for designing 1-D IIR digital filters by Chebyshev polynomials ($s = p$, Laplace variable).

eral units, are preferred. The evolution trend is, also for large computer systems, to have several *fast computing units* interconnected among them and with large capacity memories by means of fast *multiple data buses*. In particular for matrix operations, as required in image processing, *fast array processors* are more frequently used. As speed indication, while we have now computer systems working at 10–100 MIPS (millions of instructions per second), it is predicted to arrive in a decade to 1000 MIPS and more.

5.2. Hardware implementation

In the hardware implementation, in general suitable integrated circuits or other devices are used to perform the main processing operations, which consist mainly in a suitable number of multiplications and additions with memory capabilities. Indeed, the more important building blocks of a hardware digital processing system are the following ones: memory cells, adders, multipliers, and a programming unit which controls the sequence of operations [1, 2, 9].

The *technology* of *integrated digital circuits* had in last years a very fast development, in terms of complexity and working frequencies, with large scale integration (LSI) and very large scale integration (VLSI) realizations.

Arithmetic circuits are now produced, which have multiplication times (16–24 bits) in the order of about 100 nanoseconds or less, allowing the implementation of fast processing units. At the same time memories are available with capacities which range from thousands of bits to thousands of bytes, with access times ranging from a few nanoseconds to some hundreds of nanoseconds. At lower clock frequencies, MOS technology (e.g. CMOS) allow the integration (using VLSI implementation) of very complex signal processing primitives, which can be used in *pipeline* or *parallel structures* as building blocks of complex digital processing systems.

Indeed, in addition to the more conventional hardware components, today's VLSI technology offers a new powerful device: the *digital signal processor* (DSP) [4]. As shown in fig. 6, the DSP is basically a programmable VLSI circuit consisting of memory devices, arithmetic and control units and input/output (I/O) devices, all integrated on a single chip. Through the appropriate software, the DSP is capable of performing a wide variety of processing operations, including digital transformations, digital filtering and data reduction. Present DSP's are not as fast as

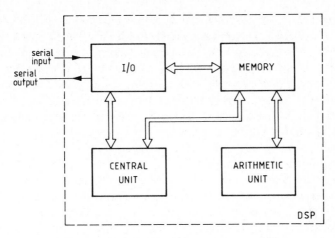

Fig. 6. Structure of a digital signal processor (DSP).

specialized digital hardware circuits, but their processing speed is expected to increase rapidly in the near future.

Microprocessor technology offers another solution, actually widely utilized, for digital processing implementation. The structure of a microprocessor-based system is schematically the same as that shown in Fig. 6; the main difference is that a few integrated circuits are now required to implement the system in place of the unique chip of a DSP. Microprocessors evolved in last years with internal word-lengths from 8 bits to 16 bits and now 32 bits, permitting higher precision digital processing operations, with also higher processing speeds.

Indeed the availability of a *limited number of bits*, in hardware implementations, for representing the input–output data, the processing operators (as the coefficients in digital filters) and the intermediate processing results (in particular regarding the multiplications), constitutes a very important aspect in designing and implementing a specific hardware system [1]. Passing from an infinite precision representation to a finite precision representation, with the connected specific *quantization*, *truncation* and *rounding*, gives origin to specific *errors* (or equivalent *noise*) in the processing results (output data) to be carefully considered and evaluated; for instance, it can be proved that the addition of one bit in the arithmetic unit increases the signal-to-noise ratio (SNR) by approximately 6 dB [1].

As an evolutionary trend in the hardware devices, we can outline that, while now silicon devices are available with delays for one elementary circuit (as a logical gate) in the order of nanoseconds, it is expected to have in one decade from now delays in the order of 100 picoseconds or less. By means of the Josephson effect or Ga–As devices, delays of 10–30 picoseconds are expected. Regarding the number of elementary circuits in a chip (a DSP or microprocessor), from about 10^5 elementary circuits presently implemented it is expected in a decade to arrive to 10^6 and more. These advances regard also peripherals, which are necessary to implement digital processing systems in a complete form. For instance, digital tape recorders are under development (HDDT–high density digital tape), with capacities ten times higher than the ones now available (greater than 10^{11} bits). By using optical technologies, optical records (*optical disks*), obtained by means of laser beams, are available with capacities up to 10^{10} bits/record and with transfer rates of about 10 Mbits/s, while higher capacity records (up to 10^{11}) are under development. Good quality units for signal and image presentation are also produced with interesting evolution trends. From displays having 256 × 256 or 512 × 512 resolution we have now displays with more than 1000 × 1000 pixels, and higher resolutions are expected in the next years.

6. Applications

The application of digital signal processing methods and systems has widely expanded in the last years, becoming very important in several fields such as speech processing, communications, radar–sonar systems, biomedicine, remote sensing, moving object recognition and robotics. In the following, some application aspects regarding these fields are given with typical processing examples.

One of the more significant applications is surely represented by digital processing of *speech signals*. Three main areas of application of digital filtering to speech processing are: a digital model for speech production, in which a time-varying digital filter reproduces the human vocal tract; filtering; band-pass analysis and spectral estimation to define the speech spectrum (as for communications) or to obtain a short-time spectral measurement; *vocoders* (analysis and synthesis) to compress speech signals reducing the amount of data for their representation and transmission (in particular obtaining a lower bandwidth) [1]. Figure 7 shows an example of speech band-pass analysis.

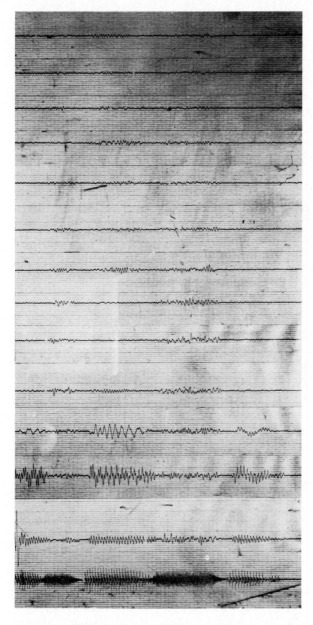

Fig. 7. Example of compressed band-pass analysis of a speech signal: 13 bands (250 Hz each) of the low-frequency speech spectrum (0,3250 Hz) are shown from the bottom to the top, corresponding to the word "processor"; the bottom trace is the input speech signal.

In *communications*, the utility of digital filtering is significant both at the transmitter terminal to define the signal spectrum and at the receiver terminal to reduce the noise introduced by the communication channel. With reference to this last aspect, *digital tone receivers* (as band-pass filters or spectral estimators) [10] and adaptive filtering (as *adaptive equalizers*) constitute important units of the receiving terminal. Data reduction systems are also used to reduce the amount of data and the required bandwidth: as the vocoders for speech signals, compression units are utilized for the transmission of static or time-varying images (TV images), in particular by using 2-D fast digital transformations (FFT, FWT) [4]. For TV images, transmission rates less than 10 Mbits/s can be obtained, which are further reduced in special communications (such as teleconferences) to a few Mbits/s.

In *radar-sonar systems*, digital filtering is very useful to reduce the noise at the receiving unit, extracting the useful signal (echo from the target). In particular, in radar systems *digital moving target indicator* (MTI) *filters*, *digital delay-line cancellers* (to eliminate fixed targets components in Doppler information extraction) and *digital matched*

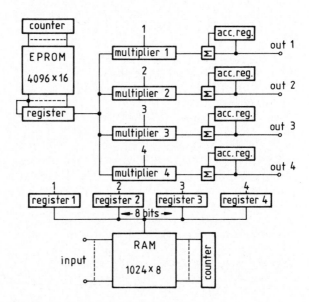

Fig. 8. Block diagram of a hardware digital signal processor performing FIR digital filtering (up to 16 bands) on 4 ECG or EEG signals.

filters for chirp signals (linearly frequency-modulated signals) are actually employed [1].

The digital processing methods and systems are of high interest for processing *biomedical signals and images*. Biomedical signals, such as electro-cardiograms (ECG) and electro-encephalograms (EEG) can be processed by means of digital filters to extract specific frequency compo-

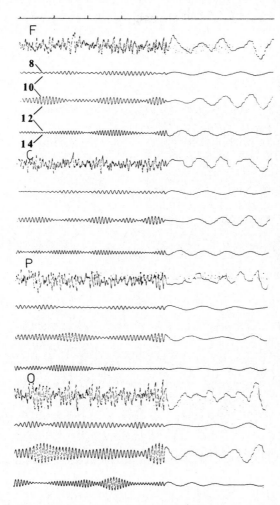

Fig. 9. Example of 3-bands analysis on 4 EEG signals (F, C, P, O).

nents, perform spectral estimation (through a band-pass analysis) or to reduce the noise. An example of a *hardware digital signal processor* performing FIR digital filtering for ECG or EEG signals is shown in fig. 8: 4 input data (8 bits) are sent (at the bottom of the figure) to an input data memory (RAM containing 1024 input data); consequently these input data are sent, through 4 registers, to an arithmetic unit having low-cost serial-parallel multipliers and serial accumulators arranged in 4 parallel processing lines; the coefficients (256 words of 16 bits) of linear phase FIR digital filters taken from an EPROM memory (containing up to 16 complete sets of coefficients, corresponding to 4096 words of 16 bits) are also sent to the arithmetic unit. The processor can indeed

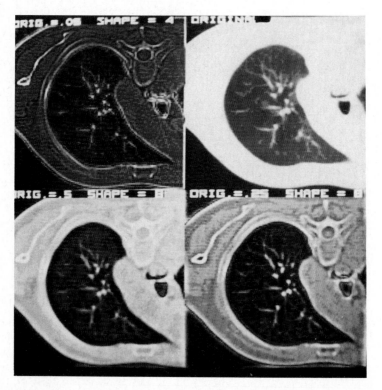

Fig. 10. Example of processing a computer tomography image (top right) by means of a 2-D FIR digital filter of parabolic type: three types of parabolic filters are used producing different enhancement results.

perform in real-time up to 16 band-pass filterings on each signal sampled at the input [11]. An example of 3-bands analysis on 4 EEG signals (F, C, P, O) is reported in fig. 9. 2-D digital filters can be very useful to process biomedical images, in particular to obtain higher quality images (enhancement). An example of processing a computer tomography image by means of a 2-D FIR digital filter of *parabolic type* (assuring good enhancement effects) is shown in fig. 10: three types of parabolic filters are used producing different enhancement results [12].

In the field of *remote sensing* many images and maps are currently collected by platforms aboard aircrafts and satellites through different sensors (optical cameras, multi-spectral-scanners, microwave radiometers, side-looking-radars, etc.). These images and maps are in general to

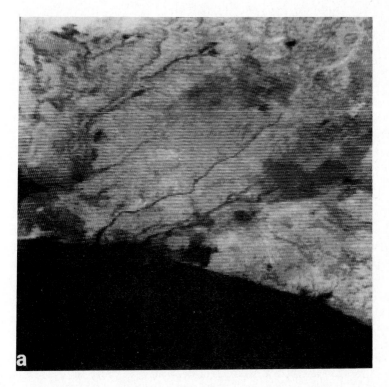

Fig. 11. Example of the application of a 2-D FIR digital filter of high-pass type to a LANDSAT-C image (North Africa): (a) original image; (b, overleaf) filtered image.

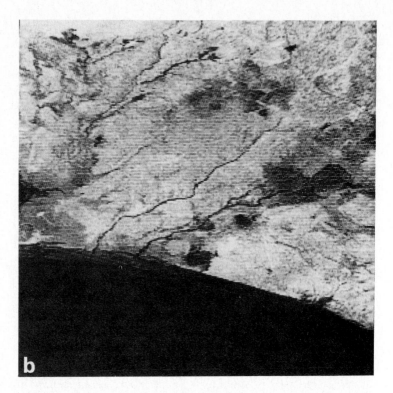

Fig. 11. continued.

be processed to improve their quality (geometric and sensor corrections, noise reduction, enhancement, ...) and to obtain final useful results (extraction of specific regions and land–sea areas as for agriculture investigations or water resource monitoring). 2-D digital filters can be applied to smooth the image data (by means of low-pass filtering), to perform a space frequency correction or to obtain enhancement (by means of high-pass, band-pass or parabolic filtering), extracting also edges and boundaries. Data compression can be further applied, as through digital transformations (FFT, FWT) to reduce the amount of data for more efficient storage or transmission. Figure 11 shows an example of the application of a 2-D FIR digital filter of high-pass type to a LANDSAT-C image (North Africa) [13].

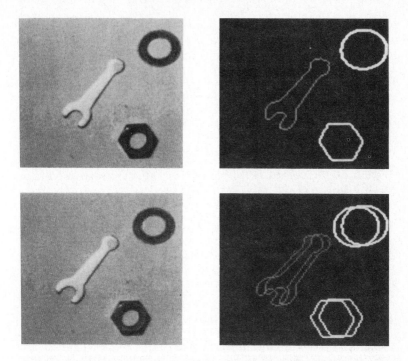

Fig. 12. Example of moving object recognition, using pre-processing digital filtering and boundary FFT matching: two consecutive frames are shown with input digitized images at the left and recognized objects at the right (each object is recognized with a different colour, here appearing as a different gray-level).

Finally, in the field of *moving object recognition and robotics*, data measuring the position of the objects or images taken on the objects can be filtered by means of 1-D and 2-D digital filters, to reduce acquisition noise and to enhance the objects in the background. High-pass or derivative filterings can further extract the edges defining the objects of interest. Figure 12 shows an example of moving object recognition, where pre-processing digital filtering is applied and the object recognition is performed through the matching of the FFT of the *boundary function* of the objects (1-D function represented by the distance of the boundary points from the centroid) with memorized FFT values (only the module of the FFT is used to assure the recognition of the objects showing any *roto-translation*) [14].

References

[1] V. Cappellini, A.G. Constantinides and P. Emiliani, Digital Filters and Their Applications (Academic Press, London, 1980).

[2] V. Cappellini and P.L. Emiliani, Two-dimensional digital systems and applications: the state of the art, in: Multidimensional Systems, ed. S. Tzafestas (Marcel Dekker Inc., New York, 1985) ch. 1.

[3] W.K. Pratt, Digital Image Processing (Wiley, New York, 1978).

[4] G. Benelli, V. Cappellini and E. Del Re, Data compression techniques, in: Data Compression and Error Control Techniques with Applications, ed. V. Cappellini (Academic Press, New York, 1985) ch. 3.

[5] J.F. Kaiser, Digital filters, in: System Analysis by Digital Computer, eds F.F. Kuo and J.F. Kaiser (Wiley, New York, 1966) p. 218.

[6] W. Hilberg and P.G. Rothe, Inf. & Control 18 (1971) 103.

[7] A.G. Constantinides, Electron. Lett. 3 (1967) 487.

[8] J.M. Costa and A.N. Venetsanopoulos, IEEE Trans. Acoust. Speech & Signal Process. ASSP-22 (1974) 432.

[9] V. Cappellini and E. Del Re, Hardware implementation of data compression and error control techniques, in: Data Compression and Error Control Techniques with Applications, ed. V. Cappellini (Academic Press, New York, 1985) ch. 8.

[10] V. Cappellini and E. Del Re, IEEE J. on Selected Areas in Communications SAC-2 (1984) 339.

[11] V. Cappellini and P.L. Emiliani, A special digital processor for ECG–EEG filtering and analysis, in: MEDINFO-83, 1983, eds. J. Van Bemmel, C. Ball and M. Wigertz (North-Holland, Amsterdam, 1983) p. 682.

[12] V. Cappellini and A. N. Venetsanopoulos, Some high-efficiency two-dimensional digital filters with application to biomedical image processing, in: CAR'85, 1985, eds. H.V. Lemke, M.L. Rhodes, C.C. Jaffee and R. Felix (Springer-Verlag, Berlin Heidelberg, 1985) p. 485.

[13] V. Cappellini, Int. J. of Remote Sensing 1 (1980) 175.

[14] V. Cappellini and A. Del Bimbo, Digital processing of time varying images, in: Issues in Acoustic Signal/Image Processing and Recognition, S. Miniato, Italy, 1983, ed. C.H. Chen (Springer-Verlag, Berlin, Heidelberg, 1983) p. 283.

COURSE 7

ARRAY PROCESSING

Johann F. BÖHME

Lehrstuhl für Signaltheorie
Ruhr-Universität
4630 Bochum, Fed. Rep. Germany

J.L. Lacoume, T.S. Durrani and R. Stora, eds.
Les Houches, Session XLV, 1985
Traitement du signal / Signal processing
© *Elsevier Science Publishers B.V., 1987*

Contents

List of symbols*

p	position vector of a sensor
t	time
ω	frequency
k	vector wavenumber
ξ, ξ_m	slowness vectors
v	velocity
λ	wavelength
τ_n, τ_{nm}	time delays
$x(t, p)$	wavefield
$Z(\omega, k)$	random function with uncorrelated increments
$f_x(\omega, k)$	frequency–wavenumber spectrum of the wavefield
$c_x(t, p)$	covariance function of the wavefield
N	number of sensors
$x(t)$	array output
$C_x(\omega)$	spectral matrix of the array output
$s(t)$	signal of a source
$f_s(\omega)$	spectrum of a signal
$h(t, p)$	impulse response of a filter
$H(\omega, k)$	frequency–wavenumber response
$\hat{s}(t)$	estimated signal
a_n	weights
$B(\omega\xi - k)$	array response pattern
$\|B(\omega\xi - k)\|^2$	beam pattern
d	distance between sensors
θ	bearing
K	number of data pieces
$x^k(t)$	piece of array output
$X^k(\omega)$	Fourier transform

---

*As far as possible in order of appearance.

$w(t)$	window
$\hat{\mathbf{C}}_x(\omega)$	estimated spectral matrix
$\hat{f}_x(\omega, \mathbf{k})$	estimated spectrum
M	number of signals
$\mathbf{d}(\mathbf{k}), \mathbf{d}, \mathbf{d}_m, \ldots$	steering vectors
ζ_m	signal (spectral) powers
ν	noise (spectral) power
$\mathbf{I}$	unit matrix
$\mathbf{C}_u$	spectral matrix of noise
$\phi_x(\boldsymbol{\xi})$	slowness vector spectrum
μ_n, λ_n	eigenvalues of $\mathbf{C}_x, \hat{\mathbf{C}}_x$
$\boldsymbol{v}_n, \boldsymbol{u}_n$	eigenvectors of $\mathbf{C}_x, \hat{\mathbf{C}}_x$
$\hat{\phi}$	high-resolution diagrams
$\tilde{\zeta}$	parameters of postulated sources
$\hat{\zeta}_C, \hat{\zeta}_{LS}, \hat{\zeta}_A, \hat{\zeta}_I$	estimates of $\tilde{\zeta}$
$\mathbf{H}, \mathbf{H}(\omega), \mathbf{H}(\boldsymbol{\xi})$	matrix of steering vectors
$\mathbf{1}$	vector of ones
$\mathbf{S}^k$	Fourier-transformed signals
$\mathbf{C}_S$	spectral matrix of signals
$\boldsymbol{\xi}$	vector of waveparameters
$\boldsymbol{\zeta}$	vector of signal spectral parameters
$\boldsymbol{\theta}$	vector of all parameters
$\hat{L}$	likelihood function
$\mathbf{D}, \hat{\mathbf{D}}$	signal matrix, its estimate
$\hat{\mathbf{C}}_s, \hat{\nu}, \boldsymbol{\xi}$	MLE of $\mathbf{C}_s, \nu, \boldsymbol{\xi}$
$\mathbf{P}$	projection matrix
$\hat{q}, \hat{Q}$	criteria for MLE of $\boldsymbol{\xi}$
$\underline{L}$	conditional likelihood function
$\bar{\mathbf{S}}^k, \bar{\nu}, \bar{\boldsymbol{\xi}}$	CMLE of $\mathbf{S}^k, \nu, \boldsymbol{\xi}$
$q, \bar{Q}$	criteria for CMLE of $\boldsymbol{\xi}$
$\mathbf{X}_l^k, \ldots$	short notation of $\mathbf{X}^k(\omega_l), \ldots$
$\text{AIC}(p)$	Akaike criterion
$\hat{\mathbf{G}}, \bar{\mathbf{G}}, \mathbf{F}$	matrices for asymptotic expressions
$\tilde{\mathbf{F}}^{-1}$	approximate Hessian matrix
$\boldsymbol{\xi}^0, \boldsymbol{\xi}^e, \boldsymbol{\xi}^L$, etc.	different estimates
$\theta_m, \rho_m, \text{SNR}_m$	bearings, ranges, signal-to-noise ratios
$\text{E}, \text{Var}, \text{d}\omega, \text{d}\,k,$	mathematical standard symbols
$\partial x, \nabla, \text{Tr},$	
$\text{Det}, \text{O}(x),$	
$\log, \sin, \ldots$	

1. Introduction

This course deals with some basic methods of signal processing for the outputs of an array of sensors in a wavefield. The purpose of this signal processing is to obtain insight into the structure of the waves traversing the array. For example, location parameters of a source generating signals which are transmitted by the waves and properties of these signals have to be measured. Typical applications are the passive sonar, where the sensors are hydrophones in the ocean and the signals are ship noise, and geophysical work with seismometers for measuring earthquakes. Signal processing in this sense is called array processing.

A variety of array processing methods have been extensively studied by researchers in different fields, e.g. in acoustics, radio astronomy, geophysics, electrical engineering and statistics. It is impractical here to survey the literature in all these fields. For example, Knight et al. (1981) presented a tutorial paper on digital signal processing for sonar with a reference list containing 253 items. In this course, we only consider some fundamentals of array processing for passive methods, i.e. targets or objects of interest generate the signals. Simple mathematical models for waves and sensors are used to illustrate concepts, both classical and more recent array-processing methods. When the wavefields are suitably defined as stochastic processes, then array processing becomes a statistical problem. Appropriate statistical methods are therefore applied to predict the behavior of known array-processing techniques and also to design new ones.

Because this course can only present a very limited view on the topic, we ignore the historical perspective and give only some references, mainly in connection with applications in underwater acoustics. Articles by Baggeroer (1978) and Knight et al. (1981) can be used as extensive introductory papers. The monographs by Monzingo and Miller (1980) and Haykin (1985) are recommended. Proceedings of NATO-Advanced Study Institutes on Underwater Acoustics and Signal Processing (e.g. Urban 1985) give good overviews on the state of the art.

Journals with the most important contributions in the field are *Journal of the Acoustical Society of America*, *IEEE Transactions on Acoustics, Speech and Signal Processing*, *and IEEE Journal on Oceanic Engineering*.

An outline of this course follows. In section 2, we first define elementary waves and the parameters of interest, and we discuss the role of time delays. Homogeneous wavefields as suitable stochastic processes consisting of plane coherent waves are introduced. Filtering of wavefields is the next point of interest. Then, we study the first classical method called phased array beamforming which means that time delays are compensated when estimating the signals of a source. Because homogeneous wavefields are stationary processes, the second moments of the wavefields are characterized by a frequency–wavenumber spectrum. The estimation of this spectrum yields the second method which is known as classical beamforming. The properties of both methods are investigated. A disadvantage of the classical methods is the limited ability to resolve the locations of sources close together. Therefore, in section 3, so-called high-resolution methods are discussed. High-resolution diagrams or peak estimates are investigated first. The second possibility is the use of more sophisticated techniques for estimating the signal spectral powers if a large number of source positions is hypothesized. The methods can also be combined. The wave models used up to this point assume uncorrelated signals from different sources. Multipath propagation, for example, cannot be handled with these models. We therefore investigate a different approach to the problem in section 4. The wave model in the frequency domain in this section is parametric. Maximum likelihood and least-squares techniques are applied to estimate the parameters of interest. Asymptotic properties of the estimates and approximations are investigated. Numerical experiments indicate what could happen in applications. In conclusion, some remarks on generalizations, current problems and models in the time domain are given.

2. Wavefields and classical methods

2.1. Elementary waves and time delays

The basis of the investigations of this section is the concept of plane coherent waves. This model describes the waves measured by an observer and generated by a source which is situated far from the observer in an ideal, infinite medium (or ocean). The most simple wave in this sense is an elementary wave.

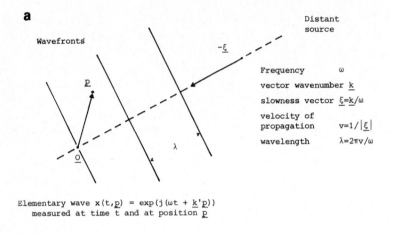

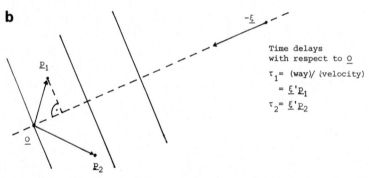

Fig. 1. Wavefronts of elementary waves: (a) parameters of elementary waves, (b) time delays.

An elementary wave measured at a position described by a vector p in Cartesian coordinates and at time t is a complex exponential $\exp[j(\omega t + k'p)]$. In this term, j is the imaginary unit, $\omega = 2\pi f$ is the frequency, k is the vector of wavenumbers or, shortly, the vector wavenumber, and k' denotes the transpose of the column vector k. We note $k = \omega\xi$, where ξ is called the slowness vector, since $v = 1/|\xi|$ is the velocity of propagation of the wavefront in the direction described by the unit vector $-\xi/|\xi|$. In fig. 1a, wavefronts (set of points in the plane with similar phases of the wave for a fixed time) are depicted together with the kind of propagation indicated by the slowness vector ξ. The distance between the wavefronts is the wavelength $\lambda = 2\pi v/\omega$.

In fig. 1b, we have a similar situation, however, with two points of observation described by position vectors p_1 and p_2. The depicted wavefront will propagate from the right to the left and will traverse first position p_2, then p_1 and finally the origin denoted by O. We now ask for the time difference τ_1 which can be measured between traversing the position p_1 and the origin O of a wavefront. τ_2 is correspondingly defined with p_2. The τ_n ($n = 1, 2$) are called time delays. Because the projection of the position vector on the direction of propagation describes the way of the wavefront, we have

$$\tau_n = \xi' p_n = (|p_n|/v) \cos(\xi, p_n) \quad (n = 1, 2). \tag{1}$$

Let us assume that the time delays can be measured. If we know the positions in addition, we can compute the slowness vector of the elementary waves. This means that the direction of the source and the velocity of propagation of the waves can be determined. We assume the vectors p_1 and p_2 to be linearly independent and to be a base of a skew coordinate system. Simple vector calculus yields

$$\xi = \left[\left(|p_2|^2 \tau_1 - p_1' p_2 \tau_2 \right) p_1 + \left(|p_1|^2 \tau_2 - p_1' p_2 \tau_1 \right) p_2 \right]$$

$$\times \left[|p_1|^2 |p_2|^2 - \left(p_1' p_2 \right)^2 \right]^{-1} \tag{2}$$

in the plane. We obtain a similar expression for three-dimensional vectors if three vectors p_1, p_2, p_3 in a general position are available and three time delays τ_1, τ_2, τ_3, respectively, can be measured. Exchanging the role of position vectors and slowness vectors, we can compute a position vector if time delays for elementary waves with different slowness vectors are known. These arguments show that the measurement of time delays is a fundamental problem in array processing.

In applications, however, we never find a simple situation as above. We have to consider the possibility that there could be more than one source in different directions, that waves with different velocities of propagation transmit different signals and that the signals are chromatic with different strengths. The signals are usually additively disturbed by noise from the sensors (sensor noise) and by many weak and not resolvable sources (ambient noise). In this section, we do not consider other effects such as reflections of waves by rigid bodies, etc., or dispersivity of the medium. One problem now is how the time delays, which contain information about the directions of the sources, and the velocities of waves can be determined. G.C. Carter (course 9, this volume) investigates this problem.

We ask in this section more generally for signal processing methods for estimating, e.g.
- the distribution of sources over the directions,
- the distribution of velocities per direction, and
- the distribution of frequencies per direction and velocity.

The methods to be discussed are nonparametric, i.e., the models cannot be characterized by a finite set of parameters.

As an example for applications we look at a passive sonar problem. A surface ship tracks a towed array by a tow cable in the depth. This array is a chain of hydrophones. The sources of interest are submarines emanating machinery noise, flow noise, etc. The hydrophones measure these signals disturbed by a lot of other marine noise. The signals of the submarines have to be detected, a problem discussed by L. Scharf (course 2, this volume). The submarines have to be localized, classified by careful frequency analysis of the signals, tracked, etc. For more details, we refer to Baggeroer (1978) and Knight et al. (1981), and also to Carter (course 9) and Moura (course 3) in this volume.

2.2. Homogeneous wavefields

We are interested in a simple model of wavefields which allows us to investigate the problems indicated in the preceding paragraphs. We assume the wavefield to be a superposition of elementary waves, where arbitrary combinations of frequencies and wavenumbers are tolerated. As with a stationary stochastic process (cf. B. Picinbono, course 1, this volume) which can be seen as a superposition of uncorrelated oscillations, $\exp(j\omega t)$, we assume the wavefield measured at position p and at time t as a superposition of uncorrelated elementary waves,

$$x(t, p) = \int \exp[j(\omega t + k'p)] \, d Z(\omega, k). \tag{3}$$

The integral is understood as a Stieltjes integral in the sense of mean squares and is taken over all frequencies ω and all vector wavenumbers k. $Z(\omega, k)$ is a random function with expectation $EZ(\omega, k) = 0$ and uncorrelated increments $d Z(\omega, k)$. The latter means that the covariances formally satisfy

$$\text{Cov}[d Z(\omega, k), d Z(\nu, g)]$$
$$= \delta(\omega - \nu, k - g) f_x(\omega, k) \, d\omega \, d k/(2\pi)^4, \tag{4}$$

where the vectors have three components, δ denotes the four-dimensional delta-distribution, and $f_x(\omega, k)$ is called the frequency–wavenumber spectrum of the wavefield $x(t, p)$. A wavefield with these properties is called homogeneous.

Equation (3) can be interpreted as a Cramer representation of a (wide sense) stationary random function $x(t, p)$. We have indeed $Ex(t, p) = 0$ and a covariance function

$$c_x(t, p) = Ex(t + s, p + q) x(s, q)$$

$$= \int \exp[j(\omega t + k'p)] f_x(\omega, k) \, d\omega \, d k/(2\pi)^4, \qquad (5)$$

which can be inferred from eq. (4). The spectrum $f_x(\omega, k)$ is then the four-dimensional Fourier transform of the covariance function $c_x(t, p)$. We refer to Yaglom (1962), Adler (1981) and Capon (1969) for specific terminology, formal methods, and applications to array-processing problems.

We now describe the outputs of N sensors of an array. The positions of the sensors are the vectors p_n ($n = 1, \ldots, N$). The sensors are ideal receivers and do not influence the wavefield. The sensor outputs are then given by sampling the wavefield, $x_n(t) = x(t, p_n)$ ($n = 1, \ldots, N$). The array output is, in a short notation, $x(t) = (x_1(t), \ldots, x_N(t))'$. The random function $x(t)$ is of course a (wide sense) stationary vector process and we can ask for its spectral (density) matrix $C_x(\omega)$. The elements of $C_x(\omega)$ are

$$C_x(\omega)_{ni} = \int \exp(-j\omega t) \, c_x(t, p_n - p_i) \, dt$$

$$= \int \exp[jk'(p_n - p_i)] f_x(\omega, k) \, d k/(2\pi)^3, \qquad (6)$$

which follows from eq. (5).

As an example, we note a wavefield consisting of waves with a fixed slowness vector ξ generated by a single source,

$$x(t, p) = \int \exp[j(\omega t + \omega\xi'p)] \, d \tilde{Z}(\omega)$$

$$= \int \exp[j\omega(t + \xi'p)] \, d \tilde{Z}(\omega) = s(t + \xi'p), \qquad (7)$$

where we only integrate over ω. The wavefield is completely described by a delayed version of the signal $s(t)$ of the source received at the origin.

The elements of the spectral matrix $\mathbf{C}_x(\omega)$ are

$$\mathbf{C}_x(\omega)_{ni} = \exp[j\omega\boldsymbol{\xi}'(\boldsymbol{p}_n - \boldsymbol{p}_i)]f_s(\omega), \tag{8}$$

where $f_s(\omega)$ is the spectrum of the stationary process $s(t)$. $\mathbf{C}_x(\omega)$ is therefore a matrix of rank 1. The frequency–wavenumber spectrum of $x(t, \boldsymbol{p})$ is discrete, namely,

$$f_x(\omega, \boldsymbol{k}) = f_s(\omega)\delta(\boldsymbol{k} - \omega\boldsymbol{\xi}). \tag{9}$$

Wavefields can be filtered. If $h(t, \boldsymbol{p})$ is the impulse response of a linear time- and space-invariant system, the filtered wavefield is

$$y(t, \boldsymbol{p}) = \int x(\bar{t}, \boldsymbol{q}) h(t - \bar{t}, \boldsymbol{p} - \boldsymbol{q})\,\mathrm{d}\bar{t}\,\mathrm{d}\boldsymbol{q}. \tag{10}$$

The frequency–wavenumber response of the system is

$$H(\omega, \boldsymbol{k}) = \int \exp[-j(\omega t + \boldsymbol{k}'\boldsymbol{p})] h(t, \boldsymbol{p})\,\mathrm{d}t\,\mathrm{d}\boldsymbol{p}. \tag{11}$$

As with stationary processes, we find the spectrum of $y(t, \boldsymbol{p})$,

$$f_y(\omega, \boldsymbol{k}) = |H(\omega, \boldsymbol{k})|^2 f_x(\omega, \boldsymbol{k}). \tag{12}$$

2.3. Compensation of time delays

After preparing a model, a simple array-processing method is to be discussed. We ask for the signal of a source which is transmitted by waves with a fixed slowness vector $\boldsymbol{\xi}$. We assume the positions $\boldsymbol{p}_1, \ldots, \boldsymbol{p}_N$ of the sensors. From section 2.1, eq. (1) we know the time delays with respect to the origin for these waves, $\tau_n = \boldsymbol{\xi}'\boldsymbol{p}_n$ ($n = 1, \ldots, N$). Looking again at fig. 1b, a method for estimating the signal is obvious. If the sensor outputs $x_n(t)$ are delayed by τ_n to $x_n(t - \tau_n)$, the signal has the same phase in all delayed outputs. Summing up these outputs means to filter the signal out of the set of other signals and noise. More generally, the signal is estimated by compensating of time delays and shading (weighting),

$$\hat{s}(t) = \sum_{n=1}^{N} a_n x_n(t - \tau_n), \quad \tau_n = \boldsymbol{\xi}'\boldsymbol{p}_n. \tag{13}$$

This method is called phased array beamforming, and $\hat{s}(t)$ is the beam signal.

The array of sensors samples the wavefield $x(t, p)$. Phased array beamforming is a filter operation. For, we note

$$\hat{s}(t) = y(t, O) = \int x(\bar{t}, q) h(t - \bar{t}, O - q) d\bar{t} dq, \qquad (14)$$

where

$$h(t, q) = \sum_{n=1}^{N} a_n \delta(t - \xi' p_n, q + p_n), \qquad (15)$$

$$H(\omega, k) = \sum_{n=1}^{N} a_n \exp[-j(\omega\xi - k)' p_n] = B(\omega\xi - k). \qquad (16)$$

$B(\omega\xi - k)$ as a function of k is called the array response pattern. Because of

$$f_y(\omega, k) = |B(\omega\xi - k)|^2 f_x(\omega, k), \qquad (17)$$

the (frequency-) spectrum of $\hat{s}(t)$ can be found by integration,

$$f_{\hat{s}}(\omega) = \int f_y(\omega, k) d k / (2\pi)^3$$

$$= \int |B(\omega\xi - k)|^2 f_x(\omega, k) d k / (2\pi)^3$$

$$= \sum_{n, i=1}^{N} a_n a_i^* \exp[-j\omega\xi'(p_n - p_i)] C_x(\omega)_{ni}. \qquad (18)$$

Equation (18) is a consequence of eqs. (5), (6), (15) and (16). The asterisk means the conjugate (-transpose) operation if complex numbers are under consideration.

Let us interpret the spectrum $f_{\hat{s}}(\omega)$ in eq. (18). It is a smoothed version of the frequency–wavenumber spectrum of the wavefield. Smoothing is done over the vector wavenumbers with a kernel centered at the vector wavenumber $\omega\xi$ of interest. However, the kernel is the so-called beam pattern $|B(\omega\xi - k)|^2$ which is, in general, not well-behaved. From eq. (16) follows that it is a finite sum of periodic functions, and complicated overlapping has to be expected. As an example let us discuss a line array which is an ideal model for a towed array. The sensors are equally spaced on a straight line, for example the first coordinate axis, with a distance d such that $p_n = d(n - 1)e$, and $e = (1, 0, 0)'$. We have, with constant weights $a_n = 1/N$ and $g =$

$(g_1, g_2, g_3)'$,

$$|B(g)|^2 = \left| \frac{1}{N} \sum_{n=1}^{N} \exp[-j(n-1)g_1 d] \right|^2. \tag{19}$$

This function is rotationally symmetric around the array axis given by the unit vector e. It is furthermore periodic in $g'ed = g_1 d$ with a period 2π, i.e., periodic in the wavenumber in e-direction with period $2\pi/d$, or periodic in the wavelength in e-direction with period d. For fixed ω, d, $v = 1/|\xi|$ and polar coordinates with respect to e, the function is periodic in $\sin\theta$ with a period λ/d. Here λ is the wavelength in the direction of propagation and θ the bearing, i.e., $\pi/2 - \theta$ is the angle between the axis and the direction of the source. In fig. 2, the last case is illustrated for a seven-element array and the most advantageous case, where $d = \lambda/2$. The gain

$$10 \log |B(\omega\xi - k)|^2 = 10 \log \left| \frac{1}{7} \sum_{m=0}^{6} \exp[-j\pi m(\sin\theta_0 - \sin\theta)] \right|^2 \tag{20}$$

is depicted as a function of bearing θ, where θ_0 is the bearing corresponding to ξ, i.e. to the source of interest.

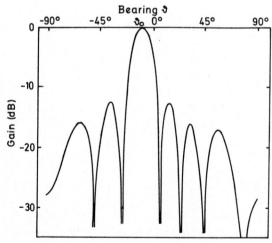

Fig. 2. Beam pattern of a seven-element line array with sensors equispaced by one half of the wavelength, steered to bearing θ_0.

Modifications and extensions of phased array beamforming are used in applications, for example, in sonar systems. We mention null-steering processors and split-beam techniques (cf. Baggeroer 1978).

2.4. Spectrum estimation

The investigations in sections 2.2 and 2.3 indicate the key role of the frequency–wavenumber spectrum $f_x(\omega, k)$ for the description of the properties of a wavefield $x(t, p)$. In our model, $f_x(\omega, k)$ describes the distribution of discrete sources over the directions, the density of the distribution of non-resolvable sources, e.g., ambient noise, the distribution of velocities of propagation per direction and the spectrum of the sum of signals per direction and velocity. The spectrum $f_x(\omega, k)$ contains the information about the wavefields in which we are interested. We should, therefore, try to estimate $f_x(\omega, k)$. It is well known (cf. Capon 1969), that we can similarly perform spectrum estimation as for time series (cf. B. Picinbono, course 1, this volume).

We assume K successively measured pieces of the array output which are denoted by $x^k(t) = (x_1^k(t), \ldots, x_N^k(t))'$ $(k = 1, \ldots, K)$. Finite Fourier transforms with a normalized window $w(t)$ are first applied,

$$X_n^k(\omega) = \int w(t) \, x_n^k(t) \exp(-j\omega t) \, dt$$

$$(k = 1, \ldots, K; n = 1, \ldots, N). \quad (21)$$

By use of the vector $X^k(\omega) = (X_1^k(\omega), \ldots, X_N^k(\omega))'$, we estimate the spectral matrix $C_x(\omega)$ of the array output $x(t)$ by

$$\hat{C}_x(\omega) = \frac{1}{K} \sum_{k=1}^K X^k(\omega) \, X^k(\omega)^*. \quad (22)$$

The frequency–wavenumber spectrum is finally estimated by a double Fourier transform of the elements $\hat{C}_x(\omega)_{ni}$ of $\hat{C}_x(\omega)$,

$$\hat{f}_x(\omega, k) = \sum_{n, i=1}^N a_n a_i^* \hat{C}_x(\omega)_{ni} \exp[-jk(p_n - p_i)]. \quad (23)$$

If we define the shaded steering vector for the vector wavenumber k by

$$d(k) = (a_1 \exp(-jk'p_1), \ldots, a_N \exp(-jk'p_N))', \quad (24)$$

eq. (23) can be written as

$$\hat{f}_x(\omega, k) = d(k)^* \hat{C}_x(\omega) \, d(k), \quad (25)$$

which is frequently called classical beamforming. A third expression for this estimate is found if we first apply a finite four-dimensional Fourier transform

$$X^k(\omega, \mathbf{k}) = \sum_{n=1}^{N} a_n \int w(t)\, x_n^k(t) \exp[-\mathrm{j}(\omega t + \mathbf{k}'\mathbf{p}_n)]\, \mathrm{d}t, \qquad (26)$$

and then compute

$$\hat{f}_x(\omega, \mathbf{k}) = \frac{1}{K} \sum_{k=1}^{K} |X^k(\omega, \mathbf{k})|^2. \qquad (27)$$

Next, we have to discuss statistical properties of the estimates. Some of them are known, e.g. from Capon (1969), more details can be found in work by Brillinger (1981). If we define the spectral window $W(\omega) = \int w(t)\exp(-\mathrm{j}\omega t)\,\mathrm{d}t$, the mean (expected value) of the estimate (23) is calculated for a given ω and $\mathbf{k} = \omega\boldsymbol{\xi}$ to be

$$\mathrm{E}\hat{f}_x(\omega, \omega\boldsymbol{\xi}) = \int |W(\omega - \nu)|^2 |B(\omega\boldsymbol{\xi} - \mathbf{g})|^2 f_x(\nu, \mathbf{g})\, \mathrm{d}\nu\, \mathrm{d}\mathbf{g}/(2\pi)^4. \qquad (28)$$

The mean is also a smoothed version of the true frequency–wavenumber spectrum. The integration is executed over frequencies and wavenumbers. The kernel contains, besides the absolutely squared spectral window, the beam pattern which is the absolutely squared array-response pattern from eq. (16). The problems caused by this smoothing are similar to those discussed in connection with eq. (19). The variance of $\hat{f}_x(\omega, \mathbf{k})$ can be expressed in a short notation by

$$\mathrm{Var}\, \hat{f}_x(\omega, \mathbf{k}) = \mathrm{O}(1/K), \qquad (29)$$

which means that it approaches to 0 not slower than $1/K$ for $K \to \infty$. Distributional properties, especially for the random variables in eq. (21) are discussed in section 4.

There is an interesting connection between phased array beamforming and classical beamforming. If we try to estimate the spectrum $f_{\hat{s}}(\omega)$ of the beamsignal $\hat{s}(t)$, cf. eqs. (18) and (13), respectively, in the same manner, we obtain for successive pieces $\hat{s}^k(t)$ of the beamsignal,

$$\hat{S}^k(\omega) = \int w(t)\, \hat{s}^k(t) \exp(-\mathrm{j}\omega t)\, \mathrm{d}t \quad (k = 1, \ldots, K), \qquad (30)$$

$$\hat{f}_{\hat{s}}(\omega) = \frac{1}{K} \sum_{k=1}^{K} |\hat{S}^k(\omega)|^2. \qquad (31)$$

The mean is

$$E\hat{f}_{\hat{s}}(\omega) = \int |W(\omega - \nu)|^2 |B(\nu\xi - g)|^2 f_x(\nu, g)\, d\nu\, dg/(2\pi)^4.$$

(32)

The variance behaves as $1/K$ too. Expression (31) is nearly the same as eq. (28). The only difference is that $\omega\xi$ is replaced by $\nu\xi$. An interpretation in connection with spectrum estimation is that phased array beamforming and classical beamforming can only result in similar results in special cases, for example, if the wavefield is narrowband with respect to frequencies and wavenumbers. Incidentally, related results can be derived for spectra which are evolutionary with time (cf. Böhme 1979).

In fig. 3, the results of a numerical example are depicted. A line array with 15 sensors spaced, for a fixed frequency ω, with a distance $d = \lambda/2$ is assumed. We further assume a wave model in the plane with signals from discrete sources disturbed only by independent sensor noise. The situation can be described generally by means of the spectral matrix of the array output,

$$C_x(\omega) = \sum_{m=1}^{M} \zeta_m(\omega)\, d_m d_m^* + \nu(\omega)\, I,$$

(33)

where $\zeta_m(\omega)$ is the spectral power of a signal m measured at the origin,

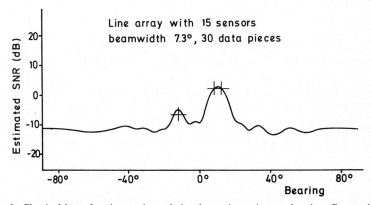

Fig. 3. Classical beamforming: estimated signal-to-noise ratio over bearing. Crosses indicate exact data of signals.

d_m is the steering vector (see eq. 24) with weights $N^{-1/2}$ and vector wavenumber $k = \omega \xi_m$. The slowness vector ξ_m characterizes the waves transmitting the signal. Finally, $\nu(\omega)$ is the spectral noise power and $\mathbf{I}$ the unit matrix. In our experiment, we have $N = 15$ sensors, $M = 3$ signals, and the constant velocity of propagation $v = 1/|\xi|$ such that the bearing θ is the only parameter of the direction vector $\xi v = (\sin \theta, \cos \theta)$. The bearings of the signals are $\theta_1 = -12°$, $\theta_2 = 8°$, $\theta_3 = 12°$ with a signal-to-noise ratio $\mathrm{SNR}_m = 10 \log(\xi_m/\nu) = -6\,\mathrm{dB}, 3\,\mathrm{dB}, 3\,\mathrm{dB}$, respectively. The data are pseudo-random vectors $X^k(\omega)$ which are complex normally distributed with zero mean and a covariance matrix $\mathbf{C}_x(\omega)$ corresponding to the model chosen. A number of $K = 30$ data pieces (or degrees of freedom, DOF) are used. In fig. 3, the diagram of $10 \log(d * \hat{\mathbf{C}}_x d/\nu - 1)$ as a function of the bearing θ is depicted, where eqs. (22) and (25) are used. The crosses indicate the exact data of the signals.

Figure 3 illustrates a typical property of classical beamforming. Sources close together in bearing cannot be resolved, for example, if the difference in bearing is smaller than the beamwidth. The beamwidth is defined by the 3 dB-points of the main lobe of the beampattern, cf. fig. 2, and is about $[51(\lambda/d)/(N - 1)]°$ for a line array. This property can also be concluded if we analyze

$$\mathrm{E}\hat{f}_x = \mathrm{E}d * \hat{\mathbf{C}}_x d \approx d * \mathbf{C}_x d = \sum_{m=1}^{M} \zeta_m |d_m^* d_m|^2 + \nu. \tag{34}$$

The last topic of this section is the numerical calculation of the different beamforming formulas. The phased array beamformer (eq. 13) usually has, e.g. for surveyance purposes, to be computed for a large number of possible slowness vectors ξ. The resulting time delays are approximated by an integer-multiple of a small delay. Then, a hard-wired tapped-delay line multichannel processor is used, for example, in some sonar systems. If only narrowband signals are expected, the computation can be approximately executed in the frequency domain via fast Fourier transforms (FFT), cf. Dudgeon (1977) as an introduction. To compute classical beamforming, two formulas, (23) and (27), can be used. If $\hat{f}_x(\omega, \omega \xi)$ is to be computed for many frequencies and slowness vectors, eq. (27) is most efficient via the use of a four-dimensional FFT followed by suitable interpolations. The dimension of the FFT can be reduced depending on the dimension of the array. For example, a line array only requires a two-dimensional FFT. The FFT's are computed by special purpose computers in many sonar systems. From a statistical point of view, however, it is more suitable to compute the four-dimensional FFT and then to smooth the spectral values over all wavenumbers which are overlapped by the beampattern. This method decreases the variance of the estimate.

Up to now we assumed a stationary wave model consisting of plane waves. For applications as a passive sonar, the sources are located not very distant from the array such that the wavefronts are spherical. The time delays then depend on directions and ranges. Though the beamformers discussed in this section were designed for direction estimation for plane wave arrivals, they can be used for range estimation in addition. For example, the classical beamformer (eq. 25) is then a rough description of the distribution of spectral power over directions and ranges. The accuracy of a range estimation is however very low. Generally, range estimation is a difficult problem which is discussed by Carter (course 9, this volume) and also in section 4.5 of this course.

3. High-resolution methods

3.1. High-resolution diagrams

To overcome the bad resolution of classical beamforming, many ideas and partly successful methods are known from the literature. In this section, we discuss methods resulting in diagrams which are comparable with the one in fig. 3; however, which should have sharper peaks at points indicating sources. They are called high-resolution diagrams or peak estimates.

We begin with a model for the spectral matrix of the array output, a little bit more general than eq. (33),

$$C_x = \sum_{m=1}^{M} \zeta_m d_m d_m^* + C_u. \tag{35}$$

We omit the notation of ω because it is fixed in the following discussion. The first term of the sum describes the spectral matrix induced by the signals and the second, C_u, that of all noise. Because of $k = \omega \xi$, we now denote the steering vectors (eq. 24) by $d(\xi)$. Let us remember eqs. (6) and (9), and that noise is also a homogeneous wavefield. Thus, eq. (35) has a spectral representation,

$$C_x = \int \left[\sum_{m=1}^{M} \zeta_m \delta(\xi - \xi_m) + \phi_u(\xi) \right] d(\xi) d(\xi)^* \, d\xi. \tag{36}$$

The expression in brackets [·] can be interpreted as a slowness-vector spectrum $\phi_x(\xi)$ of C_x. This spectrum has one part with δ-peaks describing the signals and a second part $\phi_u(\xi)$ describing noise and which is assumed to be a smooth function of ξ. Generally, $\phi_u(\xi)$ is not uniquely determined. For example, with $C_u = I$ the three-dimensional Fourier transform $\Phi_u(\eta)$ of $\phi_u(\xi)$ over ξ must only satisfy $\Phi_u(-\omega(p_n - p_i)) = \delta_{ni}$ $(n, i = 1, \ldots, N)$, where δ_{ni} is Kronecker's delta.

For estimating the parameters ξ_m of the sources if $\hat{C}_x$ is given, every diagram over ξ computed from $\hat{C}_x$ and showing only peaks in the neighborhood of the true signal parameters is suited. Such diagrams are high-resolution diagrams or peak estimates. In general, however, high-resolution diagrams cannot be used to estimate the distribution of power over ξ.

We remark that related spectral representations can be found in special situations. If, for example, $v = 1/|\xi|$ is constant and known, we take polar coordinates for ξ and obtain angle spectra, in the space over azimuth and elevation and in the plane over bearing.

Most of the known peak estimates can be motivated by some properties of model (33) if the number M of signals is smaller than N, the number of sensors. Let μ_n and v_n $(n = 1, \ldots, N)$ be the eigenvalues and corresponding orthonormal eigenvectors of the non-negative (definite) matrix C_x, i.e. $C_x v_n = \mu_n v_n$ and $v_n^* v_i = \delta_{ni}$, where $\mu_1 \geq \mu_2 \geq \cdots \geq \mu_N \geq 0$. Then, we have $\mu_{M+1} = \cdots = \mu_N = v$ and, if $\mu_M > \mu_{M+1}$ that the eigenvectors v_n $(n = M + 1, \ldots, N)$ are orthogonal to all steering vectors $d_m = d(\xi_m)$ $(m = 1, \ldots, M)$. The eigenvalues λ_n and eigenvectors u_n of the estimate $\hat{C}_x$ have, asymptotically for an increasing number K of DOF, similar properties (cf. Brillinger 1981, Böhme 1983a, Bienvenue and Kopp 1983). Therefore, they are frequently used to define peak estimates.

We use them too and consider first that $\hat{C}_x$, as every Hermitian matrix, can be written in the form

$$\hat{C}_x = \sum_{n=1}^{N} \lambda_n u_n u_n^*, \tag{37}$$

using its eigenvalues $\lambda_1 \geq \lambda_2 \geq \cdots \geq \lambda_N$ and corresponding eigenvectors $u_1, \ldots, u_N$. To complete, we first formulate classical beamforming (eq. 25) by means of this eigensystem, where $d = d(\xi)$,

$$\hat{f}_x = d^* \hat{C}_x d = \sum_{n=1}^{N} \lambda_n |u_n^* d|^2. \tag{38}$$

Capon (1969) introduced a high-resolution diagram which is also called adaptive beamforming,

$$\hat{\phi} = \left(d * C_x^{-1} d \right)^{-1} = \left(\sum_{n=1}^{N} \lambda_n^{-1} |u_n^* d|^2 \right)^{-1}. \tag{39}$$

In this formula, we assume that the inverses exist. The following methods apply the special properties of the eigensystem of C_x discussed in the preceding paragraph. They are sometimes called signal subspace, eigenstructure methods or orthogonal beamforming. Owsley (1973) and Liggett (1973) investigated

$$\hat{\phi} = |u_n^* d|^2 \quad (n = 1, 2, \ldots) \tag{40}$$

for the strongest eigenvalues $\lambda_1, \lambda_2, \ldots$. If the number M of signals is known, the following diagram can be suitable,

$$\hat{\phi} = \sum_{m=1}^{M} |u_m^* d|^2. \tag{41}$$

The orthogonality of the eigenvectors v_n $(n = M + 1, \ldots, N)$ to the steering vectors d_m of the signals motivate methods applying the approximate nulls of $|u_n^* d|^2$ with respect to the arguments ξ for $n > M$. The diagram

$$\hat{\phi} = |u_N^* d|^{-2} \tag{42}$$

is used by Pisarenko (1973), Bienvenue (1979), Cantoni and Godara (1980), etc. If again the number of signals is known, we can take

$$\hat{\phi} = \left(\sum_{n=M+1}^{N} |u_n^* d|^2 \right)^{-1} \tag{43}$$

and refer to the work of Schmidt (1979), Ziegenbein (1979), etc. There exists a variety of modifications and generalizations, e.g. for the multifrequency case. Some of them are discussed by Wax et al. (1984). It may be remarked that adaptive beamforming (eq. 39) behaves, for large SNR, as eq. (43) except for a scale factor. We refer to Sharman et al. (1983) and Böhme (1983a), who have investigated the behavior of the peak estimates.

Finally, the maximum entropy method is mentioned (cf. Burg 1967, Lacoss 1971), that was designed for equispaced line arrays. We can write the corresponding diagram as

$$\hat{\phi} = \left| \sum_{n=1}^{N} \lambda_n^{-1} u_{n1} u_n^* d \right|^{-2}, \tag{44}$$

where u_{n1} denotes the first component of u_n. Motivations for this method and its properties are discussed by Kay and Marple (1981). Generalizations of the method, allowing general array configuration, are discussed, for example, by Lang (1981). Connections with adaptive beamforming and with adaptive antennas are discussed by McDonough (1979).

3.2. Estimation of spectral signal powers

Let us re-examine property (34) of classical beamforming for model (33). If the parameters ξ_m of the steering vectors $d_m = d(\xi_m)$ of the sources are known, we can estimate the signal powers ζ_m by $d_m^* \hat{C}_x d_m$ ($m = 1, \ldots, M$) except for some bias. If not, we can postulate a large number L of possible parameters $\xi = \tilde{\xi}_l$ ($l = 1, \ldots, L$) with corresponding steering vectors $\tilde{d}_l = d(\tilde{\xi}_l)$ and compute all values $\tilde{d}_l^* \hat{C}_x \tilde{d}_l$, i.e., the estimates of the signal powers of the postulated sources. These numbers are, on the other hand, L points of the diagram of classical beamforming. The arguments for the M strongest local maxima are used later for estimating the parameters $\xi_1, \ldots, \xi_M$. The idea for methods with a better resolution in ξ compared with classical beamforming is now to find better estimates for the signal powers. d'Assumpcao (1980) and Böhme (1983b) investigated this problem.

Before we review two solutions, we have to look again at model (33) but now defined with the set of postulated steering vectors $\tilde{d}_l$ ($l = 1, \ldots, L$),

$$C_x = \sum_{l=1}^{L} \tilde{\zeta}_l \tilde{d}_l \tilde{d}_l^* + \tilde{\nu} I. \tag{45}$$

We now permit that L is greater than the number N of sensors. The question is whether the parameters $\tilde{\zeta}_1, \ldots, \tilde{\zeta}_L$ and $\tilde{\nu}$ are identifiable for a given C_x. For, we multiply eq. (45) from the left by $\tilde{d}_i^*$ and from the right by $\tilde{d}_i$ and obtain

$$\sum_{l=1}^{L} \tilde{\zeta}_l |\tilde{d}_i^* \tilde{d}_l|^2 + \tilde{\nu} = \tilde{d}_i^* C_x \tilde{d}_i \quad (i = 1, \ldots, L). \tag{46}$$

Using the traces of the matrices in eq. (45), we are able to replace $\tilde{\nu}$ in eq. (46) such that

$$\sum_{l=1}^{L} \tilde{\xi}_l \left(|\tilde{d}_i \tilde{d}_l|^2 - N^{-1} \right) = \tilde{d}_i^* \mathbf{C}_x \tilde{d}_i - N^{-1} \operatorname{Tr} \mathbf{C}_x \quad (i = 1, \ldots, L).$$

(47)

The parameters $\tilde{\xi}_1, \ldots, \tilde{\xi}_L$ must satisfy this linear equations system and must be non-negative. They are uniquely determined if the matrix with elements $|\tilde{d}_i^* \tilde{d}_l|^2$ is non-singular. In the other case, every solution of eq. (46) results in the same matrix given by eq. (45).

Next, we assume that only the estimate $\hat{\mathbf{C}}_x$ is known. The problem is to fit model (45) to $\hat{\mathbf{C}}_x$ in some sense which is called estimation of variance components in statistical literature (cf. Rao 1973). d'Assumpcao (1980) investigated many criteria and corresponding beamformers. The simplest one is found by a least-squares fit,

$$\varepsilon = \operatorname{Tr}\left(\left(\hat{\mathbf{C}}_x - \mathbf{C}_x \right)^2 \right),$$

(48)

which has to be minimized over the parameters. Setting the partial derivatives of ε to the parameters equal to zero gives necessary conditions for the optimum parameters $\hat{\xi}_1, \ldots, \hat{\xi}_L, \hat{\nu}$. We obtain equations similar to eq. (47), where $\tilde{\xi}_l$ and $\tilde{\nu}$ are replaced by $\hat{\xi}_l$ and $\hat{\nu}$, respectively, and $\mathbf{C}_x$ by $\hat{\mathbf{C}}_x$. The solution can be written in a closed form. If $\mathbf{H}$ is the matrix with column vectors $\tilde{d}_1, \ldots, \tilde{d}_L$, $\hat{\xi}_{\mathrm{LS}} = (\hat{\xi}_1, \ldots, \hat{\xi}_L)'$, $\hat{\xi}_{\mathrm{C}} = (\tilde{d}_1^* \hat{\mathbf{C}}_x \tilde{d}_1, \ldots, \tilde{d}_L^* \hat{\mathbf{C}}_x \tilde{d}_L)'$ and $\mathbf{1} = (1, \ldots, 1)'$, then the least-squares estimate is

$$\hat{\xi}_{\mathrm{LS}} = \left[(\mathbf{H}^* \mathbf{H})^{\square} (\mathbf{H}^* \mathbf{H})' - N^{-1} \mathbf{1}\mathbf{1}' \right]^{-1} \left[\hat{\xi}_{\mathrm{C}} - N^{-1} \operatorname{Tr} \hat{\mathbf{C}}_x \mathbf{1} \right].$$

(49)

Here, $\mathbf{A}^{\square}\mathbf{B} = (a_{ij} b_{ij})$ is the Schur direct product of two matrices. The inverse of the matrix in brackets can be replaced by a generalized inverse if this matrix is singular. Finally, noise power is estimated by

$$\hat{\nu}_{\mathrm{LS}} = N^{-1} \left(\operatorname{Tr} \hat{\mathbf{C}}_x - \hat{\xi}_{\mathrm{LS}}' \mathbf{1} \right).$$

(50)

The least-squares estimate has some interesting properties. Because $\hat{\xi}_{\mathrm{C}}$ is the vector of signal power estimates by the classical beamformer, we call it the classical beamformer estimate. Equation (49) means that the least-squares estimate is a linear transformation of the classical beamformer estimate. This transformation corrects the bias of $\hat{\xi}_{\mathrm{C}}$; the means of the components of $\hat{\xi}_{\mathrm{C}}$ are comparable to eq. (34), such that

$$\mathrm{E}\hat{\xi}_{\mathrm{C}} = \left[(\mathbf{H}^* \mathbf{H})^{\square} (\mathbf{H}^* \mathbf{H})' \right] \tilde{\xi} + \nu \mathbf{1},$$

(51)

where $\tilde{\zeta} = (\tilde{\zeta}_1, \ldots, \tilde{\zeta}_L)'$. Consequently,

$$E\hat{\zeta}_{LS} = \left[(H*H)^{\square}(H*H)' - N^{-1}\mathbf{1}\,\mathbf{1}'\right]^{-1}\left[E\hat{\zeta}_C - N^{-1}\mathrm{Tr}\,C_x\right] = \tilde{\zeta},$$
(52)

$$E\hat{\nu}_{LS} = \nu,$$
(53)

i.e., the least-squares estimates are unbiased.

The outputs of a classical beamformer are available in many sonar systems. At times it can be necessary to compute a high-resolution method via calculation of $\hat{\zeta}_{LS}$. One could try to use eq. (49); however, this is impractical because the matrix to be inverted is very large. Therefore, d'Assumpcao (1980) developed an iterative method for solving the least-squares problem of eq. (48) by use of the steepest ascent/descent. Some iterative corrections of the classical beamformer estimate can be sufficient for estimation of location parameters, etc. This method is numerically more efficient, however, convergence is slow. An example is discussed below.

Another criterion used by d'Assumpcao (1980) is as follows. Model (45) is to be chosen such that the product $\hat{C}_x^{-1}C_x$ is as close as possible to the unit matrix I. He defined the inverse fit estimate $\hat{\zeta}_I$ by the solution of the problem that the determinant $\mathrm{Det}(\hat{C}_x^{-1}C_x)$ is to be maximized over the parameters under the constraint, $\mathrm{Tr}(\hat{C}_x^{-1}C_x) = N$. Equivalently, $\mathrm{Det}\,C_x$ can be optimized under the same constraint. Necessary conditions for the optimum parameters $\hat{\zeta}_I$ are easily found via a formulation of the problem with a Lagrange multiplier. They are

$$\tilde{d}_l C_x^{-1}\tilde{d}_l = \tilde{d}_l^* \hat{C}_x^{-1}\tilde{d}_l \quad (l = 1, \ldots, L),$$
(54)

$$\mathrm{Tr}(C_x^{-1}) = \mathrm{Tr}(\hat{C}_x^{-1}).$$
(55)

The right-hand side of eq. (54) is the reciprocal of the adaptive beamformer output used to estimate the signal power for the postulated steering vector $\tilde{d}_l$. The inverse fit estimate $\hat{\zeta}_I$ is then a (non-linearly) corrected version of the adaptive beamformer estimate $\hat{\zeta}_A$ defined correspondingly. Because the equations system (54), (55) seems to have no closed solution for $L > N$, the only way to compute $\hat{\zeta}_I$ is an iterative correction of $\hat{\zeta}_A$. Such a method was developed by d'Assumpcao (1980). That author also conjectured that $\hat{\zeta}_I$ is unbiased, a prediction which contrasts with $\hat{\zeta}_A$.

The interpretations of this section motivate processing techniques frequently used in applications. First, relatively simple and robust esti-

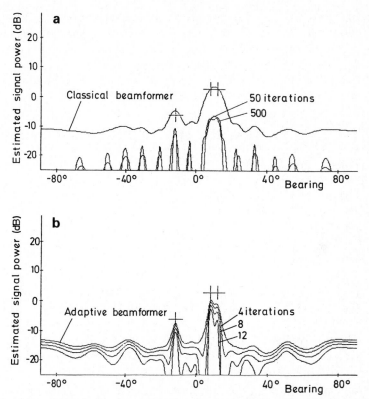

Fig. 4. Estimation of spectral signal powers: estimated signal-to-noise ratio over bearing. Crosses indicate exact data of signals. (a) Classical beamforming estimate and iterations for least-squares estimate; (b) Adaptive beamforming estimate and iterations for inverse fit estimate.

mates are computed (preprocessing) for surveyance purposes, etc. and, second, an iterative enhancement of the simple estimates is tried if some situations are of more interest (postprocessing). A simple numerical experiment was executed to indicate the behavior of the two combined processors discussed in this section. The result of this experiment is illustrated in fig. 4. The situation is the same as described in connection with fig. 3 (section 2.4), where $L = 360$ sources are postulated every 0.5° in the half plane. The classical beamformer estimate in fig. 4a is that of fig. 3. Iterative corrections to approximate the least-squares estimates are executed. The resulting estimates after 50 iterations and 500 iterations (convergence) are depicted in fig. 4a in addition. Figure 4b shows the adaptive beamformer estimate and respectively 4, 8, and 12 iterations for the inverse fit estimate.

We interpret these results and the results of similar computer experiments as follows. By use of the classical beamformer estimate, sources close together, e.g. separated by a half of the beamwidth, cannot be resolved in bearing. Many iterations for a least-squares estimate are required to obtain an estimate that allows separation of the sources. A variety of spurious peaks could mislead an observer in predicting a larger number of existing signals. The adaptive beamformer estimate in fig. 4b has just resolved the sources of interest. Some iterations for the inverse fit estimate give better results. Another experiment indicated that the application of iterations designed for the least-squares estimate to the adaptive beamformer estimate (the better initial estimate) also yields useful results.

A common property of the iteratively enhanced estimates is that they cannot be used for the estimation of the exact signal powers ζ_m in model (33). Parameters $\tilde{\zeta}_l$ of eq. (45) are estimated which are not uniquely defined in general. A further critical point of d'Assumpcao's iterations is the numerical effort per iteration. In our experiments, it is about four times that required for the parametric methods discussed in section 4.4. Those methods also start with relatively simple estimates and enhance these estimates iteratively. In a computer experiment, we used the absolute maxima of Owsley's high resolution diagrams (eq. 40) as initial estimates and enhanced them by Liggett's (1973) unitary transformations, cf. section 4.5 for more details. The accuracy of this method for bearing estimation is comparable to that of the iteratively corrected adaptive beamformer estimate.

4. Parametric methods

4.1. Introductory remarks

Homogeneous wavefields as discussed in sections 2 and 3 consist of uncorrelated plane waves. The classical beamformer (eq. 25) tries to estimate the distribution of sources in the space, the velocities of the waves and the frequency spectra of the signals. Wavefields consisting of curved waves transmitting correlated signals cannot be suitably treated with such models. Correlated signals received from different directions appear, for example, in underwater acoustics by multipath propagation.

A possibility to include such phenomena in a theory is the development of a suitable model of coherent waves and of sensor arrays. The model must be perfectly specified except for some parameters. The problem is to estimate all parameters of interest given a piece of array output of finite length. We then have a special statistical parameter estimation problem, cf. Scharf (course 2, this volume) and Rao (1973) for the statistical background.

C. Carter investigates the problem partly in course 9 in this volume on time-delay estimation. The time delays contain all information about the locations of the sources and the positions of the sensors. Via estimation of time delays, we can estimate, for example, the location parameters. However, if we have M sources and N sensors, NM time delays have to be estimated. The location parameters have then to be fitted to the time delays. In this situation, it may be more suitable to estimate the parameters of interest directly from the data. This was done for range, bearing and velocity estimation by Bangs and Schultheiss (1973), Hahn and Tretter (1973), Schultheiss and Weinstein (1981), Moura (1981), etc., and also by Carter. They all started with the likelihood ratio detector for a signal in noise with known signal and noise spectra. They formulated maximum likelihood estimates and predicted the asymptotic behavior of the estimates, for example, using the Cramer–Rao bound. In this course, we assume that signal and noise spectra are not known and have to be estimated in addition. We discuss two different maximum likelihood estimates. Both, they have the desirable property that the estimation of the location or waveparameters can be separated from that of the signal parameters and that of the noise parameters.

4.2. Data model

A conventional model is used, fig. 5 shows a simple example. Sources $m = 1, \ldots, M$ generate signals which are transmitted by a wavefield. The wavefield has known properties of propagation except for some parameters. The outputs of the sensors $n = 1, \ldots, N$ are, as in section 2.4, Fourier-transformed with a smooth normalized window of length T. For every frequency ω of interest, we have, as in eq. (21), data $X^k(\omega) = (X_1^k(\omega), \ldots, X_N^k(\omega))'$ $(k = 1, \ldots, K)$ of successive pieces of sensor outputs. Correspondingly, $S^k(\omega) = (S_1^k(\omega), \ldots, S_M^k(\omega))'$ $(k = 1, \ldots, K)$ denote the Fourier-transformed signals striking the origin. The array output is assumed to be a zero-mean stationary vector process. The propagation-reception conditions for the signals are described by a linear

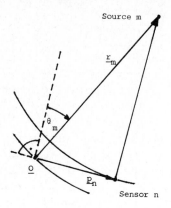

Source m

$\underline{r}_m$

θ_m

$\underline{O}$

$\underline{p}_n$

Sensor n

Time delay

with respect to $\underline{O}$

$\tau_{nm} = (|\underline{r}_m| - |\underline{r}_m - \underline{p}_n|)/v =$

$(\rho_m - (\rho_m^2 + a_n^2 - 2\rho_m a_n \sin\theta_m)^{1/2})/v$

$\rho_m = |\underline{r}_m|$ range of source m

$a_n = |\underline{p}_n|$ distance of sensor n

θ_m bearing of source m

v velocity of propagation

Fig. 7. Time delays for spheric waves in the plane.

system with a $(N \times M)$-matrix $\mathbf{H}(\omega)$ as frequency response. The columns of $\mathbf{H}(\omega)$ are the steering vectors $\boldsymbol{d}_m$ $(m = 1, \ldots, M)$. For example, the (n, m)-th element of $\mathbf{H}(\omega)$ may be $N^{-1/2}\exp(-j\omega\tau_{nm})$, where τ_{nm} is the time delay of the mth signal in the nth sensor. The array output consists of the filtered signals and is additively disturbed by independent sensor noise. The spectral matrices $\mathbf{C}_x(\omega)$ $(N \times N)$ of the array output can then be expressed by

$$\mathbf{C}_x(\omega) = \mathbf{H}(\omega)\mathbf{C}_s(\omega)\mathbf{H}(\omega)^* + \nu(\omega)\mathbf{I}, \tag{56}$$

where $\mathbf{C}_s(\omega)(M \times M)$ is the spectral matrix of the signals and $\nu(\omega)$ is the spectrum of sensor noise. If we say $\mathbf{C}_u = \nu\mathbf{I}$, model (35) is a special case of eq. (56), where $\mathbf{C}_s$ is the diagonal matrix $\mathrm{Diag}(\zeta_m)$.

Let us apply the well-known asymptotic properties of the Fourier-transformed array output if the window length T is large (cf. Brillinger 1981).

(1) $X^1(\omega), \ldots, X^K(\omega)$ are independently and identically complex-normally distributed random vectors with zero mean and covariance matrix $\mathbf{C}_x(\omega)$ as in eq. (56).

(2) The conditional distribution of data $(X^1(\omega), \ldots, X^K(\omega))$ when $(S^1(\omega), \ldots, S^K(\omega))$ is given follows: $X^1(\omega), \ldots, X^K(\omega)$ are independent

random vectors, and $X^k(\omega)$ is complex-normally distributed with mean $H(\omega) S^k(\omega)$ and covariance matrix $\nu(\omega) I$.

(3) For different discrete frequencies $\omega_1 = 2\pi n_l/T$ ($l = 1, \ldots, L$), the $X^k(\omega_l)$ ($k = 1, \ldots, K; 1, \ldots, L$) are independent random vectors.

The parameters have to be described now. The parameters of the propagation-reception conditions or, short, the wave parameters are described by the vector ξ, and we write $H(\omega) = H(\omega, \xi)$. For plane waves, as in section 2, ξ summarizes the components of all slowness vectors of the M sources. In other applications, ξ contains bearings and ranges or, possibly, some sensor positions as components. For a fixed frequency, the parameters of the spectral matrix $C_s(\omega) = C_s(\omega, \zeta)$ are summarized to a vector $\zeta = \zeta(\omega)$ which depends on ω in general. For example, the elements of the Hermitian matrix $C_s(\omega)$ may be the parameters. The spectral noise power $\nu = \nu(\omega)$ is also a parameter. Then for property (1), the parameters of $C_x(\omega) = C_x(\omega, \theta)$ are $\theta = (\zeta, \xi, \nu)$. Beginning with property (2), the components of $S^k = S^k(\omega)$ are not known and can be assumed as parameters, and we note $\theta = (S^1, \ldots, S^K, \xi, \nu)$. In addition, the number M of sources is generally unknown.

4.3. Maximum likelihood estimates

The asymptotic distributional properties (1), (2) and (3) stated in the preceding section allow us to formulate maximum likelihood estimates. The likelihood functions to be maximized are given by the logarithms of the corresponding normal probability densities of the data available. We first investigate the single frequency case and omit the notation of ω. Furthermore, the number M of signals is assumed to be known and to be less than the number N of sensors. Its estimation is discussed at the end of this section.

The application of property (1) yields (except for a scale factor and an additive constant) the likelihood function

$$\hat{L}(\theta) = -\log \text{Det}\, C_x(\theta) - \text{Tr}\big[\hat{C}_x C_x(\theta)^{-1}\big], \tag{57}$$

where the data are collected to the estimate $\hat{C}_x$ of C_x in eq. (22). An MLE $\hat{\theta}$ maximizes $\hat{L}(\theta)$ over θ and satisfies

$$\frac{\partial \hat{L}}{\partial \theta_i}\bigg|_{\hat{\theta}} = \text{Tr}\bigg(C_x^{-1} \frac{\partial C_x}{\partial \theta_i}\bigg)\bigg|_{\hat{\theta}} + \text{Tr}\bigg(C_x^{-1}\hat{C}_x C_x^{-1} \frac{\partial C_x}{\partial \theta}\bigg)\bigg|_{\hat{\theta}} = 0 \tag{58}$$

for all components θ_i of θ. Here, Tr means the trace of a matrix and the usual differential calculus for matrices is applied.

A solution of eq. (58) is, in general, a difficult task. Only in special cases a closed solution or simplifications are possible. Two of them should be discussed. We first assume that the steering vectors themselves are unknown unit vectors (which is, of course, an unrealistic assumption in applications) and furthermore that the matrix C_s is not known. Then, it is easy to show that the MLE $\hat{D}$ of the signal matrix $D = HC_sH^*$ can be constructed by the eigenvalues λ_i, decreasingly ordered, and the eigenvectors u_i of C_x:

$$\hat{D} = \sum_{i=1}^{M} (\lambda_i - \hat{v}) u_i u_i^*, \tag{59}$$

where

$$\hat{v} = \sum_{i=M+1}^{N} \lambda_i/(N - M) \tag{60}$$

is the MLE of v. This solution is not surprising because D has a similar representation with the eigensystem of the exact matrix C_x. The result once again motivates the use of the eigenvalues and eigenvectors of $\hat{C}_x$ for constructing peak estimates as in section 3.1.

The second special case is more relevant for applications and results from the assumption that all elements of C_s are unknown parameters. The idea of constructing the MLE is as follows: we first try to optimize $\hat{L}(C_s, \xi, v)$ over C_s and v, where the parameter vector ξ is fixed. If we have explicit solutions, we put them into the likelihood function and maximize the resulting criterion over ξ. The corresponding solution was indicated by Böhme (1986), assuming that H has a rank M, namely,

$$\hat{C}_s(\xi) = (H^*H)^{-1}H^*\hat{C}_xH(H^*H)^{-1}\big|_\xi - \hat{v}(\xi)(H^*H)^{-1}\big|_\xi, \tag{61}$$

and

$$\hat{v}(\xi) = \text{Tr}\big[(I - P)\hat{C}_x\big]\big|_\xi/(N - M), \tag{62}$$

where

$$P(\xi) = H(H^*H)^{-1}H^*\big|_\xi \tag{63}$$

is the projector into the linear space spanned by the columns of H, i.e. by the steering vectors. Substituting eqs. (61) and (62) in eq. (57) yields

$$\hat{L}\big(\hat{C}_s(\xi), \xi, v(\xi)\big) = -\log \text{Det}\big[P\hat{C}_xP + \hat{v}(I - P)\big]\big|_\xi - N. \tag{64}$$

This means that an MLE $\hat{\xi}$ minimizes

$$\hat{q}(\xi) = \text{Det}\big[P\hat{C}_xP + \hat{v}(I - P)\big]\big|_\xi. \tag{65}$$

The MLE of C_s and v are, consequently, $\hat{C}_s(\hat{\xi})$ and $\hat{v}(\hat{\xi})$.

Obviously, with this approach, the estimation of the wave parameters can be separated from that of the other parameters. We remark that for a fixed ξ, the estimates (61) and (62) are also the solution of a least-squares fit, where $\text{Tr}[(\hat{\mathbf{C}}_x - \mathbf{C}_x)^2]$ has to be minimized.

We now apply property (2) that characterizes a conditional distribution. We formulate a conditional maximum likelihood function (except for a scale factor and an additive constant),

$$\bar{L}(\boldsymbol{\theta}) = -N \log \nu - \frac{1}{\nu K} \sum_{k=1}^{K} |X^k - \mathbf{H}(\xi)S^k|^2, \tag{66}$$

that is to be maximized over the parameters $\boldsymbol{\theta} = (S^1, \ldots, S^k, \xi, \nu)$. The optimum parameters $\bar{\boldsymbol{\theta}} = (\bar{S}^1, \ldots, \bar{S}^K, \bar{\xi}, \bar{\nu})$ are called conditional maximum likelihood estimates (CMLE).

We can use a similar strategy as in the last special case to simplify the problem (cf. Böhme 1983b, 1985a). We first solve for the signals S^k and then for ν and ξ. The solution is

$$\bar{S}^k(\xi) = (\mathbf{H}^*\mathbf{H})^{-1}\mathbf{H}^*X^k\big|_\xi \quad (k = 1, \ldots, K), \tag{67}$$

$$\bar{\nu}(\xi) = \bar{q}(\xi)/N, \tag{68}$$

and $\bar{\xi}$ maximizes

$$\bar{L}(S^1(\xi), \ldots, S^K(\xi), \xi, \bar{\nu}(\xi)) = -N \log(\bar{q}(\xi)/N) - N \tag{69}$$

over ξ, i.e. $\bar{\xi}$ minimizes

$$\bar{q}(\xi) = \text{Tr}[(\mathbf{I} - \mathbf{P})\hat{\mathbf{C}}_x]\big|_\xi. \tag{70}$$

The CMLE's are finally $\bar{S}^1(\bar{\xi}), \ldots, \bar{S}^K(\bar{\xi})$, $\bar{\xi}$ and $\bar{\nu}(\bar{\xi})$. It may be remarked that in this approach the S^k cannot be stably estimated, which contrasts to the estimation of ξ and ν. A stable estimate of the underlying spectral matrix $\mathbf{C}_s$, however, is $\hat{\mathbf{C}}_s(\bar{\xi})$ using eq. (61).

Again, we obtain that the estimation of the wave parameters can be separated from that of the other parameters. It may be remarked that eq. (67) and minimizing of eq. (70) are also the solution of a nonlinear regression problem with the sum of squares given by the sum in eq. (66). The estimation of frequencies or of wavenumbers by nonlinear regression is a known method in the literature (cf. Hinich and Shaman 1972, Tufts and Kumaresan 1980, Nickel, 1983, and van den Bos 1981).

Let us interpret criteria (65) and (70) to be minimized for the MLE $\hat{\xi}$ and CMLE $\bar{\xi}$, respectively, of the waveparameters ξ. First, the MLE $\hat{\xi}$ has to be chosen so that the optimum parameters minimize a generalized variance of the model spectral matrix (56). The special case where only one signal corresponding to a steering vector d yields $P = dd^*$, and

$$\hat{q}(\xi) = \left[(\text{Tr}\,\hat{C}_x - d^*\hat{C}_x d)/(N-1) \right]^{N-1} d^*\hat{C}_x d. \tag{71}$$

This function is monotonically decreasing in $d^*\hat{C}_x d$, the output of a classical beamformer. The MLE $\hat{\xi}$ then maximizes the classical beamformer output. If there are more than one signal, the optimization problem is much more complicated. The CMLE $\bar{\xi}$ minimizes eq. (70). This means, in connection with eq. (68), that the estimated noise power has to be minimized. In the case of a single signal, we again obtain that the optimum $\bar{\xi}$ maximizes the classical beamformer output.

The estimates up to this point are designed for a single frequency ω. A generalization to the multifrequency case is not difficult if we apply property (3) in section 4.2. For different discrete frequencies ω_l, the random vectors $X^k(\omega_l)$ $(k = 1, \ldots, K; l = 1, \ldots, L)$ are independently and normally distributed random vectors. Using a short notation $X_l^k = X^k(\omega_l)$, $H_l(\xi) = H(\omega_l, \xi)$, etc., the likelihood function corresponding to eq. (57) is

$$\hat{L}(\theta) = -\sum_{l=1}^{L} \left\{ \log \text{Det}\,C_{xl}(\theta) + \text{Tr}\big[\hat{C}_{xl} C_{xl}^{-1}(\theta)\big] \right\}. \tag{72}$$

Here, as in the following, we assume that the wave parameters ξ do not depend on the frequency. We easily find necessary conditions for MLE's as in eq. (58). We only note the corresponding solution if all signal spectral matrices C_{sl} are unknown. The criterion (65) has to be generalized as follows:

$$\hat{Q}(\xi) = \sum_{l=1}^{L} \log \text{Det}\big[P_l \hat{C}_{xl} P_l + \hat{v}_l (I - P_l) \big] \Big|_{\xi}. \tag{73}$$

This means that an MLE $\hat{\xi}$ minimizes the geometric mean of generalized variances of the model matrices C_{xl}. The MLE's for C_{sl} and v_l are given by eq. (61) and eq. (62), respectively, if we add the subscript to all matrices and use $\xi = \hat{\xi}$.

We generalize eq. (66) in a similar way, where the parameters now are given by $\theta = (S_1^1, \ldots, S_L^K, \xi, v_1, \ldots, v_L)$. The criterion corresponding to eq. (70) is

$$\overline{Q}(\xi) = \sum_{l=1}^{L} \log \text{Tr}\big[(I - P_l)\hat{C}_{xl}\big] \Big|_{\xi}. \tag{74}$$

A CMLE $\bar{\xi}$ then minimizes the geometric mean of the estimated spectral noise powers.

Finally, we discuss the problem of estimating the number M of sources. Because we started with the principle of maximum likelihood, it seems to be suitable to estimate the number of sources by means of Akaike's (1974) criterion, that is, for a possible number p of sources,

$$\text{AIC}(p) = -2\log\left(\begin{array}{c}\text{maximum}\\\text{likelihood}\end{array}\right) + 2\left(\begin{array}{c}\text{number of free adjustable}\\\text{parameters within the model}\end{array}\right). \tag{75}$$

The criterion has to be computed for $0 \leqslant p \leqslant N - 1$. The smallest number $p = \hat{M}$ that minimizes $\text{AIC}(p)$ is an estimate of the number of sources. For example, Wax and Kailath (1983) applied this criterion for the array processing problem. Those authors also applied Rissanen's (1978) criterion which yields unbiased estimates under suitable assumptions. However, they investigated only the special case where the steering vectors are perfectly unknown unit vectors and they found, for example, for the single-frequency case

$$\text{AIC}(p) = -2K(N-p)\log\frac{\left(\prod_{i=p+1}^{N}\lambda_i\right)^{1/(N-p)}}{1/(N-p)\sum_{i=p+1}^{N}\lambda_i}$$

$$+ 2p(2N - p). \tag{76}$$

Thus, as well known from statistics (cf. James 1969), the geometric mean of the smallest eigenvalues of $\hat{C}_x$ has to be compared with the arithmetic mean.

In Böhme (1985a), Akaike's criterion was formulated in connection with eq. (66) for the single-frequency case. If $\bar{q}_p$ denotes the minimum value $\bar{q}(\bar{\xi})$ for p sources, then

$$\text{AIC}(p) = -2KN\left(-\log\bar{q}_p + \log N - 1\right) + 2[p(2K + a) + 1], \tag{77}$$

where pa is the number of waveparameters in ξ. Using eq. (74), we can easily define the criterion for the wideband case with similar notations,

$$\text{AIC}(p) = -2KLN\left(-\bar{Q}_p/L + \log N - 1\right)$$

$$+ 2[p(2KL + a) + L]. \tag{78}$$

Also in this situation, a sequence of likelihood ratio tests could be designed for estimating M, similar to that investigated by Nickel (1983) for the single-frequency case. Finally, we formulate Akaike's criterion in connection with eq. (65) if the elements of $\mathbf{C}_s$ (i.e. p^2 real parameters) are unknown,

$$\text{AIC}(p) = -2K\left(-\log \hat{q}_p - N\right) + 2\left(p^2 + pa + 1\right), \tag{79}$$

where $\hat{q}_p = \hat{q}(\hat{\xi})$ for p sources. A generalization to the multifrequency case is now obvious. A problem for further research is whether the use of Rissanen's criterion instead of Akaike's criterion results in better estimates for the cases investigated in this paper.

4.4. Asymptotic properties and approximations

In this section we indicate statistical properties of some estimates of the preceding section. For more details we refer to Brillinger (1981), Hinich and Shaman (1972) and Böhme (1985a). Starting with complex-normally distributed data and the likelihood function (57), we find, asymptotically for large degrees K of freedom, that $K^{1/2}(\hat{\theta} - \theta)$ is normally distributed with zero mean and covariance matrix $\hat{\mathbf{G}}(\theta)^{-1}$. The elements of $\hat{\mathbf{G}}(\theta)$ (assuming non-singularity) are

$$\hat{\mathbf{G}}(\theta)_{in} = \text{Tr}\left(\frac{\partial \mathbf{C}_x}{\partial \theta_i} \mathbf{C}_x^{-1} \frac{\partial \mathbf{C}_x}{\partial \theta_n} \mathbf{C}_x^{-1}\right)\bigg|_{\theta} = K\,\text{E}_\theta \frac{\partial \hat{L}}{\partial \theta_i} \frac{\partial \hat{L}}{\partial \theta_n}\bigg|_{\theta}. \tag{80}$$

Interpreting, the MLE $\hat{\theta}$ is consistent, i.e. converges in probability to the true parameters, and the asymptotic covariance matrix is given by the inverse of the Fisher information matrix. A suitable submatrix of $K^{-1}\hat{\mathbf{G}}(\theta)^{-1}$ then gives us the asymptotic covariance matrix of the MLE $\hat{\xi}$. Closed expressions for that matrix are not yet known to the author.

This is in contrast to the CMLE $\bar{\xi}$. This estimate has the property that $K^{1/2}(\bar{\xi} - \xi)$ is asymptotically, for large K, normally distributed with zero mean and covariance matrix $\mathbf{F}\bar{\mathbf{G}}\mathbf{F}|_{\mathbf{C}_s, \xi, \nu}$ where the elements of $\bar{\mathbf{G}}$ are

$$\bar{\mathbf{G}}_{in} = \text{Tr}\left(\frac{\partial \mathbf{P}}{\partial \xi_i} \mathbf{C}_x \frac{\partial \mathbf{P}}{\partial \xi_n} \mathbf{C}_x\right) = K\,\text{E} \frac{\partial \bar{q}}{\partial \xi_i} \frac{\partial \bar{q}}{\partial \xi_n}, \tag{81}$$

and the elements of $\mathbf{F}^{-1}$ (assuming non-singularity) are

$$(\mathbf{F}^{-1})_{in} = 2\,\text{Re}\,\text{Tr}\left(\mathbf{C}_s \frac{\partial \mathbf{H}^*}{\partial \xi_n}(\mathbf{I} - \mathbf{P})\frac{\partial \mathbf{H}}{\partial \xi_i}\right) = \text{E}\frac{\partial^2 \bar{q}}{\partial \xi_n \partial \xi_i}. \tag{82}$$

This property can be proved using a method developed in Dzhaparidze and Yaglom (1983). In all cases, the variances of the estimates as well as the covariances go to zero as $1/K$. If we ask for the variance of an element $\bar{\xi}_i$ of $\bar{\xi}$, for example, the range estimate for source 1, we roughly predict it by $1/K$ times the corresponding diagonal element of $\mathbf{FGF}$, where the unknown parameters are replaced by consistent estimates. For example, we can use the CMLE $\bar{\xi}$, the estimates $\bar{\mathbf{C}}_s = \hat{\mathbf{C}}_s(\bar{\xi})$ and $\bar{\nu} = \hat{\nu}(\bar{\xi})$ from eqs. (61) and (62), respectively. We could also use $\hat{\mathbf{C}}_X$ in eq. (81).

The numerical efforts required for computing the MLE's and CMLE's of the last section should be commented. The numerical solution of the likelihood equations system (58) (that is highly nonlinear) for array processing problems as discussed in this course seems to be very complicated. The essential problem of the simplified version in eqs. (61), (62) and (65) is the minimization of eq. (65) which is tractable. The more simple problem is that of eq. (70). Therefore, we only discuss it in more detail. We develop a numerical method which requires approximately the same computer time as Liggett's (1973) method that can be seen as an approximate solution of eq. (58) for uncorrelated signals. In section 4.5, results of numerical experiments with both methods are presented.

The numerical method is a modification of the well-known Gauss–Newton iteration. Let $\nabla \bar{q}$ denote the gradient of eq. (70). The Gauss–Newton iterations for the minimization of $\bar{q}(\xi)$ over ξ, i.e. the solution of $\nabla \bar{q}|_{\xi} = \mathbf{0}$, are

$$\xi^{n+1} = \xi^n - \left(\nabla\nabla'\bar{q}\right)^{-1}\nabla\bar{q}\big|_{\xi^n} \quad (n = 0,1,2,\dots). \tag{83}$$

Here, $(\nabla\bar{q})_i = -\mathrm{Tr}(\partial\mathbf{P}/\partial\xi_i \hat{\mathbf{C}}_x)$, and

$$(\nabla\nabla'\bar{q})_{ik} = -\mathrm{Tr}\left(\frac{\partial^2\mathbf{P}}{\partial\xi_i\,\partial\xi_k}\hat{\mathbf{C}}_x\right) \tag{84}$$

is the Hessian matrix. To simplify the numerical computations, we approximate the Hessian by its expected value which is $\mathbf{F}^{-1}$, cf. eq. (82). Corresponding iterations are called Fisher's scoring. However, we cannot execute them because $\mathbf{C}_s$ is unknown. We replace $\mathbf{C}_s$ by an estimate, for example, $\hat{\mathbf{C}}_s(\xi^n)$ from eq. (61) and obtain the desired method,

$$\xi^{n+1} = \xi^n - \tilde{\mathbf{F}}\nabla\overline{Q}\big|_{\xi^n} \quad (n = 0,1,2,\dots), \tag{85}$$

where (assuming non-singularity)

$$(\tilde{\mathbf{F}}^{-1})_{ik} = -2\,\mathrm{Re}\,\mathrm{Tr}\left(\hat{\mathbf{C}}_s\frac{\partial\mathbf{H}^*}{\partial\xi_k}(\mathbf{I}-\mathbf{P})\frac{\partial\mathbf{H}}{\partial\xi_i}\right)\bigg|_{\xi}. \tag{86}$$

If we have a useful initial estimate ξ^0, a few iterations may be sufficient to get a good estimate of ξ. This is a consequence of a theorem which can be shown by a technique investigated by Dzhaparidze and Yaglom (1983). This theorem states that only one correction in the sense of eqs. (83) or (85) of any consistent estimate yields an estimate having the same asymptotic behavior as the CMLE $\bar{\xi}$, if certain regularity conditions are satisfied.

A simple method to construct a suitable initial estimate was introduced by Böhme (1984). The estimates are tolerable if the column vectors of $\mathbf{H}$, i.e. the steering vectors $d_m = d(\xi_m)$, are approximately orthogonal and if the signals are not strongly correlated. The idea is to decompose $\hat{\mathbf{C}}_x$ similar to a singular-value decomposition as follows.

$\mathbf{C}_1 = \hat{\mathbf{C}}_x$, for $m = 1, \ldots, M$, compute

$$\Big[\gamma_m = d\big(\xi_m^0\big)^* \mathbf{C}_m d\big(\xi_m^0\big) = \max_{\eta} \, d(\eta)^* \mathbf{C}_m d(\eta)$$

$$v_m = \mathbf{C}_m d\big(\xi_m^0\big)$$

$$\mathbf{C}_{m+1} = \mathbf{C}_m - v_m v_m^* / \gamma_m \Big]. \tag{87}$$

Combining the components of all ξ_m^0 yields the initial estimate ξ^0 of all waveparameters. If we correct ξ^0 by some iterations (eq. 85), the result is called the enhanced estimate ξ^e.

The first estimate ξ_1^0 of this simple method is given by the absolute maximum of the classical beamformer output and indicates the strongest source. The second, ξ_2^0, does the same with a reduced spectral matrix, etc. The quality of these estimates is comparable to that of Owsley's (1973) estimates, which are given by the absolute maxima of the diagrams (40) for $n = 1, \ldots, M$. Both the simple and Owsley's estimates can well separate sources close together, however, they tend to put the source so that the corresponding steering vectors are approximately orthogonal. We finally remark that the M strongest peaks of Schmidt's (1979) diagram (43) can also be used to obtain initial estimates of the waveparameters if the sources are resolved.

For completeness, we describe Liggett's (1973) geometrically motivated method for enhancement of Owsley's diagrams (40). The signal space spanned by the exact steering vectors $d_1, \ldots, d_M$ is well estimated by the space of the eigenvectors $u_1, \ldots, u_M$ belonging to the M strongest eigenvalues λ_m of $\hat{\mathbf{C}}_x$, if the signals are not strongly correlated and have sufficient SNR's. The problem is to construct another base (not orthonormal in general) of the estimated signal space that fits the base $\zeta_1^{1/2}d_1, \ldots, \zeta_M^{1/2}d_M$ as well as possible. Here, the ζ_m are the spectral signal powers. Liggett started with the vectors $w_m = (\lambda_m - \hat{v})^{1/2}u_m$ ($m = 1, \ldots, M$), cf. eqs. (59) and (60). He constructed successively unitary transformations,

$$v_m = \sum_{i=1}^{M} a_{mi}w_i, \quad \text{with} \quad \sum_{m=1}^{M} a_{mk}^* a_{mi} = \delta_{ki}, \tag{88}$$

so that the following criterion,

$$b = \sum_{m=1}^{M} \max_{\xi_m} |v_m^* d(\xi_m)|^2 / |v_m|^2, \tag{89}$$

is maximized. The optimizing parameters ξ_m^L are the final estimates of ξ. The signal powers can be simply estimated by $\zeta_m^L = |v_m|^2$ and the noise power by $\hat{v}$ in eq. (60).

4.5. Numerical experiments

For large degrees K of freedom, the asymptotic behavior of the MLE's and CMLE's can be predicted as shown in the last section. In applications, the behavior of the approximations for moderate K is of interest. Precision and stability, and also the common behavior of the estimates have been investigated by numerical experiments, especially when resolution problems are expected. Such experiments are suitably executed on a vector computer to reduce the numerical burden of matrix operations.

Model (56) is used for a line array of 15 sensors spaced by a half wavelength. Three sources located approximately broadside generate uncorrelated signals, i.e. $C_s = \text{Diag}(\zeta_1, \zeta_2, \zeta_3)$. Unknown waveparameters are either bearings θ_m when Liggett's estimates θ_m^L are compared with the enhanced estimates θ_m^e, or bearings θ_m and ranges ρ_m otherwise. In the first experiment, the sources are assumed to be in a symmetric configuration: one strong source in the center is surrounded by two weak sources separated by 0.68 of the beamwidth. For signal-to-noise ratios $\text{SNR}_m = 10\log(\zeta_m/\nu) = 3, 12, 3\,\text{dB}$ and $-6, 3, -6\,\text{dB}$ ($m = 1, 2, 3$), and $K = 20$ and $100\,\text{DOF}$, groups of 10 matrices $\hat{C}_x$ are computed. $K\hat{C}_x$ is simulated by complex Wishart-distributed matrices with K DOF and the parameter matrix C_x. For each group, Liggett's estimates θ_m^L and ζ_m^L and the enhanced estimates ξ_m^e and the diagonal elements of eq. (61) as estimates ζ_m^e of ζ_m are computed. Correspondingly, figs. 6 and 7 illustrate estimated SNR_m (with respect to the exact noise power ν) over estimated bearings θ_m. Crosses give reference to exact source data.

J.F. Böhme

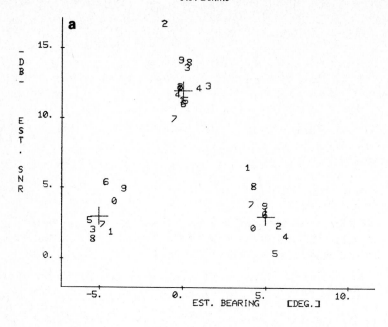

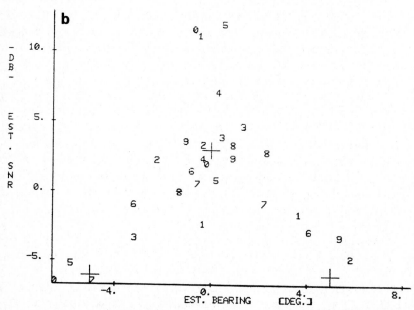

Fig. 6. Liggett's estimates: estimated signal-to-noise ratios over estimated bearings, 10 simulations. Crosses indicate exact data of signals. (a) Strong signals, 20 DOF; (b) Weak signals, 20 DOF; (c) Weak signals, 100 DOF.

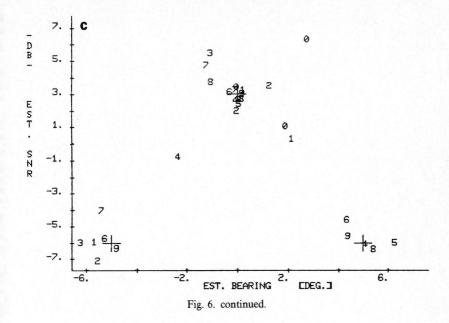

Fig. 6. continued.

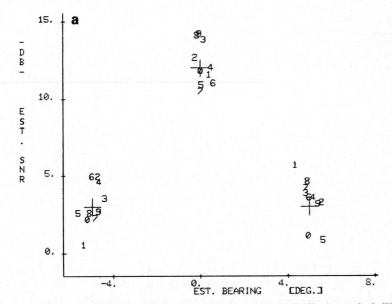

Fig. 7. Enhanced estimates as in fig. 6: (a) Strong signals, 20 DOF; (b, overleaf) Weak signals, 20 DOF; (c, overleaf) Weak signals, 100 DOF.

J.F. *Böhme*

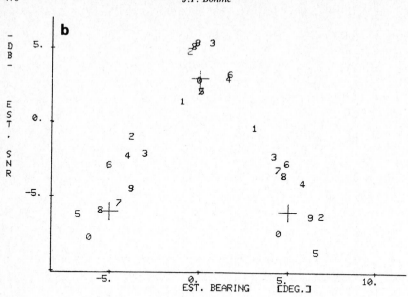

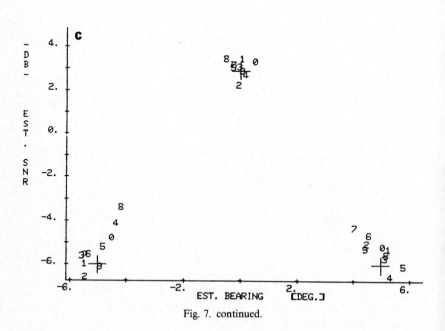

Fig. 7. continued.

Experiences of a lot of similar experiments allow to state some qualitative results:

(a) If the sources are well separated in bearing and if SNR > 0 dB for every signal, and $K = 20$, then both methods, as well as both initial estimates, estimate bearings comparatively well. For weaker signals, K has to be increased.

(b) If the sources are close together, i.e. separated by less than one beamwidth, the initial estimates separate the sources. However, they estimate with bias as predicted. Both, Liggett's estimates and the en-hanced estimates try to correct for bias. They behave approximately equivalent if signal powers are not very different. The enhanced esti-mates are better in some symmetric source configurations, where Liggett's as well as Owsley's estimates encounter problems in discovering weak sources, and also for increasing K.

The second experiment investigates the common behavior of the enhanced estimates for bearings, ranges and signal powers. Two equally

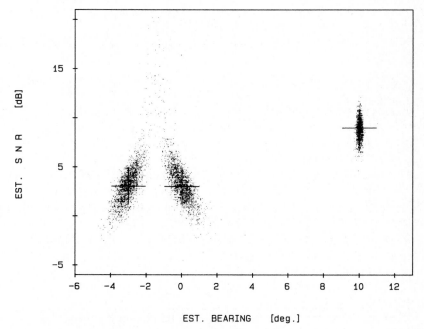

Fig. 8. Enhanced estimates: estimated signal-to-noise ratios over estimated bearings, 2048 simulations, 20 DOF. Crosses indicate exact data of signals.

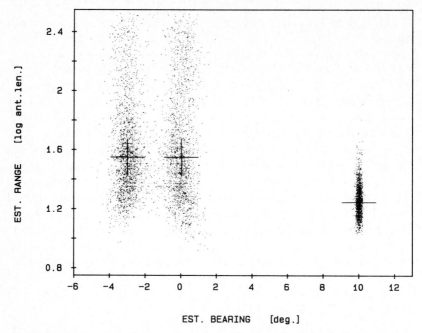

Fig. 9. See fig. 8: estimated ranges over estimated bearings. Vertical scale: log (range/antenna length).

weak sources separated in bearing by 0.41 of the beamwidth and a stronger source well separated from the others are assumed. Signal-to-noise ratios are $SNR_m = 3, 3, 9\, dB$ $(m = 1, 2, 3)$ and ranges $\rho_m = 35, 35, 18$ times the antenna length, respectively. $\hat{C}_x$ is computed as in the first experiment with $K = 20\, DOF$. All estimates and 2048 matrices $\hat{C}_x$ are computed. For range estimation, a simple plane model is used. Figure 5 illustrates the geometry and the definition of corresponding time delays. The results are indicated in the scatter diagrams in figs. 8 and 9. In fig. 8, estimated SNR_m (with respect to exact spectral noise power) over estimated bearing are depicted. Exact values given by the parameters of the model are denoted by crosses. Estimated range in multiples of array length and logarithmic scale over estimated bearing are depicted in fig. 9.

Let us summarize some results of this and similar experiments. If a source is well separated from other sources, the estimates of the wave parameters and of the signal spectral powers are precise and stable. The estimates tend to be uncorrelated and not affected by the other sources.

A normal distribution could be used as a first approximation. This property contrasts with estimations for sources close together, for example less than a half of the beamwidth. Estimations of range and bearing are nearly uncorrelated, but signal spectral powers and bearings are not. We find significant influence by the neighbour source. Also an hypothesis of normal distribution should be rejected.

4.6. Concluding remarks

Let us re-examine the main problem of array processing: the estimation of the locations of sources and that of their spectral properties, if a piece of array output is observed. In section 2 we found that phased array beamforming and classical beamforming are suitable methods for obtaining an overview, or for surveillance purposes. These methods are relatively robust, however, they have bad resolution properties. The problem to resolve two sources, for example in bearing, can be attacked by the use of high-resolution methods discussed in section 3. Generally, these methods are successful if weak model assumptions are not violated. The high-resolution methods can separate sources close together. The resulting estimates of the location parameters may be biased in this case. For correcting, more careful modelling is required and parametric methods are suitable. Initial estimates which separate the sources are iteratively enhanced. These methods can estimate precise and stable as discussed in the preceding sections. However, they can break down if the model mismatches. Therefore, interactive systems should be used in practice, for example in sonar systems. The operator should apply different methods if a situation requires a more detailed analysis.

Some comments on current research problems are added. The observation in the numerical experiments of section 4.5 that the estimates of bearings and spectral signal powers are strongly correlated for sources close together can be found also by an analysis of the asymptotic covariances of the estimates. First results in this direction are now available and will be published together with a statistical analysis of the numerical experiments by U. Sandkühler. The noise model we used in eq. (56) is unsatisfactory in many applications. For a more complex spectral matrix of noise, an alternating estimation of signal and noise parameters (cf. Hannan 1971), seems to be a reasonable method. The details require further research. The wavefields we discussed consist of coherent waves such that time delays and wave parameters do not depend on frequency. For example in a dispersive medium, the velocity of propagation of wavefronts depends on frequency. If we construct a suitable matrix **H**, we can apply slight modifications of the estimation techniques developed in sections 4.3 and 4.4. We first optimize over all frequency-dependent parameters and then optimize the geometric means of the resulting minimal criteria as in eqs. (73) and (74) over the parameters not depending on frequency.

In section 4.1, we introduced a parametric model in the frequency domain (cf. eq. 56), because of the nice asymptotic distributional properties of the Fourier-transformed array outputs. In principle, we can also use parametric models in the time domain. For rational models, asymptotic distributional properties of prediction errors could be applied to construct well-motivated parameter estimates. We will not go into the details. Signal subspace methods can also be designed for such models. We only refer to the work of Porat and Friedlander (1983) and that of Su and Morf (1983), who investigated similar methods for sampled data.

Acknowledgement

U. Sandkühler's assistance in designing, programming and executing the numerical experiments and analysing the results, is acknowledged.

References

Adler, R.J., 1981, The Geometry of Random Fields (Wiley, New York).
Akaike, H., 1974, A new look at the statistical model identification, IEEE Trans. Autom. Control **19**, 716–723.
Baggeroer, A.B., 1978, Sonar signal processing, in: Applications of Digital Signal Processing, ed. A.V. Oppenheim (Prentice Hall, Englewood Cliffs) ch. 6.
Bangs, W.J., and P.M. Schultheiss, 1973, Space–time processing for optimal parameter estimation, in: Signal Processing, eds. J.W.R. Griffiths, P.L. Stocklin and C. van Schooneveld (Academic Press, New York) pp. 577–590.
Bienvenue, G., 1979, Influence of the spatial coherence of the background noise on high-resolution passive methods, Proc. IEEE-ICASSP, Washington, 306–309.
Bienvenue, G., and L. Kopp, 1983, Optimality of high-resolution array processing using the eigensystem approach, IEEE Trans. Acoust. Speech Signal Process. **31**, 1235–1248.
Böhme, J.F., 1979, Array processing in semi-homogeneous random fields, Proc. GRETSI-Septième colloque sur le traîtement du signal et ses applications, Nice, France, 104/1–4.
Böhme, J.F., 1983a, On the stability of some high-resolution beamforming methods, Inf. Sciences **29**, 75–88.
Böhme, J.F., 1983b, On parametric methods for array processing, in: Signal Processing II: Theories and Applications, ed. H.W. Schüßler (North-Holland, Amsterdam) pp. 637–644.
Böhme, J.F., 1984, Estimation of source parameters by maximum likelihood, Proc. IEEE-ICASSP, San Diego, 7.3.1–4.
Böhme, J.F., 1985a, Source-parameter estimation by approximate maximum likelihood and nonlinear regression, IEEE J. Oceanic Eng. **10**, 206–212.

Böhme, J.F., 1986, Estimation of spectral parameters of correlated signals in wavefields, Signal Processing, Vol. 10.

Brillinger, D.R., 1981, Time Series: Data Analysis and Theory, Expanded Ed. (Holden-Day, San Francisco).

Burg, J.P., 1967, Maximum entropy spectrum analysis, 37th Annual Meeting Soc. Explor. Geophysicists, Oklahoma.

Cantoni, A., and L.C. Godara, 1980, Resolving the directions of the sources in a correlated field incidenting on an array, J. Acoust. Soc. Am. **67**, 1247–1255.

Capon, J., 1969, High-resolution frequency-wavenumber spectrum analysis, Proc. IEEE **57**, 1408–1418.

Carter, G.C., 1987, course 9, this volume.

d'Assumpcao, H.A., 1980, Some new signal processors for arrays of sensors, IEEE Trans. Inf. Theor. **26**, 441–453.

Dudgeon, D.E., 1977, Fundamentals of digital array processing, Proc. IEEE **65**, 898–904.

Dzhaparidze, K.O., and A.M. Yaglom, 1983, Spectrum parameter estimation in time series analysis, in: Developments in Statistics, Vol. 4, ed. P.R. Krishnaiah (Academic Press, New York) pp. 1–96.

Hahn, W.R., and S.A. Tretter, 1973, Optimum processing for delay-vector estimation in passive signal arrays, IEEE Trans. Inf. Theor. **19**, 608–614.

Hannan, E.J., 1971, Non-linear time series regression, J. Appl. Probab. **8**, 762–780.

Haykin, S., ed., 1985, Array Signal Processing (Prentice Hall, Englewood Cliffs).

Hinich, M.J., and P. Shaman, 1972, Parameter estimation for an r-dimensional plane wave observed with additive independent Gaussian errors, Ann. Math. Statist. **43**, 153–169.

James, A.T., 1969, Tests for equality of latent roots of the covariance matrix, in: Multivariate Analysis II, ed. P.R. Krishnaiah (Academic Press, New York) pp. 205–218.

Kay, S.M. and S.L. Marple, 1981, Spectrum analysis–a modern perspective, Proc. IEEE **6**, 1380–1419.

Knight, W.C., R.G. Pridham and S.M. Kay, 1981, Digital signal processing for sonar, Proc. IEEE **69**, 1451–1506.

Lacoss, R., 1971, Data adaptive spectral analysis methods, Geophysics **36**, 661–675.

Lang, S.W., 1981, Spectral estimation for sensor arrays, Ph.D. dissertation (Massachusetts Institute of Technology, Cambridge, MA).

Liggett, W.S., 1973, Passive sonar: fitting models to multiple time series, in: Signal Processing, eds. J.W.R. Griffiths, P.L. Stocklin and C. van Schooneveld (Academic Press, New York) pp. 327–345.

McDonough, R.N., 1979, Application of the maximum likelihood method and the maximum entropy method to array processing, in: Nonlinear Methods of Spectral Analysis, ed. S. Haykin (Springer, Berlin).

Monzingo, R.A., and T.W. Miller, 1980, Introduction to Adaptive Arrays (Wiley, New York).

Moura, J.M.F., 1981, The hybrid algorithm: a solution to aquisition and tracking, J. Acoust. Soc. Am. **69**, 1663–1672.

Moura, J.M.F., 1987, course 3, this volume.

Nickel, U., 1983, Super-resolution by spectral line fitting, in: Signal Processing II: Theories and applications, ed. H.W. Schüßler (North-Holland, Amsterdam) pp. 645–648.

Owsley, N.L., 1973, Spectral signal set extraction, in: Signal Processing, eds. J.W.R. Griffiths, P.L. Stocklin and C. van Schooneveld (Academic Press, New York) pp. 469–475.

Picinbono, B., 1987, course 1, this volume.

Pisarenko, V.F., 1973, The retrieval of harmonics from covariance functions, Geophys. J. R. Astron. Soc. **33**, 347–366.

Porat, B., and B. Friedlander, 1983, Estimation of spatial and spectral parameters of multiple sources, IEEE Trans. Inf. Theor. **29**, 412–425.

Rao, C.R., 1973, Linear Statistical Inference and its Applications (Wiley, New York).

Rissanen, J., 1978, Modeling by shortest data description, Automatica **14**, 465–471.

Scharf, L., 1987, course 2, this volume.

Schmidt, R.O., 1979, Multiple emitter location and signal parameter estimation, Proc. RADC Spectrum Estimation Workshop, Rome, NY, 243–258.

Schultheiss, P.M., and E. Weinstein, 1981, Lower bounds on the localization errors of a moving source observed by a passive array, IEEE Trans. Acoust. Speech Signal Proc. **29**, 600–607.

Sharman, K., M. Wax, and T. Kailath, 1983, Asymptotic statistics of covariance eigenstructure spectral analysis methods, subm. to IEEE Trans. Acoust. Speech Signal Proc.

Su, G., and M. Morf, 1983, The signal subspace approach for multiple wide-band emitter location, IEEE Trans. Acoust. Speech Signal Proc. **31**, 1502–1522.

Tufts, D.W. and R. Kumaresan, 1980, Improved spectral resolution II, Proc. IEEE-ICASSP, Denver, pp. 592–597.

Urban, H., ed., 1985, Adaptive Methods in Underwater Acoustics (Reidel, Dordrecht).

van den Bos, A., 1981, Recent developments in least-squares model fitting, in: Underwater Acoustics and Signal Processing, ed. L. Bjørnø (Reidel, Dordrecht) pp. 455–469.

Wax, M., and T. Kailath, 1983, Determining the number of signals by information theoretic criteria, Proc. IEEE Acoust. Speech Signal Proc. Spectrum Estimation Workshop II, 192–196.

Wax, M., T.J. Shan, and T. Kailath, 1984, Spatio-temporal spectral analysis by eigenstructure methods, IEEE Trans. Acoust. Speech Signal Proc. **32**, 817–827.

Yaglom, A.M., 1962, An Introduction to the Theory of Stationary Random Functions (Prentice Hall, Englewood Cliffs).

Ziegenbein, J., 1979, Spectral analysis using the Karhunen–Loeve transform, Proc. IEEE-ICASSP, Washington, 182–185.

COURSE 8

ADAPTIVE SIGNAL PROCESSING

Tariq S. DURRANI BSc MSc PhD CEng FIEE

Department of Electronic and Electrical Engineering
University of Strathclyde
204 George Street, Glasgow G1 1XW, Scotland

J.L. Lacoume, T.S. Durrani and R. Stora, eds.
Les Houches, Session XLV, 1985
Traitement du signal / Signal processing
© Elsevier Science Publishers B.V., 1987

Contents

1. Introduction

Architectures and algorithms for signal processing have their origins in the well-springs of approximation theory and optimisation theory. They range from structures which are fixed in size and format, such as digital filters, or fixed length FFTs, to flexible and programmable forms such as adaptive systems. While the first two instances are discrete-time/discrete-element approximations to continuous functional operations, the last is essentially an implementation of architectures aimed at optimising a prespecified performance criterion. Adaptive systems have been variously referred to as self-optimising systems, learning systems, updating systems, model reference systems, and so on. Irrespective of their name, adaptive systems, within the context of this course, will refer to algorithms or architectures which can change their characteristics with time; they involve an element of 'training' during which period the structural parameters change to meet some predetermined performance index, thus programability is an integral aspect of such systems. The systems involve an iterative procedure to achieve the desired performance objective, and it is during the iteration period or 'training' period that the system is said to be 'adapting' or 'learning' or 'self-optimising'. In this respect the systems are manifestly time-varying.

The origins of adaptive systems can be traced back to Newton; the root-finding algorithm for a polynomial is the first ever implementation of the steepest-descent technique. Similarly the minimisation via the Newton–Raphson technique is another manifestation of an adaptive computing scheme. In this course, I do not propose to trace the history of adaptive systems from ancient times to the state of the art—rather I propose to concentrate on techniques which are now classed as conventional, such as the steepest-descent method and its variants, and the more recent developments such as recursive least-squares and lattice algorithms, and then to round-off the course by a discussion on some new directions such as the potentially more exciting knowledge-based systems.

Fig. 1. Signal processing system.

The fundamental precept of signal processing is depicted in fig. 1, where an operation is performed on the input data to improve the output in order to characterise the input information. This may be considered as either improving the signal to noise ratio at the output to identify the input signal characteristics or to facilitate the parametrisation of the underlying process, or the system which is generating the input data. The former category includes techniques of noise filtering and noise cancelling, while the latter covers aspects of system identification and system equalisation.

Classical signal processing is concerned with the use of structures with fixed elements and formats, such as matched filters or digital filters, for implementing the processing system. The approach relies on significant a priori information about the input data, or equivalently a lack of concern about the input data; for instance for matched filtering it is assumed that the signal of interest is of a known form which can be matched. Equivalently in digital filtering or FFT analysis, the form of the underlying information contained is secondary to the requirement of suppressing noise in the filter stop bands, or, as in the FFT, decomposing the input data into an a priori chosen set of frequency bands or ordinates. In adaptive signal processing, on the other hand, the processing system is considered to have a time-varying structure which is modified as more data becomes available (or with every iteration) in a manner which allows the system performance to improve with increasing input data, and in this respect the system is *data-dependent*.

The most general structure for an adaptive system is given in fig. 2, where the system response is a time-varying function

$$H(z, nT) = \frac{A(z, nT)}{B(z, nT)} = \frac{\sum_{p=0}^{N-1} a_p(nT) z^{-p}}{\sum_{q=0}^{M-1} b_q(nT) z^{-q}} \tag{1}$$

such that the coefficients $\{a_p, b_p\}$ for any arbitrary order M, N are time-varying parameters; their value at any time instant nT being given by $\{a_p(nT), b_p(nT)\}$. It is the modification of these coefficients and the attendant filter response which underlies the updating technique, and the consequent behaviour of the adaptive system. Starting with arbitrary initial values, the parameters need to be modified to achieve desired performance characteristics, so it is worthwhile discussing the criterion for optimality of the adaptive systems.

2. Criterion for optimality and adaptive system model

The adaptive system in fig. 2 has to process the input data to yield a desired signal; in this respect the system should be as 'close' as possible to an ideal. The measure of closeness can be defined in terms of some objective criterion. The criterion should have the following properties:

Useful measure of performance;

Mathematically tractable;

Unique solution;

Ease of computation;

Little prior information.

It is imperative that the optimality criterion should represent a quantifiable measure of the 'distance' between the characteristics of the actual system and the desired (optimum). The criterion should obviously lend itself to mathematical analysis, and while computational burden is of secondary importance these days, nevertheless, it is useful to look for algorithms which do not involve massive computing requirements, especially if real time processing is of interest. Also, it is essential that the criterion leads to a unique solution as the eventual system realisation should be unique. Thus multi-modality optimising indexes are out of the question.

A criterion which is simple, quadratic, and leads to unique solutions is the Minimum Mean Square Error (MSE) criterion. In the main, it has the advantage of requiring only linear operations on the data, and only second-order statistics of the input. It can be shown that for Gaussian input data the criterion leads to a solution identical to those based on probability measures such as maximum a posteriori probability or maximum likelihood.

Recently some criteria have been proposed which are based on higher-order moments of the error (Walach and Widrow 1984). Here,

Fig. 2. Time-varying (adaptive) system.

instead of using the L^2 norm for the error, a norm of L^4 or even $L^{2p}(\forall p)$ is considered, and preliminary results seem to indicate better iterative performance than algorithms based on the L^2 norm.

Using the proposed criterion, the configuration of fig. 2 for adaptive signal processing is modified to appear as that given in fig. 3.

The basic requirement for adaptive signal processing thus consists of the input data (or a primary input sequence), the desired signal (or, the reference input) and a technique for minimising the mean square error between the output of the adaptive process and the desired signal. This requirement implies that the process coefficients need to be chosen such that the output from the adaptive processor is a replica of the reference input in the mean square sense, after the appropriate training period during which the coefficients are modified to minimise the mean square error.

Referring to the general structure of eq. (1) and to fig. 3, setting b_0 to unity arbitrarily, and dropping the dependence on time, we obtain the output

$$y_k = \sum_{q=1}^{M-1} b_q y_{k-q} + \sum_{p=0}^{N-1} a_p x_{k-p}. \tag{2}$$

Using vector notation

$$\boldsymbol{A}^{\mathrm{T}} = [a_0, a_1, \ldots, a_{N-1}],$$

$$\boldsymbol{B}^{\mathrm{T}} = [b_1, \ldots, b_{M-1}],$$

$$\boldsymbol{X}_k^{\mathrm{T}} = [x_k, x_{k-1}, \ldots, x_{k-N+1}],$$

$$\boldsymbol{Y}_{k-1}^{\mathrm{T}} = [y_{k-1}, y_{k-2}, \ldots, y_{k-M+1}],$$

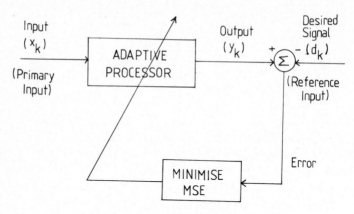

Fig. 3. Adaptive system.

we may rewrite eq. (2) as

$$y_k = B^T Y_{k-1} + A^T X_k. \tag{3}$$

The mean square error between y_k and the reference input d_k is given by

$$J = E[\varepsilon_k^2] = E[(d_k - y_k)^2], \tag{4}$$

where E denotes the expectation operator. Here both the primary and reference inputs are considered as stochastic. Substituting eq. (3) into eq. (4) yields

$$
\begin{aligned}
J &= E\left[\left(d_k - A^T X_k - B^T Y_{k-1}\right)^2\right] \\
&= R_{dd}(0) - 2A^T r_{xd} - 2B^T r_{yd} + A^T R_{xx} A + B^T R_{yy} B \\
&\quad + 2A^T R_{xy} B,
\end{aligned} \tag{5}
$$

where

$R_{dd}(0) = E[d_k^2]$ = mean square value of reference input;

$r_{xd} = E[d_k X_k]$ = cross-correlation vector between primary and reference inputs;

r_{yd} = cross-correlation vector between system output and reference input;

R_{xx} = autocorrelation matrix of input data $\{X_k\}$;

R_{yy} = autocorrelation matrix of system output $\{y_k\}$;

R_{xy} = cross-correlation matrix $\triangleq E[X_k Y_{k-1}^T]$.

The unknown coefficients $\{A, B\}$ can be determined from eq. (5) by minimising J. Thus

$$\frac{\partial J}{\partial A} = 0 = -2r_{xd} + 2R_{xx}A - \frac{\partial}{\partial A}\left(B^T r_{yd}\right) + \frac{\partial}{\partial A}\left(B^T R_{yy}B\right)$$

$$+ 2\frac{\partial}{\partial A}\left(A^T R_{xy}B\right), \tag{6}$$

$$\frac{\partial J}{\partial B} = 0 = -2\frac{\partial}{\partial B}\left(B^T r_{yd}\right) + \frac{\partial}{\partial B}\left(B^T R_{yy}B\right) + 2\frac{\partial}{\partial B}\left(A^T R_{xy}B\right). \tag{7}$$

For completeness, any pth element in eq. (7) may be written as

$$\frac{\partial J}{\partial b_p} = 0 = -2r_{yd}(p) - 2\sum_{l=1}^{M-1} b_l r_{yd}(p-l) - 2\sum_{l=1}^{M-1} b_l \frac{\partial r_{yd}(p-l)}{\partial b_p}$$

$$+ 2\sum_{l,s=1}^{M-1} b_l b_s \frac{\partial r_{yd}(l-s)}{\partial b_p} + 2\sum_{m=0}^{N-1} a_m r_{xy}(m-p)$$

$$+ 2\sum_{l,m=0}^{M-1\,N-1} a_m \frac{\partial r_{xy}(m-l)}{\partial b_p} b_l. \tag{8}$$

Here $r_{yd}(p) = p$th element of vector r_{yd}.

From above, the last two terms in eqs. (6) and (7) arise from the fact that any variation in $\{A, B\}$ affects the output $\{y_k\}$ due to feedback.

While eqs. (6) and (7) are not intractable, they represent a nonlinear set of equations whose solution is extremely cumbersome, and more importantly, could lead to several (non-unique) values for the coefficients $\{A, B\}$, corresponding to local minima for the performance criterion J. Some of the solutions may represent non-realisable elements. Due to these problems, it is usual to consider the structure of the adaptive system as that consisting of zeros only, i.e. the coefficients $\{b_k\}$ are all taken as zero (except for $b_0 = 1$). Under these circumstances, the adaptive system is modified to

$$H(z, kT) = \sum_{p=0}^{N-1} a_p(kT)z^{-p}, \tag{9}$$

which implies an FIR (Finite Impulse Response) structure (see fig. 4) and leads to

$$J = R_{dd}(0) - 2A^T r_{xd} + A^T R_{xx}A. \tag{10}$$

Following from this, the global (steady-state) solution to the adaptive system may be realised from eq. (6) as

$$r_{xd} = R_{xx}A \tag{11a}$$

or

$$A = R_{xx}^{-1}r_{xd}. \tag{11b}$$

Equation (11a) is called the Wiener–Hopf equation, and its solution (11b) represents the Wiener filter. It has all the advantages of mathematical tractability, uniqueness, stability (due to the FIR structure), and for the global solution of eq. (10) requires:

(a) Evaluation of the input data $\{x_k\}$ autocorrelation.
(b) Evaluation of the cross-correlation between the input $\{x_k\}$ and the reference input $\{d_k\} \rightarrow r_{xd}$.
(c) Setting up of the auto-correlation matrix R_{xx}.
(d) Inversion of R_{xx} (with the use of fast algorithms for Toeplitz matrix inversion, this is not too cumbersome).
(e) Multiplication between R_{xx}^{-1} and r_{xd} to yield the coefficient vector A.

The overall approach is essentially a batch processing approach requiring storage of data to compute the above, and it represents a tedious computing effort.

Note, from eq. (10)

$$J_{\min} = R_{dd}(0) - r_{xd}^{T}R_{xx}^{-1}r_{xd}. \tag{12}$$

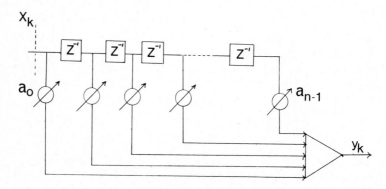

Fig. 4. FIR structure for adaptive filter.

Furthermore, as

$$\frac{\partial J}{\partial A} = 2E\left[\varepsilon_k \frac{\partial \varepsilon_k}{\partial A}\right] = -2E[\varepsilon_k X_k],$$

then for the minimum

$$\frac{\partial J}{\partial A} = 0 = E[\varepsilon_k X_k]. \tag{13}$$

This equation embodies the Orthogonality principle which implies that the optimum solution always leads to an orthogonal relationship between the data and the error.

3. Conventional adaptive signal processing algorithms

The above analysis yields the Wiener filter as the optimum solution to the minimum mean square error problem. In adaptive signal processing the requirement is to develop an iterative (adaptive) procedure to obtain the global optimal solution. The aim of the procedure is, starting with an arbitrary set of values for the coefficients $\{A\}$, to improve their values as more and more data become available in order to converge towards the optimum solution of eq. (11). In the following sections a number of algorithms are discussed which allow such iterative solutions.

3.1. Steepest-descent method

The steepest-descent method ensures that a matrix inversion in eq. (10) is avoided by the use of an iterative scheme based on a gradient-following method. From eq. (11)

$$\frac{\partial J}{\partial A} = -2r_{xd} + 2R_{xx}A. \tag{14}$$

This leads to an algorithm, known as the *steepest-descent* algorithm, which is characterised by the equation

$$A(k+1) = A(k) - \mu \frac{\partial J}{\partial A}\bigg|_k \tag{15}$$

with an arbitrary initial value of $A(0)$.

Here $A(k)$ is the coefficient vector of the kth iteration cycle of the algorithm, μ is a constant gain which controls the convergence rate of the algorithm, and is called the *Step Length*, while $(\partial J/\partial A)|_k$ denotes the value of the gradient of the optimising criterion J for $A = A(k)$ in eq. (14). The algorithm thus includes a correction term at each iteration which is proportional to the gradient of J, and represents a step taken in the direction of the gradient of J where the step size is controlled by μ. From eq. (14), the steepest descent algorithm is given by

$$A(k + 1) = (I - 2\mu R_{xx}) A(k) + 2\mu r_{xd}. \tag{16}$$

From eqs. (15) and (16) it is obvious that the algorithm converges to a steady-state value when the gradient goes to zero, i.e.

$$R_{xx}A^0 = r_{xd} \tag{17}$$

which yields the Wiener–Hopf solution. A^0 represents the optimal solution, and hence eq. (16) is an iterative scheme for solving the Wiener–Hopf equations.

3.2. Convergence aspects

An important consideration in any iterative scheme is the rate with which the algorithm converges to the optimum solution. Referring to eq. (16), if A^0 represents the optimum solution as given by eq. (11), then defining the error vector at each iteration as

$$A(k) - A^0 = \tilde{A}(k)$$

we obtain

$$\tilde{A}(k + 1) = (I - 2\mu R_{xx}) \tilde{A}(k). \tag{18}$$

To see more precisely the behaviour of the error vector, eq. (18) may be decoupled into N independent equations by applying a similarity transformation. Since the covariance matrix R_{xx} is symmetric and positive definite it can be decomposed into

$$R_{xx} = UDU^T, \tag{19}$$

where U is an orthonormal matrix of eigenvectors of R_{xx}, and D is a diagonal matrix containing its eigenvalues, such that

$$D = \text{Diag}\{\lambda_1, \lambda_2, \ldots, \lambda_N\}. \tag{20}$$

Utilising eq. (19) in eq. (18) leads to

$$U^T\tilde{A}(k + 1) = [I - 2\mu D]U^T\tilde{A}(k). \tag{21}$$

Considering the similarity-transformed error vector,

$$U^T\tilde{A}(k) = V(k),$$

any pth element of $V(k)$ may be expressed as

$$v_p(k + 1) = \left[1 - 2\mu\lambda_p\right]v_p(k) \quad \text{for any } p. \tag{22}$$

Equation (22) will converge,

$$0 < \left|1 - 2\mu\lambda_p\right| < 1, \tag{23}$$

for all eigenvalues $\{\lambda_p\}$ of D. The condition is most certainly satisfied if

$$0 < \mu < \frac{1}{\lambda_{\max}}, \tag{24}$$

where $\lambda_{\max}$ is the largest eigenvalue of R_{xx}, and represents the worst-case requirement for eq. (22). Equation (24) is an extremely important relationship as it yields the bounds within which the step-length parameter should lie for the iterative algorithm to converge.

An indication of the rate of convergence of the iterative algorithm can be obtained from eq. (22). More specifically the rate at which the various modes $\{v_p(n)\}$ decay to zero can be derived from eq. (24):

$$1 - 2\mu\lambda_p = e^{1/\tau_p} \approx 1 - \frac{1}{\tau_p}$$

or

$$\tau_p \simeq \frac{1}{2\mu\lambda_p} \tag{25}$$

The longest time constant involved in the error equation (24) is

$$\tau_{\max} \approx \frac{1}{2\mu\lambda_{\min}},$$

where $\lambda_{\min}$ is the smallest eigenvalue of the covariance matrix R_{xx}. Referring to the bound on μ, we see that

$$\tau_{\max} > \frac{\lambda_{\max}}{2\lambda_{\min}}. \tag{26}$$

This leads to the conclusion that the convergence rate of the steepest-descent algorithm is dictated by the eigenvalue spread of the data covariance matrix. The ratio $\lambda_{\max}/\lambda_{\min}$ is called the condition number of the matrix, and eq. (26) implies that the larger the condition number of the data covariance matrix, the longer will it take the steepest-descent method to converge to the optimum solution.

3.3. Least-Mean-Squares (LMS) algorithm

The steepest-descent algorithm of section 3.1 requires prior knowledge of the data covariance matrix R_{xx} and the cross-correlation vector r_{xd}. In practice neither is available a priori. A number of algorithms have been developed which allow joint estimation of these quantities, and iterative solutions for the optimal Wiener filters of eq. (16). The most popular of these is called the Least-Mean-Squares (LMS) algorithm first proposed by Widrow and Hoff (1960) in the context of adaptive signal processing. In the LMS algorithm, the matrix quantities R_{xx} and r_{xd} appearing in eq. (16) are replaced by their instantaneous estimates $X_k X_k^T$ and $X_k d_k$ respectively. This is a common procedure in stochastic approximation theory. This leads to the LMS algorithm as

$$A(k+1) = \left(I - 2\mu X_k X_k^T\right)A(k) + 2\mu d_k X_k$$

$$= A(k) + 2\mu e_k X_k, \tag{27}$$

where the instantaneous error is

$$e_k = d_k - A^T(k)X_k = d_k - y_k, \tag{28}$$

$\{y_k\}$ being the filter output.

This algorithm has several attractive features: it is simple and easy to implement, it requires no a priori information or data storage, and it lends itself to real-time processing. On other hand, it is relatively sensitive to the choice of the step length or convergence factor (μ), and can be relatively slow to converge thus requiring a long training sequence to compute the coefficients. The implementation of the algorithm thus involves:

Arbitrary initial choice of vector $A(0)$;

Training sequence (desired or reference input);

Computation of eq. (27) with each new data-point till steady-state values for $A(k)$ prevail.

The algorithm allows processing or tracking of nonstationary data, provided the underlying characteristics of the data change slowly to allow the coefficients to converge over the quasi-stationary segments of the data. Thus the adaptive processor involves two stages of operations: a training period to evaluate the coefficients, and a tracking (or processing period) when the desired or reference signal is removed and the processor operates on the primary input only.

3.4. Stochastic-gradient-descent algorithm and modified-least-squares algorithm

In the *stochastic-gradient-descent algorithm* an estimate of the gradient of the mean square error criterion is evaluated over a finite data length and then substituted into the iterative scheme of eq. (15). The coefficients are maintained constant over this time interval. Hence,

$$A((n + 1)K) = A(nK) - \mu \nabla e(nK),$$

$$\nabla e(nK) = \frac{1}{K} \sum_{q=0}^{K-1} e(nK + q)X(nK + q),$$

$$A(nK + q) = A(nK) \quad q = 1, 2, \ldots, (K - 1), \tag{29}$$

and

$$e(nK) = d_{nK} - x_{nK},$$

where K is the averaging time interval, and ∇e is the stochastic gradient of the time-averaged mean square error

$$\overline{e^2(nk)} = \frac{1}{K} \sum_{q=0}^{K-1} e^2(nK + q). \tag{30}$$

This is also an estimate (and a more robust one than the instantaneous estimate employed in the LMS algorithm) of the gradient of the mean square error since

$$E[\nabla e(nK)] = E[e(nK)X(nK)] \quad \text{and} \quad E[\overline{e^2(nK)}] = J.$$

This algorithm has the advantage of using a more realistic estimator for the gradient of the mean square error than the LMS algorithm, with a consequently increased computational cost. Note that for $K = 1$, the algorithm reduces to the LMS algorithm. Gardner (1984) has studied the properties of this algorithm in some detail.

The *Modified LMS (MLMS) algorithm* is a variation on the steepest-descent and LMS algorithms, and was first proposed by Griffiths (1969). The algorithm tackles the problem where no reference signal $\{d_k\}$ is available, but requires that the cross-correlation vector r_{xd} is known. In this case the MLMS algorithm is given by

$$A(k + 1) = (\mathbf{I} - 2\mu X_k X_k^{\mathrm{T}})A(k) + 2\mu r_{xd}, \tag{31}$$

where the autocorrelation matrix is replaced by its instantaneous estimate. The assumption of prior knowledge about r_{xd} is valid practically, in that while one may not know the desired signal a priori, some classification of its desired characteristics is usually known. It has been found in practice that even crude estimates (or first-order approximations) of r_{xd} lead to rapid convergence of the algorithm.

3.5. Convergence analysis for the LMS algorithm

The LMS algorithm has been studied extensively (see for instance Gardner 1984, Bershad and Qu 1984, Feuer and Weinstein 1985). The iterative scheme of the algorithm does not lend itself to analysis for all classes of data, and almost all results to date (with the exception of Bershad and Qu's work for the frequency-domain (scalar) adaptive filter), have been developed on the assumption that the reference input data consists of a sequence of statistically independent random vectors (and in some cases a Gaussian distribution for the data is also inferred). Nevertheless, it has been shown experimentally (see Gitlin and Weinstein 1979) that for a sufficiently small step size (μ), i.e. slow convergence rates, the results based on the independence assumption closely agree with the experimental results. In this section we would like to study the convergence properties of the algorithm in terms of convergence in the mean and mean square sense, and introduce some new results. Referring to the LMS algorithm, the updating vector is given by eq. (27).

Defining A^0 as the optimal (Wiener–Hopf) solution we note the instantaneous error $\tilde{A}(k) = A(k) - A^0$, which on inserting into eqs. (27) and (28) leads to

$$\tilde{A}(k+1) = \left(\mathbf{I} - 2\mu X_k X_k^{\mathrm{T}}\right)\tilde{A}(k) + 2\mu\varepsilon(k)X_k, \qquad (32)$$

where

$$\varepsilon(k) = d_k - A^{0\mathrm{T}}X_k. \qquad (33)$$

Defining $V_k = U^{\mathrm{T}}\tilde{A}(k)$, where $UDU^{\mathrm{T}} = E[XX_k^{\mathrm{T}}] = R_{xx}$,

$$V_{k+1} = \left(\mathbf{I} - 2\mu U^{\mathrm{T}}X_k X_k^{\mathrm{T}}U\right)V_k + 2\mu\varepsilon(k)U^{\mathrm{T}}X_k. \qquad (34)$$

Assuming that the data vector is independent of the coefficient vector, and defining

$$M_k = E[V_k] \quad \text{and} \quad P_k = E\left[V_k V_k^{\mathrm{T}}\right] \qquad (35)$$

we obtain from eq. (34)

$$M_{k+1} = (I - 2\mu D)M_k \tag{36}$$

as $E[\varepsilon(k)X_k] = 0$. Here D is the diagonal matrix of eigenvalues of R_{xx}. Equation (36) is identical to eq. (21), and thus convergence in the mean is assured if μ satisfies the condition given in eq. (24).

From eq. (34) defining

$$Z_k = U^T X_k, \tag{37}$$

$$P_{k+1} = P_k - 2\mu \{ E[Z_k Z_k^T V_k V_k^T] + E[V_k V_k^T Z_k Z_k^T] \}$$
$$+ 3\mu^2 E[Z_k Z_k^T V_k V_k^T Z_k Z_k^T] + 3\mu^2 \sigma_e^2 E[Z_k Z_k^T]. \tag{38}$$

Here $\sigma_e^2 = E[\varepsilon^2(k)]$ is given in eq. (12) as $J_{\min}$. Again exploiting the independence of X_k or Z_k, and noting that $E[Z_k Z_k^T] = U^T R U = D$, we obtain for eq. (38)

$$P_{k+1} = P_k - 2\mu(DP_k + P_k D) + 4\mu^2 \Gamma_k + 4\mu^2 \sigma_e^2 D, \tag{39}$$

where

$$\Gamma_k = E[Z_k Z_k^T V_k V_k^T Z_k Z_k^T]. \tag{40}$$

This is a sixth-order joint moment. Detailed expressions for this are beyond the scope of this presentation, but it can be shown (Horowitz and Senne 1981) that

$$\Gamma_k = 2DP_k D + \mathrm{Tr}[DP_k]D \tag{41}$$

In earlier work Widrow et al. (1975) assumed that

$$\Gamma_k = E[Z_k Z_k^T] E[V_k V_k^T] E[Z_k Z_k^T] = DP_k D. \tag{42}$$

Using eq. (42) in eq. (39) leads to a Ricatti-type equation given by

$$P_{k+1} = P_k - 2\mu(DP_k + P_k D) + 4\mu^2 DP_k D + 4\mu^2 \sigma_e^2 D. \tag{43}$$

This yields a steady state solution as $k \to \infty$, as

$$P_\infty = \mathrm{Diag}[\, p_1, \ldots, p_l, \ldots, p_N]$$

which is a diagonal matrix with any lth element as

$$p_l = \frac{\mu \sigma_e^2}{1 - \mu \lambda_l}, \tag{44}$$

where $D = \mathrm{Diag}[\lambda_1, \ldots, \lambda_l, \ldots, \lambda_N]$ is the eigenvalue matrix of R_{xx}. It may be seen that eq. (43) leads to a convergent result if the step length μ obeys the bound given in eq. (23).

If the exact expression (41) for Γ_k is employed, then after tedious computation, we obtain

$$P_\infty = \frac{\mu\sigma_e^2}{1 - f(\mu)} \text{Diag}[p_1, p_2, \ldots, p_l, \ldots, p_N] \tag{45}$$

where

$$p_l = \frac{1}{1 - 2\mu\lambda_l} \quad \text{and} \quad f(u) = \mu \sum_{k=1}^{N} \frac{\lambda_k}{1 - 2\mu\lambda_k}. \tag{46}$$

In recent work it has been shown that using eq. (45), a tighter bound can be established for μ, i.e.

$$0 < \mu < \frac{1}{3\lambda_{\max}} \tag{47}$$

provided

$$f(\mu) < 1.$$

3.5.1. Misadjustment

It is important to determine the variance of the output of the adaptive filter. This is given by

$$e(k) = d_k - A^{\mathrm{T}}(k)X_k \tag{48}$$

$$= d_k - \left(A^0 + \tilde{A}^{\mathrm{T}}(k)\right)X_k$$

$$= \varepsilon(k) - \tilde{A}^{\mathrm{T}}(k)X_k, \tag{49}$$

where $\varepsilon(k)$ is as defined in eq. (33). Using eq. (37),

$$e(k) = \varepsilon(k) - V_k^{\mathrm{T}}Z_k$$

$$\therefore E\left[e^2(k)\right] = E\left[\varepsilon^2(k)\right] + E\left[V_k^{\mathrm{T}}Z_kZ_k^{\mathrm{T}}V_k\right]$$

$$= \sigma_e^2 + \text{Tr}[DP_k] \tag{50}$$

where $\sigma_e^2 = J_{\min}$. Using the stationary solution of eq. (43) we obtain

$$E\left[e^2(\infty)\right] = \sigma_e^2 + \mu\sigma_e^2 + \mu\sigma_e^2 \sum_{k-1}^{N} \frac{\lambda_k}{1 - \mu\lambda_k}. \tag{51}$$

From this we find the misadjustment, which is a measure of the effect of the noisiness of the coefficients on the output error variance:

$$M = \frac{E\left[e^2(\infty)\right] - \sigma_e^2}{\sigma_e^2} = \mu \sum_k \frac{\lambda_k}{1 - \mu\lambda_k}. \tag{52}$$

For small values of μ,

$$M \simeq \mu \sum \lambda_k = \mu \operatorname{Tr}[R_{xx}] \tag{53}$$

which is an important measure of the performance of the adaptive filter. From the exact solution, using eq. (45), we obtain the misadjustment as

$$M = \frac{1}{\sigma_e^2} \sum_i \lambda_i p_i. \tag{54}$$

The bound on μ requires information about λ_{max}, which in practice is replaced by using the fact that

$$\operatorname{Tr}[R_{xx}] = \sum_{i=1}^{N} \lambda_i > \lambda_{max}.$$

Hence $NR(0) = \operatorname{Tr}[R_{xx}]\lambda_{max}$, where $R(0) = $ mean square power of the input signal. Then μ is chosen as

$$\mu < \frac{1}{NR(0)}. \tag{55}$$

Using the exact solution (45) for P_∞, the misadjustment for the LMS algorithm is given by

$$M = \frac{1}{\sigma_e^2} \operatorname{Tr}(D\Gamma_\infty) = \frac{f(\mu)}{1 - f(\mu)}, \tag{56}$$

where $f(\mu)$ is as defined in eq. (46).

The misadjustment can thus be minimised by minimising $f(\mu)$. Note that the range for $f(\mu)$ is

$$0 < f(\mu) < 1.$$

However, it can be shown that the rate of convergence of the mean square error to its steady state can be measured by the following sum:

$$I = \sum_{k=0}^{\infty} E\left[e^2(k)\right] - E\left[e^2(\infty)\right].$$

This may be solved by using eq. (39) and eq. (50), and after intensive mathematical manipulation this leads to

$$I = \frac{1}{4} \frac{\sum_{i=1}^{N} \gamma_{0i}/(1 - 2\mu\lambda_i)}{\mu(1 - f(\mu))}.$$

(57)

Here $\gamma_{0i} = i$th element of the initial covariance matrix $\{P_0\}$. Note that the misadjustment may be controlled by choosing μ, (such that $f(\mu) \rightarrow 0$), but this leads to a slow rate of convergence (see eq. 57). From eq. (50), a value of μ can be chosen that results in the fastest possible convergence rate.

Setting $\partial I/\partial \mu = 0$, and considering the simplified case of $\lambda_i = \lambda$ $(i = 1, 2, \ldots, N)$, this leads to

$$I = \frac{1}{4\mu} \frac{\sum_{i=1}^{N} \gamma_{0i}}{(1 - \mu(N + 2)\lambda)},$$

(58)

which depends upon the initial value $\{\gamma_{0i}\}$. Considering $\{\gamma_{0i}\}$ to be independent of μ, eq. (58) can be minimised by the choice of

$$\mu^* = \frac{1}{2(N + 2)\lambda}.$$

Note that here we can replace λ by its average value,

$$\lambda_{av} = \frac{1}{N}\left(\sum_{k=1}^{N} \lambda_k\right) = R_{xx}(0) = \text{input signal power,}$$

then,

$$\mu^* = \frac{1}{2(N + 2)R_{xx}(0)}$$

(59)

which is a closer bound than given in eq. (46). Substituting this relationship into the expression for the misadjustment, and employing the simplifying assumption given above, we arrive at

$$M_* = \frac{N}{(N + 2)}.$$

(60)

Note that from Γ_k it may be observed that $(2\mu\lambda_i)$ is reminiscent of the poles of the function $f(\mu)$, thus a time constant for the process can be defined as (see also eq. 25)

$$\tau_{av} = \frac{1}{2\mu\lambda_{av}},$$

and from above this is $(N + 2)$, so we see that the algorithm would on average converge in $(N + 2)$ steps, where N is the number of unknown coefficients.

$$M_* = \left(\frac{N}{\tau_{av}} \right). \tag{61}$$

This equation relates the misadjustment factor to the average settling time of the mean square error and the number of filter coefficients.

4. Recursive least-squares techniques

Recursive Least-Squares (RLS) techniques are finding renewed interest with the ready availability of fast algorithms (Ljung et al. 1978); with increasing computing speeds, and more importantly, with the current emphasis on parallel architectures and systolic/wavefront processors even computationally intensive algorithms are being used increasingly. In essence, the RLS technique is a sequential method for minimising the sum of squares. In the context of adaptive signal processing, the objective is to minimise (refer to fig. 5)

$$J = \text{Min} \sum_{k=1}^{N} e^2(k), \tag{62}$$

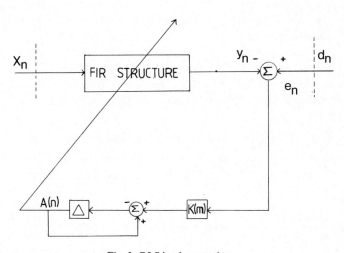

Fig. 5. RLS implementation.

where the error is

$$e(k) = d_k - \sum_{s=0}^{N-1} a_s x_{k-s}. \tag{63}$$

Here the FIR model for the adaptive system is being used. Concatenating the error $\{e(k)\}$ for various time instants, we obtain

$$\begin{bmatrix} e_0 \\ e_1 \\ \vdots \\ e_n \end{bmatrix} = \begin{bmatrix} d_0 \\ d_1 \\ \vdots \\ d_n \end{bmatrix} - \begin{bmatrix} x_0 & 0 & \cdots & 0 \\ x_1 & x_0 & \cdots & 0 \\ \vdots & \vdots & & \vdots \\ x_n & x_{n-1} & \cdots & x_{n-N+1} \end{bmatrix} \begin{bmatrix} a_0 \\ a_1 \\ \vdots \\ a_{N-1} \end{bmatrix} \tag{64}$$

or

$$e(n) = d(n) - X(n)A. \tag{65}$$

Using this notation, the minimisation problem is to determine

$$J = \operatorname{Min} e^T(n)e(n). \tag{66}$$

The well-known minimum norm solution to this is given by

$$A(n) = [X^T(n)X(n)]^{-1}X^T(n)d(n). \tag{67}$$

This solution is computationally intensive, requiring a matrix inversion for each new value of data. The RLS algorithm avoids this inversion, and allows the coefficient vector to be updated with each new data point. It has all the advantages of a recursive structure, being fixed in size and format (once the choice of the number of coefficients has been made) and requires no storage of the data (either the primary input $\{x_k\}$ or the reference input $\{d_k\}$). Note that the form of eq. (62) is called the "prewindowed" form where the data $x_k = 0$ for $k < 0$.

Defining

$$P(n) = [X^T(n)X(n)]^{-1}, \tag{68}$$

then with a new data point we have

$$P(n+1) = [X^T(n+1)X(n+1)]^{-1}.$$

From eqs. (62) and (63), $X(n+1)$ may be partitioned as

$$X^T(n+1) = [X^T(n)\Phi(n+1)], \tag{69}$$

where $\Phi(n + 1)$ reflects the new data vector, i.e.

$$\Phi^{\mathrm{T}}(n + 1) = [x_{n+1}, x_n, \ldots, x_{n-N}].$$

Then

$$P(n + 1) = P(n) - P(n)\Phi(n + 1)$$
$$\times [1 + \Phi^{\mathrm{T}}(n + 1)P(n)\,\Phi(n + 1)]^{-1}$$
$$\times \Phi^{\mathrm{T}}(n + 1)P(n). \tag{70}$$

Note that

$$P(n + 1)\Phi(n + 1) = P(n)\Phi(n + 1)$$
$$\times [1 + \Phi^{\mathrm{T}}(n + 1)P(n)\Phi(n + 1)]^{-1},$$

which we define as

$$K(n + 1) = P(n + 1)\Phi(n + 1). \tag{71}$$

Now considering the updated coefficient vector from eq. (65)

$$A(n + 1) = P(n + 1)X^{\mathrm{T}}(n + 1)d(n + 1), \tag{72}$$

where

$$d^{\mathrm{T}}(n + 1) = [d^{\mathrm{T}}(n)d_{n+1}], \tag{73}$$

d_{n+1} being the latest reference input value. Inserting eq. (70) into eq. (71) and using eqs. (67) and (72), we obtain

$$A(n + 1) = P(n + 1)[X^{\mathrm{T}}(n)d(n) + \Phi(n + 1)d_{n+1}]$$
$$= [\mathbf{I} - K(n + 1)\Phi^{\mathrm{T}}(n + 1)]$$
$$\times P(n)X^{\mathrm{T}}(n)d(n) + P(n + 1)\Phi(n + 1)d_{n+1}$$
$$= [\mathbf{I} - K(n + 1)\Phi^{\mathrm{T}}(n + 1)]A(n) + K(n + 1)d_{n+1}$$

or

$$A(n + 1) = A(n) + K(n + 1)[d_{n+1} - \Phi^{\mathrm{T}}(n + 1)A(n)]. \tag{74}$$

This is the recursive least squares estimator for the coefficients A. The vector K is usually referred to as the Kalman gain. Note that $\Phi^{\mathrm{T}}(n + 1)A(n) = y_{n+1}$, the filter output (see fig. 3) and eq. (62). Hence

$$A(n + 1) = A(n) + K(n + 1)\tilde{e}(n + 1). \tag{75}$$

The updating technique thus involves a correction on the coefficients based on the error generated by the receipt of the latest information $\{d_{n+1}, \text{ and } x_{n+1}\}$.

The algorithm may be initiated by choosing $P(0) = \alpha \mathbf{I}$ where α is a large positive value, such as $100 \cdot r_{xx}(0)$, and the coefficient vector $A(0) = \mathbf{0}$, for simplicity.

Referring to eq. (74) and taking expectations, we may approximate the above to (provided we consider independent data vectors)

$$E[A(n + 1)] = E[A(n)] + E[P(n + 1)][\Phi(n + 1) d_{n+1}]$$
$$- E[\Phi(n + 1)\Phi^{\mathrm{T}}(n + 1)] E[A(n + 1)].$$

Noting that $\Phi(n + 1)$ is the input data vector and $P(n)$ is a measure of the inverse of the data correlation matrix $[X^{\mathrm{T}}(n)X(n)]$, the above leads to

$$E[A(n + 1)] \approx E[A(n)] + E[(X^{\mathrm{T}}(n)X(n))^{-1}]$$
$$\times [r_{xd} - R_{xx}E[A(n)]]. \tag{76}$$

Thus in the mean the algorithm leads to a sequential Wiener–Hopf solution. Further, as $[X^{\mathrm{T}}(n)X(n)]^{-1}$ is the Hessian matrix for the least-squares criterion J, the RLS technique is a manifestation of the Newton–Raphson method for minimising the mean square error, with its attendant accelerated speed of convergence. In theory, for an N-element coefficient vector, the algorithm should converge in N iterations. Note that the initial rate of convergence of the algorithm is dictated by the initial value chosen for $P(0)$. For large $P(0)$ this effect damps out rapidly.

5. Adaptive lattices

Lattice configurations form an important class of architectures for signal processing. They possess regularity of structures comprising identical stages (sections), which have orthogonal properties, and involve bounded coefficients. Lattices are thus inherently stable. These properties make them particularly attractive for adaptive processing. Lattices arise in the autoregressive modelling of input data, and provide outputs which are the residuals of the models. Figure 6 illustrates an M-stage lattice. It consists of two channels corresponding to the outputs $\{f_m(k)\}$ and $\{g_m(k)\}$ which are referred to as the forward and backward residuals at various stages $\{m\}$. Each stage of the lattice comprises a pair of adders and (for modelling second-order stationary data) a reflection coefficient multiplier $\{K_m\}$ and a delay element. The governing equations for each

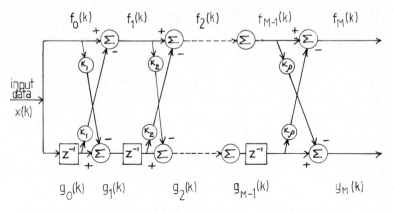

Fig. 6. Lattice structure.

stage of the lattice are

$$f_m(k) = f_{m-1}(k) - K_m g_{m-1}(k-1), \tag{77}$$

$$g_m(k) = g_{m-1}(k-1) - K_m f_{m-1}(k), \tag{78}$$

$$f_0(k) = g_0(k) = x(k), \qquad m = 1, 2, \ldots, M,$$

where $\{x(k)\}$ is the input data sequence, and $\{f_m(k)\}$ and $\{g_m(k)\}$ are the forward and backward residuals at the mth lattice stage, for the kth time instant. K_m is the reflection coefficient for the mth stage.

Defining the forward and backward transfer function between any mth stage output and the input as $A_m(z) = F_m(z)/X(z)$, and $B_m(z) = G_m(z)/X(z)$, where $F_m(z)$, $G_m(z)$ and $X(z)$ are the z-transforms of f_m, g_m and x respectively, and letting

$$A_m(z) = \sum_{p=0}^{m} a_m(p)z^{-p}, \qquad B_m(z) = \sum_{p=0}^{m} b_m(p)z^{-p}, \qquad a_0 = 1$$

for all m, then

$$f_m(k) = x(k) + \sum_{p=1}^{m} a_m(p)x(k-p). \tag{79}$$

Here the second term is the negative of the forward predictor of $x(k)$. Hence $f_m(k)$ is the mth-order forward residual, and $\{a_m(k)\}$ are the coefficients of the m-length one step ahead forward predictor. Using eqs.

(77) and (78) it may be shown that

$$g_m(k) = x(k - m) + \sum_{p=1}^{M} a_m(p)x(k - m + p). \qquad (80)$$

Thus $g_m(k)$ is the backward residual as the second term is the backward predictor of $x(k - m)$. The prediction coefficients are computed from the reflection coefficients as

$$a_m(m) = K_m,$$
$$a_m(s) = a_{m-1}(s) - K_m a_{m-1}(m - s), \quad s = 1, 2, \ldots, (M - 1). \qquad (81)$$

For stationary stochastic data, the reflection coefficient at each stage is evaluated by minimising both the forward and the backward residuals, i.e. by the following criterion:

$$J_m = E\left[f_m^2(k)\right] + E\left[g_m^2(k)\right]$$
$$= E\left[\left(f_{m-1}(k) - K_m g_{m-1}(k - 1)\right)^2\right.$$
$$\left. + \left(g_{m-1}(k - 1) - K_m f_{m-1}(k)\right)^2\right] \qquad (82)$$

which leads to

$$K_m = 2\frac{E\left[f_{m-1}(k)g_{m-1}(k - 1)\right]}{E\left[f_{m-1}^2(k)\right] + E\left[g_{m-1}^2(k)\right]}. \qquad (83)$$

It is easy to show that

$$E[g_p(k)\, g_q(k)] = 0, \qquad p = q,$$

and

$$E[f_m(k)\, x(k - j)] = 0, \qquad 1 \leqslant j \leqslant p,$$
$$E[g_m(k - 1)\, x(k - 1)] = 0, \qquad 1 \leqslant j \leqslant p. \qquad (84)$$

The last two follow from the orthogonal properties of residuals (mean square errors), and eq. (84) represents the orthogonalisation inherent in the lattice. As each stage involves a minimisation with respect to only one (reflection) coefficient, the global mean square prediction error minimisation problem is reduced to that of a stage by stage minimisation.

A number of methods have been proposed for estimating K_m, including an adaptive procedure by Griffiths (1978), which involves iterative estimation of the reflection coefficients by means of a gradient technique.

The method is susceptible to the usual problems associated with fixed-step-length algorithms in terms of convergence rates, misadjustment factors and choice of step length. A recursive least-squares technique has been proposed by Lee et al. (1981) and by Shensa (1981), amongst others. The computational burden is excessive, and algorithms which yield reflection coefficients which are the harmonic mean of the ratio between the forward and backward residual cross-correlations and residual variances, involve complicated stage by stage updating. A simple algorithm is included here (see Durrani and Murukutla, 1976) which gives an exact least-squares solution for the estimation of $\{\nabla_m\}$.

Consider the cost function required to design an M-stage lattice, for n data points as

$$J_m(n) = \frac{1}{2n} \sum_{t=1}^{n} \left[f_{m-1}(t) - K_m(n) g_{m-1}(t-1) \right]^2$$

$$+ \left[g_{m-1}(t-1) - K_m(n) f_{m-1}(t) \right]^2, \tag{85}$$

where $K_m(n)$ is an estimate of the mth reflection coefficient. It is obvious that $E[J_m(n)] = E[J_m]$ of eq. (82).

For minimisation, $\partial J_m(n)/\partial K_m(n) = 0$ yields

$$\frac{1}{n} \sum_{t=1}^{n} \left[f_{m-1}^2(t) + g_{m-1}^2(t-1) \right] K_m(n)$$

$$= \frac{2}{n} \sum_{t=1}^{n} f_{m-1}(t) g_{m-1}(t-1). \tag{86}$$

From this it may be seen that $E[K_m(n)] = E[K_m]$ of eq. (82). Defining

$$S_m(n) = \sum_{t=1}^{n} f_{m-1}^2(t) + g_{m-1}^2(t-1) \tag{87}$$

we obtain for eq. (86)

$$2 \sum_{t=1}^{n-1} f_{m-1}(t) g_{m-1}(t-1) = S_m(n-1) K_m(n-1). \tag{88}$$

Using eqs. (88) and (87) leads to (here the same term is added and subtracted)

$$S_m(n)K_m(n)$$

$$= S_m(n-1)K_m(n-1) + 2f_{m-1}(n)g_{m-1}(n-1)$$

$$+ \left[K_m(n-1) - K_m(n-1)\right]\left[f_{m-1}^2(n) + g_{m-1}^2(n-1)\right]$$

$$= S_m(n)K_m(n-1) + f_{m-1}(n)\left[g_{m-1}(n-1) - K_m(n-1)f_{m-1}(n)\right]$$

$$+ g_{m-1}(n-1)\left[f_{m-1}(n) - K_m(n-1)g_{m-1}(n-1)\right]. \tag{89}$$

Using the recursions in eq. (77) and (78), this leads to

$$K_m(n) = K_m(n-1) + S_m^{-1}(n)$$

$$\times \left[f_{m-1}(n)\tilde{g}_m(n-1) + \tilde{f}_m(n)g_{m-1}(n-1)\right]. \tag{90}$$

Note the differences in time updating between $f_m(n)$ and $\tilde{f}_m(n)$ and $g_m(n)$ and $\tilde{g}_m(n)$. Finally from the basic definition of $S_m(n)$, it is easy to show that

$$S_m^{-1}(n) = S_m^{-1}(n-1) - \frac{\left\{S_m^{-1}(n-1)\right\}^2\left\{g_{m-1}^2(n-1) + f_{m-1}^2(n)\right\}}{1 + S_m^{-1}(n-1)\left\{g_{m-1}^2(n-1) + f_{m-1}^2(n)\right\}}$$

$$\tag{91}$$

with initial values being $f_0(n) = g_0(n) = x(n)$. Equations (90) and (91) establish the time update equations for computing the reflection coefficients with increasing data lengths, $\{f_m^2(n)\}$ yields the forward mean square error at each stage and can be calculated iteratively, and further exploited to yield the optimal number of lattice stages. The order updating can be achieved either by using eq. (81) or by the method of Le Roux and Gueguen (1977) which is an implementation of the Schur algorithm, and for Toeplitz matrices allows order update of reflection coefficients without the need for computing the prediction coefficients $\{a_m, b_m\}$.

The algorithm lends itself to multichannel implementation. However other recursive least-squares algorithms are available for tackling multi-dimensional data (see Lee et al. 1981). Lattices play an important role in data equalisation as adaptive predictors, and for the modelling of speech. They are particularly useful in signal modelling and spectral analysis, and more recently have become so for signal detection (Durrani and Arslanian 1984), and with attendant lattice–ladder techniques for noise cancelling and radar signal processing (Corredera et al. 1985).

This is a rich area of research, and it should lead to important innovations and applications for signal processing, in terms of new algorithms as well as new architectures.

6. New directions in adaptive systems

Adaptive systems are essentially self-learning systems. They come into their own in applications which are concerned with input data characteristics which range from stationary stochastic data to data with slowly time-varying characteristics, through to non-stationary data. The aim of processing the data is to identify some underlying scenario, such as signal frequencies or strengths, directionality of incident wavefield, spectral characteristics of input data, and so on. In practice, an adaptive processing scheme would be application-specific, and the corresponding adaptive filter characteristics (coefficients or system response) would be domain-specific. Thus these coefficients can be considered to span a finite space specific to the application. The adaptive filtering problem is to determine the most likely vector from the space which would match the particulars (characteristics) of the input data to provide the 'optimal' filter for the current set of requirements.

Given that the adaptive scheme is application-specific, two new directions of work open up from the above discussion: (i) use of a code book of vectors, or (ii) use of a knowledge base. The first is based on exploiting the availability of a large set of vectors, predetermined either through prior experimentation or by a gradual assimilation of information, by storing (coefficients) vectors generated by conventional computation techniques. The prior information is used to make up a code book or coefficient (data-)base corresponding to possible situations, and the adaptive filter is operated by using the input data to call up an 'address' from the code book or data base to output the filter parameters relevant to the current situation. The 'learning' effort is involved in generating the code book, while the real-time computing effect is concerned with determining the relevant address. The latter may be achieved by using a simple 'distance' measure. The technique has been used very successfully in the compression-coding of speech. Several new methods have recently been proposed for the rapid scanning and sorting of a large data base, and thus the method offers speed as the only time required is to set up a 'distance' measure for the code book address and then to scan the code book for the 'closest' address (set of filter coefficients). The cost–speed trade-off is in the storage requirements for the data base.

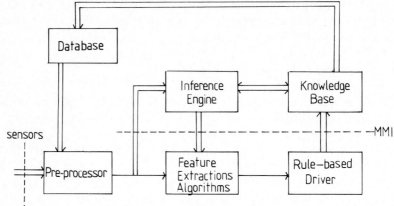

Fig. 7. Expert system for adaptive signal processing.

The second approach, which is concerned with the use of a knowledge base, is potentially much more powerful and comprehensive, and offers the prospect of marrying symbolic processing with signal processing and allows the use of both declarative programming and procedural programming.

Such a scheme would be constructed around an expert system shell for adaptive processing which will have as elements a knowledge base, an inference engine to drive the knowledge base, and a run-time generator. The scheme is depicted in fig. 7. the expert system would be embedded within a common processing framework. The mathematical processing (using procedural programming) will be concerned with the 'pre'-processing of data such as noise filtering or the (adaptive) filtering of the data to improve the input sequence from the data. This data would then be analysed, and relevant features extracted using conventional feature detection (or pattern recognition) algorithms. Initially a rule-driven knowledge base would be developed based on tests and expert knowledge. In the field, the knowledge base would feed the data base (or the equivalent of the proposed vector code-book) to drive the pre-processor (for instance, for specific SNR it would generate the appropriate filter weights). The processed data would be used to drive the inference engine which may call upon the knowledge base to determine the appropriate features (such as signal frequency, power, bearing) by calling upon the appropriate algorithms. The knowledge base would be replenished, via the rule driver. The proposed scheme has the value of allowing the most appropriate algorithm to be made available for the processing, plus ensuring that desired features are extracted and the data interpreted in the most suitable form. Such a scheme lifts conventional signal processing into a domain which allows ease of processing, detection and control, and facilitates interpretation and decision making. The computational costs are obviously significant. However, the facilities would, in general, outweigh these considerations. Reference to similar concepts is made in the Proceedings edited by Nii (1984), particularly for achieving signal-to-symbol transformation, which allows the combined use of low-level signal processing and knowledge representation and reasoning techniques for processing sonar data from linear arrays.

References

Bershad, N.J., and L.Z. Qu, 1984, IEEE Trans. Acoust., Speech & Signal Process. **ASSP-32**, 659–700.

Corredera, C., T.S. Durrani and A.S. Arslanian, 1985, Proc. Institute of Mathematics Conference, Bath UK (September 1985), ed. T.S. Durrani, to be published by Oxford University Press.

Durrani, T.S., and A.S. Arslanian, 1982, Proc. IEEE Int. Conf. on Acoustics, Speech and Signal Processing, Paris (April 1982) pp. 1021–1024.

Durrani, T.S., and A.S. Arslanian, 1984, Proc. IEEE Int. Conf. on Acoustics, Speech and Signal Processing, San Diego, CA (March 1984) pp. 47.4.1–47.4.4.

Durrani, T.S., and N.L.M. Murukutla, 1979, IEE Electr. Lett. **15**, 831–833.

Feuer, A., and E. Weinstein, 1985, IEEE Trans. Acoust., Speech & Signal Process. **ASSP-33**, 222–230.

Gardner, W.A., 1984, Signal Processing 6, 113–133.

Gitlin, R.D., and S.B. Weinstein 1979, Bell. Syst. Tech. J. **58**, 301–321.

Griffiths, L.J., 1969, Proc. IEEE **57**, 1696–1702.

Griffiths, L.J., 1978, Proc. IEEE Int. Conf. on Acoustics, Speech and Signal Processing, Tulsa, OK (April 1978) pp. 87–90.

Horowitz, L.L., and K.D. Senne, 1981, IEEE Trans. Acoust., Speech & Signal Process. **ASSP-29**, 722–736.

Le Roux, J., and C. Gueguen, 1977, IEEE Trans. Acoust., Speech & Signal Process. **ASSP-25**, 257–259.

Lee, D.T.L., et al., 1981, IEEE Trans. Acoust., Speech & Signal Process. **ASSP-29**, 627–641.

Ljung, L., M. Morf and D. Falconer, 1978, Int. J. Control **27**, 1–19.

Nii, P.N., ed. 1984, Proc. IEEE Int. Conf. on Acoustics, Speech and Signal Processing, San Diego, CA (March 1984) pp. 39A3.1–4.

Shensa, M.J., 1981, IEEE Trans. Autom. Control **AC-26**, 695–702.

Walach, E., and B. Widrow, 1984, IEEE Trans. Inf. Theory **IT-30**, 275–283.

Widrow, B., and M.J. Hoff, 1960, IRE WESCON Convention Record, Part 4, pp. 96–104.

Widrow, B., et al., 1975, Proc. IEEE **63**, 1692–1716.

COURSE 9

COHERENCE AND TIME DELAY ESTIMATION*

G. Clifford CARTER

Naval Underwater Systems Center
New London, CT 06320, USA

J.L. Lacoume, T.S. Durrani and R. Stora, eds.
Les Houches, Session XLV, 1985
Traitement du signal / Signal processing
© *Elsevier Science Publishers B.V., 1987*

Contents

Introduction

This course presents a tutorial review of work in coherence and time delay estimation. References to much of the relevant work in these two fields are included. It is a summary of work done by the author and several co-authors over more than a decade. A review of coherence research and development is presented. A derivation of the ML estimator for time delay is presented together with an interpretation of that estimator as a special member of a class of generalized cross correlators. The performance of the estimator is given for both high and low signal-to-noise ratio cases. The proposed correlator is implemented and stimulated with synthetic data. The results are compared with performance predictions and are found to be in good agreement. The paper is organized into 5 sections. Section 1 is a review of coherence. Section 2 is a review of the generalized framework for coherence estimation. Section 3 is a summary of statistics of the MSC estimator and contains subsections discussing the probability density function, experimental results, bias, receiver operating characteristics and confidence bounds. Section 4 discusses time delay estimation. Finally, section 5 discusses the focused time delay beamformer form of passive ranging.

1. Coherence

This section discusses application of the coherence function. Much of the material in this section is a summary of the work by Carter and Knapp (1975). The coherence function between two wide-sense stationary random processes x and y is equal to the cross-power spectrum $G_{xy}(f)$ divided by the square root of the product of the two auto-power spectra. Specifically, the complex coherence is defined by

$$\gamma_{xy}(f) = \frac{G_{xy}(f)}{\sqrt{G_{xx}(f)G_{yy}(f)}}, \tag{1.1}$$

where f denotes the frequency of interest. The coherence is a normalized

518

cross-spectral density function; in particular, the normalization constrains eq. (1.1) so that the magnitude-squared coherence (MSC), defined by

$$C_{xy}(f) \triangleq \left| \gamma_{xy}(f) \right|^2, \tag{1.2a}$$

lies in the range

$$0 \leqslant C_{xy}(f) \leqslant 1, \quad \nabla f. \tag{1.2b}$$

Throughout the text we use C and $|\gamma|^2$ interchangeably.

The coherence function has uses in numerous areas, including system identification, measurement of signal-to-noise ratio (SNR), and determination of time delay. The coherence—in particular, magnitude-squared coherence (MSC)—can only be put to use when its value can be accurately estimated. Indeed it is highly desirable to understand the statistics of the estimator. Therefore, this section addresses interpretations of the coherence function and following sections address procedures for properly estimating it as well as statistics of the estimator.

One interesting interpretation of coherence—particularly MSC—is that it is a measure of the relative linearity of two processes. To illustrate this, consider fig. 1 in which a sample function, $y(t)$, of an arbitrary stationary random process consists of the response, $y_0(t)$, of a linear filter plus an error component, $e(t)$. When the linear filter is chosen to minimize the mean-square value of $e(t)$, i.e., the area under the error spectrum, then $y_0(t)$ becomes that part of $y(t)$ linearly related to $x(t)$. The spectral characteristics of $e(t)$ are given by

$$\begin{aligned} G_{ee}(f) = G_{yy}(f) + G_{xx}(f)|H(f)|^2 \\ - H(f)G_{xy}^*(f) - H^*(f)G_{xy}(f), \end{aligned} \tag{1.3}$$

where the asterisk indicates complex conjugation and $H(f)$ is the filter

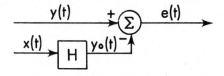

Fig. 1. Model of error resulting from linearly filtering $x(t)$ to match any desired signal $y(t)$.

transfer function. The error spectrum is given by

$$G_{ee}(f) = G_{xx}(f)\left|H(f) - \frac{G_{xy}(f)}{G_{xx}(f)}\right|^2 + G_{yy}(f)\left[1 - C_{xy}(f)\right].$$

$$(1.4)$$

Hence, the optimum filter is given by

$$H_0(f) = \frac{G_{xy}(f)}{G_{xx}(f)}.$$

$$(1.5)$$

Note that the coherence is related to the optimum linear filter according to

$$\gamma_{xy}(f) = H_0(f)\sqrt{\frac{G_{xx}(f)}{G_{yy}(f)}}$$

$$(1.6a)$$

and

$$C_{xy}(f) = |H_0(f)|^2\frac{G_{xx}(f)}{G_{yy}(f)}.$$

$$(1.6b)$$

These results apply regardless of the source of $y(t)$. When the linear filter is optimum in the mean-square sense, the error is uncorrelated with $x(t)$, i.e.,

$$G_{xy_0}(f) = H_0(f)G_{xx}(f) = G_{xy}(f).$$

$$(1.7)$$

Furthermore, the minimum value of $G_{ee}(f)$ is given by

$$G_{ee}(f) = G_{yy}(f)\left[1 - C_{xy}(f)\right]$$

$$(1.8)$$

and

$$G_{y_0y_0}(f) = |H_0(f)|^2G_{xx}(f) = G_{yy}(f)C_{xy}(f).$$

$$(1.9)$$

From the identity

$$G_{yy}(f) = C_{xy}(f)G_{yy}(f) + \left[1 - C_{xy}(f)\right]G_{yy}(f)$$

$$(1.10a)$$

it follows that

$$G_{yy}(f) = G_{y_0y_0}(f) + G_{ee}(f),$$

$$(1.10b)$$

indicating that the MSC is the proportion of $G_{yy}(f)$ contained in the linear component of $y(t)$, and $1 - C_{xy}(f)$ is the proportion of $G_{yy}(f)$ contained in the error, or the nonlinear component of $y(t)$.

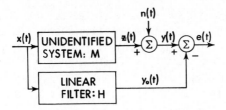

Fig. 2. Model of error resulting from linear approximation of an unidentified system.

These results can be applied to fig. 2 directly. For example, in the special case of fig. 2, when system M is linear and noise $n(t)$ is uncorrelated with $x(t)$, and the linear filter H is optimal, then $y_0(t) = z(t)$ and $e(t) = n(t)$. Thus, the noise power spectrum

$$G_{nn}(f) = G_{yy}(f)\left[1 - C_{xy}(f)\right]. \tag{1.11}$$

This is an intuitively satisfying result since if the MSC is unity, there is no noise, whereas if the MSC is zero, the output is all noise. For linear systems, additive noise uncorrelated with the input reduces the MSC according to the ratio of $G_{nn}(f)$ to $G_{yy}(f)$.

The power spectrum from the output of an arbitrary system can always be viewed in terms of its two components $G_{yy}(f)$ times $C_{xy}(f)$, and $G_{yy}(f)$ times $[1 - C_{xy}(f)]$, regardless of how $y(t)$ is produced. It is interesting to note that the ratio of these components can be viewed as a signal-to-noise ratio (SNR); when the system is a simple delay and the excitation is $s(t)$ with power spectrum $G_{ss}(f)$, then it follows that

$$\frac{G_{ss}(f)}{G_{nn}(f)} = \frac{C_{xy}(f)}{1 - C_{xy}(f)}, \tag{1.12}$$

or the linear-to-nonlinear ratio, depending on the application. For practical nonlinear systems, the identification of the optimum linear component is not as obvious as one might suspect. For example, in the system without noise described by $y(t) = x^3(t) + bx(t)$, the optimal linear part is not $bx(t)$. To clarify this point, it will be demonstrated that for a limited class of inputs and a limited class of nonlinearities, analytic expressions for the optimal linear part can be obtained. This offers interesting insight into both the general system identification problem and the coherence interpretation problem. First, the nonlinearity is constrained to have no memory and no noise, i.e. $y = f(x)$. Second, the input processes are constrained to be separable in the Nuttall (1958) sense. A description of separable processes is beyond the scope of this course. However, it can be shown that a Gaussian process possesses these properties and, hence, is a separable process.

Under the no-memory nonlinearity and separable process constraints, it has been proven that when $n(t) = 0$, the cross correlation between $x(t)$ and $y(t)$ at delay τ is given by

$$R_{xy}(\tau) = R_{xx}(\tau)K, \tag{1.13a}$$

where

$$K = \frac{1}{\sigma^2} \int f(x)(x - \mu)p(x)\,\mathrm{d}x, \tag{1.13b}$$

and $p(x)$ is the first-order probability density function of $x(t)$, $y = f(x)$ is a complete description of the nonlinear function without noise, μ is the mean of $x(t)$, and σ^2 is the variance of $x(t)$. Notice that K does not depend on frequency, but only on the first-order probability density and the nonlinearity. It follows directly that, for no-memory nonlinearities excited by separable processes,

$$\gamma_{xy}(f) = K\sqrt{\frac{G_{xx}(f)}{G_{yy}(f)}}\ . \tag{1.14}$$

Comparison of eq. (1.14) with eq. (1.6) shows that the constant K is the optimum linear filter in the mean-squared sense.

As an example, suppose $x(t)$ is Gaussian zero-mean with variance σ^2; then

$$K = \frac{1}{\sigma^2} \int_{-\infty}^{\infty} f(x)x \frac{1}{\sqrt{2\pi\sigma^2}} \exp\left(\frac{-x^2}{\sigma^2}\right) \mathrm{d}x. \tag{1.15}$$

Whenever $p(x)$ is even and $f(x)$ is an even function, $K = 0$ so that the coherence is zero. However, when $f(x)$ is an odd function, K does not necessarily vanish even though the unidentified system is nonlinear. For example, when $f(x) = x^3(t) + bx(t)$, application of eq. (1.15) yields $K = 3\sigma^2 + b$. Therefore, the optimal linear part of $x^3(t) + bx(t)$ is not $bx(t)$, but rather $y_0(t) = (b + 3\sigma^2)x(t)$ for a zero-mean Gaussian process with variance σ^2. For $b = 0$, it follows that $K \neq 0$ and $C_{xy}(f) \neq 0$, provided $G_{xx}(f) \neq 0$. However, if $b = -3\sigma^2$, then $K = 0$ and $C_{xy}(f) = 0$. Thus, the coherence may still be zero even though the nonlinearity is not even. A computer simulation of the example with $\sigma^2 = 1/2$

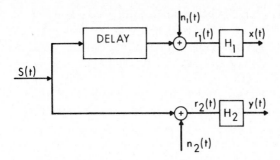

Fig. 3. Model of directional signal corrupted with additive noise and processed.

and $b = -3/2$ was conducted, and the results agreed with the theory. This result can be independently verified by calculating $R_{xy}(\tau) = E\{x(t)[x^3(t + \tau) + bx(t + \tau)]\}$, which for Gaussian processes is $3\sigma^2 R_{xx}(\tau) + bR_{xx}(\tau)$. Therefore, the MSC equals zero if $b = -3\sigma^2$, and there is no power in the optimum linear part of the nonlinearity $f(x) = x^3(t) - 3\sigma^2 x(t)$.

For situations like those shown in fig. 2, the coherence measures which proportion of an unidentified system output is "linear". Thus, the MSC provides a comparison of the proportion of system power that is linear with the proportion that is nonlinear in exactly the same way in which the SNR was measured for the output of a linear system corrupted by additive noise. However, in other system configurations, such as that shown in fig. 3, where noise and signal have a different model, different relationships hold. Figure 3 is analogous to the physical situation in which signal $s(t)$ from a distant source is received at two geographically separated sensors. Each observed signal is corrupted by additive stationary noise and is linearly filtered. When $n_1(t)$ and $n_2(t)$ are uncorrelated, but have the same power spectra $G_{nn}(f)$, the SNR is

$$\frac{G_{ss}(f)}{G_{nn}(f)} = \frac{\sqrt{C_{xy}(f)}}{1 - \sqrt{C_{xy}(f)}}, \tag{1.16}$$

which differs from eq. (1.12). Later we will see whether eq. (1.12) or eq. (1.16) plays a fundamental role in time delay estimation. Specific references to related work can be found in the article by Carter and Knapp (1975).

2. Generalized framework for coherence estimation

The purpose of this section is to review a generalized framework for power spectral estimation and to show how three estimation methods fit into this framework.

In the generalized framework of Nuttall and Carter (1980) we are concerned with both auto- and cross-spectral estimation; hence, we consider two discrete random processes. As is often the case in practice, we are limited to a single time-limited realization (TLR) of each random process. Within our generalized framework for power spectral estimation, we *first* partition each TLR into N segments, where the segments may be overlapped. *Second*, each segment is multiplied by a time-weighting function (the weighting function may be unity everywhere within the segment, as for example rectangular weighting, or it may be smooth, as for example Hanning weighting). *Third*, the discrete Fourier coefficients (DFC) are computed for each weighted segment via an appropriate algorithm such as the FFT after each segment has been appropriately appended with zeros. *Fourth*, the DFC's for one TLR segment are multiplied by the complex conjugate of the DFC's for the other segment (or same segment for auto spectra). *Fifth*, the complex products are averaged over the N available segments (one segment if $N = 1$). *Sixth*, the resultant spectral estimates are Fourier transformed into the correlation or lag domain, where they are multiplied by a lag-weighting function (which may be unity). *Finally*, the results are transformed back into the frequency domain. (Alternatively, the last two steps can be replaced by a convolution in the frequency domain. Depending on the extent of the frequency domain convolution, the former alternative may be computationally preferable over this latter.) Mathematical details can be found in the paper by Nuttall and Carter (1982).

We now point out how three spectral analysis techniques fit into this generalized framework. All three achieve virtually the same mean and variance performance. First, the Blackman and Tukey (1958) (BT) method allows for only one segment with rectangular time weighting over the entire record (from each TLR), and it applies a smooth lag-weighting function in the correlation domain, which goes to zero well before the end of the data record. (We note that, historically, the BT approach estimated the correlation function directly rather than first transforming into the frequency domain, but since it is faster than the original time domain BT method, it is a viable and equivalent alternative approach.) By adjusting how quickly the lag weighting goes to zero, resolution and

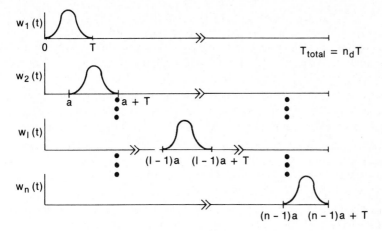

Fig. 4. Overlapped weighting functions.

stability can be compromised. For example, if the weighting goes to zero quickly the spectral estimates will have coarse resolution and good stability compared with a lag weighting that does not go to zero quickly. The exact shape of the weighting will influence the exact shape of the sidelobes and the main lobe in the frequency domain.

The second method that falls within this framework is the weighted overlapped-segment averaging or WOSA technique. In the WOSA method we apply a smooth multiplicative time weighting to each of a large number of segments, and average the DFC products from these overlapped segments to obtain a final spectral estimate, without employing additional lag weighting. See for example fig. 4 for a graphical portrayal of a family of overlapped time-weighting functions. The time weighting is typically a smooth weighting, such as the Hanning weighting in the WOSA method, because it yields good sidelobe behavior. Overlap is important in the WOSA method in order to realize maximum stability (that is, minimum variance) of the spectral estimate. The WOSA method is a statistically sound method widely in use today.

A third technique that falls within the generalized framework is the lag reshaping method (see e.g. Nuttall and Carter 1982). The lag reshaping method recognizes that the number of available data points may be so large as to preclude the normal BT method in practical situations. We segment the data without overlapping and apply unit gain rectangular time weighting to each segment (this rectangular weighting requires no time weighting multiplications). Later, we will undo the bad sidelobe effects that this rectangular time weighting initially causes, and gain additional stability. Note the segment-averaged power spectrum will be transformed into the correlation (or lag) domain, where a smooth multiplicative lag-weighting function will be applied before transforming back into the frequency domain. The smooth lag weighting will be the product of two lag weightings one for the desired window and one for lag reshaping. This lag-domain "reshaping" is an acceptable method for almost completely undoing the bad sidelobe effects of rectangular time weighting.

All three of these techniques that fall within the generalized framework have good statistical properties. The Blackman and Tukey (1958) method attains minimum variance spectral estimates and is the benchmark against which other techniques have been measured. The key assumptions for a statistical investigation of any spectral analysis technique are stationarity, Gaussian random processes, and a large product of observation time and desired resolution bandwidth. Since we are interested in spectral estimates, second-order stationarity is required. Since one must investigate variances of second-order quantities for stability determination, fourth-order moments of the random processes are required; hence, for mathematical tractability, the Gaussian assumption is needed. And in many practical cases, to have meaningful spectral estimates with any method, large observation-time resolution-bandwidth products are required. These assumptions are the practical essentials of any mathematical analysis of spectral estimation methods.

Under these assumptions, Nuttall (1971) has shown that the WOSA method can achieve the same stability, in particular, the same number of equivalent degrees of freedom (EDF) as the BT spectral estimation method for both auto- and cross-spectral estimation if the proper overlap is used for each weighting, when both methods operate on the same amount of data and are constrained to the same frequency resolution. For many practical time weightings, most of the maximum EDF (minimum variance) can be attained by a computationally reasonable amount of overlap. For example, with Hanning weighting, 92 percent of the maximum EDF can be realized with 50 percent overlap. And for Parzen (cubic) weighting, 93 percent of the maximum EDF can be realized with 62.5 percent overlap. Furthermore, the number of FFT's required is virtually independent of the particular time weighting employed (with its optimum overlap), but depends only upon the observation-time resolution-bandwidth product.

For more complicated spectral measures, such as coherence, the analysis of WOSA becomes unwieldy, and one is driven to simulation. In particular, Carter et al. (1973a) empirically investigated the effect of overlap on the variance of the coherence estimate via the WOSA method. There was a pronounced improvement (about a factor of two in variance reduction) with overlap as opposed to no overlap. In another experiment, Carter and Knapp (1975) compared the use of Hanning and rectangular weightings for estimating coherence between a flat broadband input to a second-order digital filter and its output. It was demonstrated that smooth weighting functions are required to obtain good coherence estimates with the WOSA method. (Recall that the WOSA method does not use additional lag shaping.)

The WOSA method with proper overlap can attain the EDF of the BT method, when both methods operate on the same amount of data and are constrained to have the same frequency resolution. Further, for good time weightings, reasonable amounts of overlap achieve most of the available EDF.

Based on analytic work by Nuttall and Carter (1982), the lag reshaping method can virtually attain the EDF of the BT method and yield very good sidelobes through the use of unusual lag weighting. It appears highly certain that the lag reshaping method requires fewer computations (perhaps by a factor of two) than the WOSA method in practice, and therefore deserves serious consideration as a replacement for the widely used WOSA method.

3. Statistics of MSC estimates obtained via the weighted overlapped segment averaging (WOSA) method

3.1. Introduction

Much of the historical work on the statistics of the MSC estimates centers on the WOSA method; by proper interpretation of variables, these results also apply to the lag reshaping method. Recall that the WOSA method consists of obtaining two finite-time series from the random processes being investigated. Each time series is partitioned into equal length segments and sampled at equally spaced data points. The segments are overlapped. However, the statistics are analytically developed for non-overlapped segments. Empirical results are presented for overlapped segments. Samples from each segment are multiplied by a weighting function, and the FFT of the weighted sequence is performed.

Then the Fourier coefficients for each weighted segment are used to estimate the auto- and cross-power spectral densities. The spectral density estimates thus obtained are used to form the MSC estimate.

Spectral resolution of the estimates varies inversely within the segment length T. Proper weighting or "windowing" of the T-second segment is also helpful in achieving good sidelobe reduction. On the other hand, for independent segments with ideal windowing, the bias and the variance of the MSC estimate vary inversely with the number of segments n. Therefore, to generate a good estimate with limited data, one may be faced with conflicting requirements on n and T. Segment overlapping can be used to increase both n and T. When the segments are disjoint, that is non-overlapping, we call the number of segments n_d. As the percentage of overlap increases, however, the computational requirements increases rapidly, while the improvement stabilizes owing to the greater correlation between data segments (see Carter et al. 1973a).

3.2. Probability density for the estimate of the MSC

The first-order probability density and distribution functions for the estimate of MSC, given the true value of MSC, and the number of independent segments n_d are given in table 1. Recall $|\gamma|^2 = C$. Equa-

Table 1
Probability density and distribution functions

Function	Expression
Density function	$p(|\hat{\gamma}|^2 \| n_d, |\gamma|^2) = (n_d - 1)(1 - |\gamma|^2)^{n_d}(1 - |\hat{\gamma}|^2)^{n_d - 2}$
	$\times {}_2F_1(n_d, n_d; 1; |\gamma|^2 |\hat{\gamma}|^2), \quad 0 \leqslant |\gamma|^2 |\hat{\gamma}|^2 < 1 \qquad (T1.1)$
	$= (n_d - 1)(1 - |\gamma|^2)^{n_d}(1 - |\hat{\gamma}|^2)^{n_d - 2}$
	$\times (1 - |\gamma|^2 |\hat{\gamma}|^2)^{1 - 2n_d} {}_2F_1(1 - n_d, 1 - n_d; 1; |\gamma|^2 |\hat{\gamma}|^2) \quad (T1.2)$
	$= (n_d - 1)\left[\dfrac{(1 - |\gamma|^2)(1 - |\hat{\gamma}|^2)}{(1 - |\gamma|^2 |\hat{\gamma}|^2)^2} \right]^{n_d}$
	$\times \dfrac{(1 - |\gamma|^2 |\hat{\gamma}|^2)}{(1 - |\hat{\gamma}|^2)^2} {}_2F_1(1 - n_d, 1 - n_d; 1; |\gamma|^2 |\hat{\gamma}|^2). \qquad (T1.3)$
Distribution function	$P(|\hat{\gamma}|^2 \| n_d, |\gamma|^2) = |\hat{\gamma}|^2 \left(\dfrac{1 - |\gamma|^2}{1 - |\gamma|^2 |\hat{\gamma}|^2} \right)^{n_d} \sum_{k=0}^{n_d - 2} \left(\dfrac{1 - |\hat{\gamma}|^2}{1 - |\gamma|^2 |\hat{\gamma}|^2} \right)^k$
	$\times {}_2F_1(-k, 1 - n_d; 1; |\gamma|^2 |\hat{\gamma}|^2). \qquad (T1.4)$

tions (T1.2) and (T1.3) in table 1 are useful because the $_2F_1$ hypergeometric function is an $(n_d - 1)$-order polynomial.

Figures 5 and 6 illustrate the probability density and distribution functions for several cases, as computed using eqs. (T1.2) and (T1.4) from table 1. It is evident from fig. 6 that the bias and variance of the MSC estimate decrease when n_d is increased.

The bias and variance of the MSC estimate can be evaluated using a general expression for the mth moment of the MSC estimate. (See Carter et al. 1973a.)

Bias and variance expressions obtained are summarized in table 2. Approximations (T2.3) and (T2.4) are the result of truncating the series (T2.1) and (T2.2). Equations (T2.5) through (T2.7) then follow for large n_d; they indicate that the MSC estimate is asymptotically unbiased, and that for large n_d the following is true.

(1) The bias is greatest, $1/n_d$, when the MSC equals zero and smallest, 0, when the MSC equals unity.

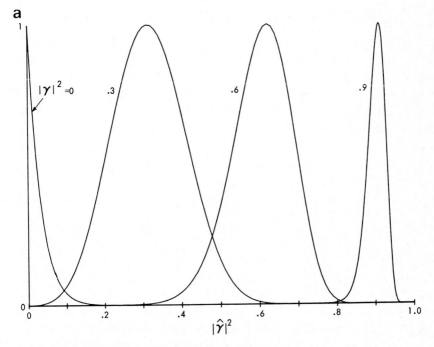

Fig. 5. (a) Probability density functions (functions have been normalized by maximum values, which are 31.0, 4.13, 5.23, and 17.5). (b, overleaf) Distribution functions.

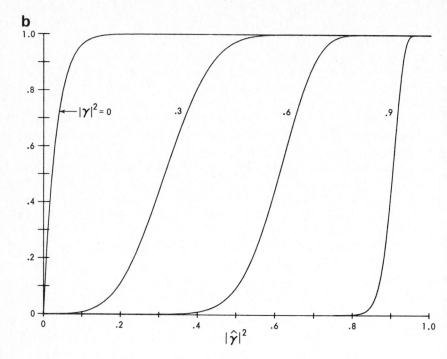

Fig. 5. Continued.

(2) The variance is zero when the MSC equals unity, and greatest, $(2/3)^3/n_d$, when the MSC equals one third.

(3) The mean-square error from the true value is equal to the variance, provided the MSC is not zero.

Figures 7 and 8, respectively, show the bias and variance as functions of n_d and $|\gamma|^2$. For values of n_d in the range from 32 to 64, expressions (T2.5) through (T2.7) are good approximations; however, the curves in figs. 7 and 8 were obtained using the exact formulas, (T2.1) and (T2.2). Peaks in the variance curves when the MSC equals one third are evident. We note, however, that there is an additional bias when our FFT size is too small. This second type of bias can be extremely important. It is the subject of many publications, including the books by Koopmans (1974) and Brillinger (1975), and is discussed in section 3.4.

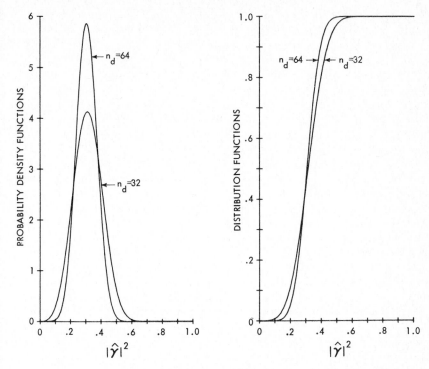

Fig. 6. Probability density functions (left) and distribution functions (right) of $|\hat{\gamma}|^2$ for $|\gamma|^2 = 0.3$.

3.3. Experimental investigation of overlap effects

An experimental study has been made of the effect of overlap of data on the MSC estimate. The analytical results presented earlier relate only to the case of independent segments, that is, the case of zero overlap. Intuitively, the application of non-overlapped smooth weighting functions does not make the most efficient use of the data when forming the MSC estimate. The experiment described herein examines this inefficiency, in terms of bias and variance of the MSC estimate as a function of different amounts of overlap.

The method of evaluating overlap is straightforward in concept. Data are generated with an accurately pre-specified value of MSC that is independent of frequency. Then the sample mean and variance of the

Table 2
Bias (B) and variance (V) expressions

Function	Expression	
Exact	$B = E[\lvert\hat{\gamma}\rvert^2 \mid n_d, \lvert\hat{\gamma}\rvert^2] - \lvert\gamma\rvert^2 = \dfrac{(1-\lvert\gamma\rvert^2)^{n_d}}{n_d}\,{}_3F_2(2, n_d, n_d; n_d+1, 1; \lvert\gamma\rvert^2) - \lvert\gamma\rvert^2$	(T2.1)
	$\quad = \dfrac{1}{n_d} + \dfrac{n_d-1}{n_d+1}\lvert\gamma\rvert^2\,{}_2F_1(1,1; n_d+2; \lvert\gamma\rvert^2) - \lvert\gamma\rvert^2.$	
	$V = E(\lvert\hat{\gamma}\rvert^4 \mid n_d, \lvert\gamma\rvert^2) - E^2(\lvert\hat{\gamma}\rvert^2 \mid n_d, \lvert\gamma\rvert^2)$	(T2.2)
	$\quad = \dfrac{2(1-\lvert\gamma\rvert^2)^{n_d}}{n_d(n_d+1)}\,{}_3F_2(3, n_d, n_d; n_d+2, 1; \lvert\gamma\rvert^2) - \left[\dfrac{(1-\lvert\gamma\rvert^2)^{n_d}}{n_d}\,{}_3F_2(2, n_d, n_d; n_d+1, 1; \lvert\gamma\rvert^2)\right]^2.$	
Approximate	$B_0 \cong \dfrac{1}{n_d} - \dfrac{2}{n_d+1}\lvert\gamma\rvert^2 + \dfrac{1!(n_d-1)}{(n_d+1)(n_d+2)}(\lvert\gamma\rvert^2)^2 + \dfrac{(n_d-1)2!}{(n_d+1)(n_d+2)(n_d+3)}(\lvert\gamma\rvert^2)^3;$	
	$B \cong \begin{cases} B_0, & B_0 \geqslant 0, \\ 0, & B_0 < 0. \end{cases}$	(T2.3)

$$V_0 \cong \frac{(n_d - 1)}{n_d(n_d + 1)} \left[\frac{1}{n_d} + 2\frac{n_d - 2}{n_d + 2}|\gamma|^2 - 2\frac{2n_d^3 - n_d^2 - 2n_d + 3}{(n_d + 1)(n_d + 2)(n_d + 3)}(|\gamma|^2)^2 \right.$$

$$+ 2\frac{n_d^4 - 6n_d^3 - n_d^2 + 10n_d - 8}{(n_d + 1)(n_d + 2)(n_d + 3)(n_d + 4)}(|\gamma|^2)^3$$

$$+ \left. \frac{13n_d^5 - 15n_d^4 - 113n_d^3 + 27n_d^2 + 136n_d - 120}{(n_d + 1)(n_d + 2)^2(n_d + 3)(n_d + 4)(n_d + 5)}(|\gamma|^2)^4 \right]; \quad V \cong \begin{cases} V_0, & V_0 \geqslant 0, \\ 0, & V_0 < 0. \end{cases} \qquad (\text{T2.4})$$

Approximation for large n_d

$$B \cong \frac{1}{n_d^2}[1 - |\gamma|^2]^2 \qquad (\text{T2.5})$$

$$\leqslant \frac{1}{n_d}[1 - |\gamma|^2] \qquad (\text{T2.6})$$

$$V \cong \begin{cases} \dfrac{1}{n_d^2}, & |\gamma|^2 = 0, \\[2mm] \dfrac{2|\gamma|^2}{n_d}(1 - |\gamma|^2)^2, & 0 < |\gamma|^2 \leqslant 1. \end{cases} \qquad (\text{T2.7})$$

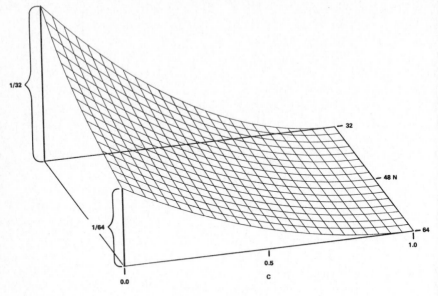

Fig. 7. Bias of $|\hat{\gamma}|^2$ versus $|\gamma|^2$ and n_d.

MSC estimate can be measured for the given overlap by averaging over frequency. Details are described in the paper by Carter et al. (1973a).

Results of the experiment are summarized in figs. 9 and 10. It is apparent from these results that the bias and variance can be reduced through overlapped processing. For example, when the MSC is 0.0, the variance of the estimator with a 50-percent overlap equals 31 percent of the variance of the estimator with no overlap. With a 50-percent overlap, the bias is 55 percent as large as without overlap. Similarly, when the MSC = 0.3 and the overlap is 50 percent, the variance is 55 percent of the non-overlapped estimator, and the bias is 50 percent as large. Observe that the bias and variance for zero overlap agree very well with eqs. (T2.5) through (T2.7) in table 2. With a 62.5-percent overlap, or greater, the bias and variance achieve values corresponding to an effective n_d of about 64 with non-overlapped processing. In the WOSA method, clearly, as the overlap increases, the computational cost must also increase. Increasing the overlap from 50 to 62.5 percent requires 32 percent more FFT's, but the variance of the MSC estimator decreases only from 80 to 95 percent of its value at a 50-percent overlap. It is doubtful, therefore, that the improvement in the WOSA method derived

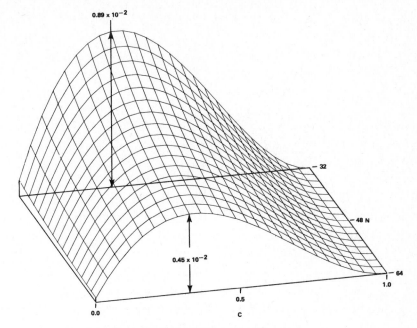

Fig. 8. Variance of $|\hat{\gamma}|^2$ versus $|\gamma|^2$ and n_d.

from using a 62.5-percent overlap, as opposed to a 50-percent overlap, will warrant the increased computational costs, except in unusual circumstances. Overlap percentages of 50 percent are quite reasonable and widely used. However, with the advent of the lag reshaping method other computational efficiencies need to be explored.

3.4. MSC bias

One type of bias, derived under simplifying assumptions including that each data segment is sufficiently long to ensure adequate spectral resolution, has been shown by Nuttall and Carter (1976) from table 2 to be

$$E[\hat{C}] - C \cong \frac{1}{n_d}(1 - C)^2\left(1 + \frac{2C}{n_d}\right), \tag{3.1}$$

where E denotes the expected value. Equation (3.1) corroborates the observation by Bendat and Piersol (1971) that more than one segment

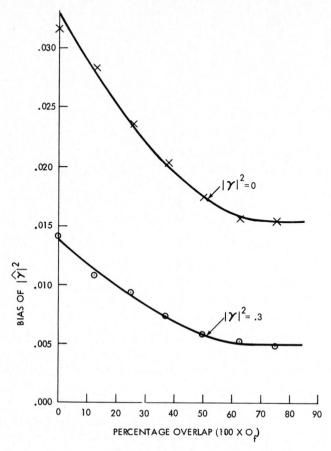

Fig. 9. Bias of $|\hat{\gamma}|^2$ when $n_d = 32$.

must be used to estimate MSC. Indeed, for $n_d = 1$, the estimated MSC equals unity regardless of the true value of MSC.

However, there is a second type of bias, described by Koopmans (1974), that can be extremely serious. This is the bias due to misalignment, or rapidly changing phase. In particular, Koopmans (1974) notes that if the phase angle of the cross power spectrum is a rapidly varying function of frequency at the frequency that the coherence is to be estimated, the estimated coherence (in particular, MSC) can be biased downward to such an extent that a strong coherence is masked. An

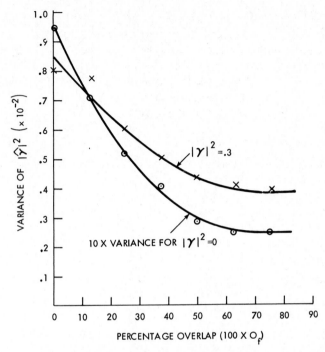

Fig. 10. Variance of $|\hat{\gamma}|^2$ when $n_d = 32$.

expression for the bias as a function of the first derivative of the phase spectrum was given by Jenkins and Watts (1968). Based on results which follow here, Koopmans' (1974) statement is correct; however, Jenkins and Watts' (1968) bias results are quantitatively incorrect for the application here. A brief derivation of the effect of misalignment akin to rapidly changing phase will be given below. These results compare favorably with analytical results by Halvorsen and Bendat (1975), with empirical results by Carter and Knapp (1975) and with empirical results to be presented in this course.

Rapidly varying phase as a function of frequency is caused by a time delay. One way to see this is to consider that the units of the slope of the phase are radians divided by radians per second, or simply seconds. The data can be realigned to compensate for a time delay. As stated by Brillinger (1975), the importance of some form of prefiltering cannot be over-emphasized, the simplest form being to lag one time series relative to the other (we note here, this is also true for time delay estimation). This procedure for coherence estimation has been suggested by others, including Akaike and Yamanouchi (1963), Jenkins and Watts (1968), and Koopmans (1974). Important to the concept of prefiltering two time series before estimating the MSC is that (unlike the estimated value of MSC) the (true value of) MSC is invariant under the linear filtering of the two series, as shown, for example, by Carter et al. (1973a), and by Koopmans (1974).

The effect of misalignment can be seen in the correspondence by Carter (1980). The results of that work show that the magnitude of the cross-power spectrum (and cross correlation) is decreased by a constant factor, depending on the ratio of the delay misalignment to the FFT time duration. Note, though, that the average phase estimate remains unaltered. Further, we note that the constant degradation factor will not appear in either of the auto-power spectral densities. Thus, the complex coherence is degraded by the same factor as the cross spectrum and the MSC is degraded by the square of this factor. That is,

$$E[\hat{C}(f)] \cong \left(1 - \frac{|D|}{T}\right)^2 C(f), \tag{3.2}$$

which agrees with the results of Halvorsen and Bendat (1975). Heuristically, this makes sense because for no delay there is no degradation and for a delay equal to, or greater than, the FFT size, the estimated MSC is zero. Note that the bias due to misalignment D, with FFT time duration T, is

$$E[\hat{C}] - C \cong \frac{-2|D|}{T} C + \left(\frac{|D|}{T}\right)^2 \tag{3.3a}$$

$$\cong \frac{-2|D|}{T} C, \qquad |D| \ll T. \tag{3.3b}$$

For example, if $|D|/T = 0.25$, the expected value of the estimated MSC from (3.3b) is about one-half of its true value. Clearly, the effect is important. Indeed empirical results bear this out.

One of the results of Carter and Knapp (1975) was the demonstration of the need to make the FFT (or equivalent transform) size larger. Empirically, large T was observed to reduce the bias in MSC estimation. This is consistent with eq. (3.3). Looking at these data again (fig. 5 of Carter and Knapp 1975), we see that the phase appears to have undergone a 1.5 radian change in 10 Hz, or a 24 ms delay was encountered in the band. Since each FFT was 500 ms, the estimated value of MSC should be about 0.91 of its true value. Indeed, this is what was indicated in fig. 7 of Carter and Knapp (1975).

One practical means of reducing the bias due to a single path misalignment is to realign the two time series under investigation before estimating the MSC. The effects of misalignment were evident in an empirical investigation in which a broadband underwater acoustic signal was transmitted through a direct path from a submerged transmitter to a submerged receiver. The source and received signals were recorded on two different tape recorders with a stable servo-lock. Subsequently, the

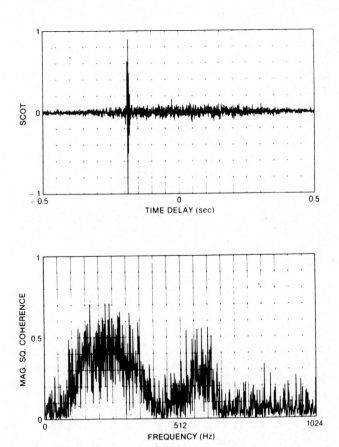

Fig. 11. SCOT estimate showing a −180 ms delay (top) and the corresponding MSC estimate of 0.45 at 250 Hz with a −180 ms delay (bottom).

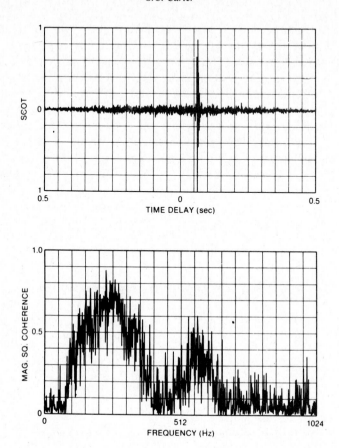

Fig. 12. SCOT estimate showing a 70 ms delay (top) and the corresponding MSC estimate of 0.7 at 250 Hz with a 70 ms delay (bottom).

two recorded signals were played back, digitized, and processed with a number of different bulk delays inserted before estimating the MSC. The bulk delays were quantized to 250 ms; the FFT size was 1.0 s.

The effect of degraded MSC estimation is evident in figs. 11 and 12, computed from 16 independent FFT's. In fig. 11, $|D| = 0.18$ and the estimated MSC appears to be about 0.45 at 250 Hz. From eq. (3.2) we predict that this estimated MSC is about 0.67 times the true value. Thus, we might expect the true value to be 0.672 (namely, 0.45/0.67). In fig. 12, introducing another 250 ms bulk delay moves the generalized cross correlation SCOT peak from −0.18 to 0.07. (See, for example, Carter et al. (1973b), Knapp and Carter (1976), and Kuhn (1978) for a discussion of the SCOT.) Now the estimated MSC is about 0.7 at 250 Hz in fig. 12. This notable increase in the MSC estimate is due to realignment. From eq. (3.2) we predict that the estimated MSC is still about 0.865 times the true value. Thus, we might expect that the true value is 0.81 (0.7/0.865). In both cases we expect that the true MSC is in about the 0.7 to 0.8 range, but our estimates varied from 0.4 to 0.7 because of misalignment when estimating the MSC.

Thus, we see that even with a large number of FFT segments, estimates of the MSC can be significantly biased downward, giving an erroneous indication of the value of the coherence. When the data are realigned and processed, estimates of the coherence are informative descriptors of the extent to which the ocean channel can be modeled by a linear time-invariant filter.

3.5. Receiver operating characteristics for a coherence detector

An algorithm for computing the receiver operating characteristics (ROC) or the probability of detection, P_D, versus the probability of false alarm, P_F, for a linearly thresholded coherence estimation detector is presented together with an example of an ROC table. More details can be found in Carter (1977b). An article by Gevins et al. (1975) presents results on using linearly thresholded coherence estimates to detect biomedical phenomena. The desire to establish a threshold, below which coherence estimates are not presented to a human decision maker, is an important issue in certain areas, such as brain wave analysis and sonar, where the volume of sensor data is large. For a fixed amount of averaging and a fixed threshold value E in the absence of a coherent source, there is still a certain probability P_F that an estimated value of coherence will exceed the threshold. Moreover, although the false alarm rate can be reduced by increasing E, to do so decreases P_D when a coherent source is present. How much it decreases P_D will depend on the strength of the coherent source, that is, the true or underlying coherence that is being estimated. We present an algorithm for computing P_D versus P_F for a specified amount of averaging and underlying coherence. Under simplifying assumptions the probability density function of $\hat{C}$, when $C = 0$, is obtained from table 1. In particular,

$$p\left(\hat{C}|n_d, C = 0\right) = (n_d - 1)(1 - C)^{(n_d - 2)}. \tag{3.4}$$

Hence, the probability of false alarm is

$$P_F = 1 - \int_0^E (n_d - 1)(1 - \hat{C})^{(n_d - 2)} \, d\hat{C}, \tag{3.5a}$$

and the threshold,

$$E = 1 - (P_F)^{1/(n_d - 1)}, \tag{3.5b}$$

that is, for a specified P_F we establish a threshold according to eq. (3.5b). Now the computationally more complex question is: what probability of detection is achieved for this threshold value E? The answer is

$$P_D = \int_E^1 p(\hat{C}|n_d, C) \, d\hat{C} = 1 - P(C \leqslant E|n_d, C), \tag{3.6}$$

where $P(\hat{C} \leqslant E/n_d, C)$ is the cumulative distribution function (CDF). The CDF is given in table 1. An example is illustrative.

For models of the form

$$x(t) = z_1(t) + n_1(t), \tag{3.7a}$$

$$y(t) = z_2(t) + n_2(t), \tag{3.7b}$$

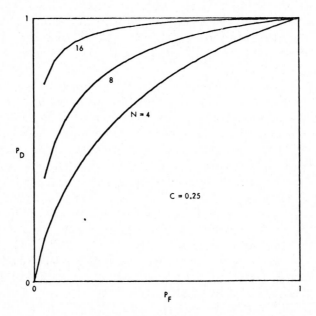

Fig. 13. ROC curves for $|\gamma|^2 = C = 0.25$, $n_d = N = 4, 8$ and 16.

where $z_1(t)$ is the output of a linear filter $H_i(f)$ excited by $s(t)$, $i = 1, 2$ and the noises are mutually uncorrelated and also uncorrelated with the signal; then it can be shown that

$$C_{xy}(f) = C_{sx}(f)C_{sy}(f), \tag{3.8}$$

that is, the coherence between two receivers is the product of the coherence between the source and each of the individual receivers for the model (3.7). Substituting results in

$$\frac{G_{z_1 z_1}(f)}{G_{n_1 n_1}(f)} \cdot \frac{G_{z_2 z_2}(f)}{G_{n_2 n_2}(f)} = \frac{C_{xy}(f)}{[1 - C_{sx}(f)][1 - C_{sy}(f)]}. \tag{3.9}$$

Now, if $C_{sx}(f) = C_{sy}(f) = [C_{xy}(f)]^{1/2}$, then it follows that

$$\left[\frac{G_{z_1 z_1}(f) G_{z_2 z_2}(f)}{G_{n_1 n_1}(f) G_{n_2 n_2}(f)} \right]^{1/2} = \frac{\sqrt{C_{xy}(f)}}{1 - \sqrt{C_{xy}(f)}}. \tag{3.10}$$

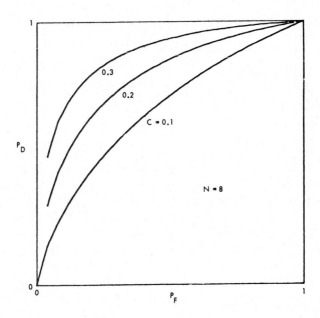

Fig. 14. ROC curves for $n_d = N = 8$, $|\gamma|^2 = C = 0.1, 0.2$ and 0.3.

```
1Ø    N=8
2Ø    N1=N-1
3Ø    N2=N-2
4Ø    A=1-N
5Ø    C=Ø.25
6Ø    PRINT "THIS RUN IS FOR N="N" AND MSC="C"
7Ø    FOR F1=Ø.Ø4 TO 1 STEP Ø.Ø4
8Ø    E=1-EXP(LOG(F1)/N1)
9Ø    Z=E*C
1ØØ   C4=(1-E)/(1-Z)
11Ø   C2=E*((1-C)/(1-Z))↑N
12Ø   S=Ø
13Ø   FOR L=Ø TO N2
14Ø   C3=C4↑L
15Ø   T=1
16Ø   F=1
17Ø   IF (L=Ø) THEN 23Ø
18Ø   FOR K=1 TO L
19Ø   K1=K-1
2ØØ   T=T*(A+K1)*(K1-L)*Z/(K*K)
21Ø   F=F+T
22Ø   NEXT K
23Ø   S=S+C3*F
24Ø   NEXT L
25Ø   P=C2*S
26Ø   FIXED 3
27Ø   PRINT E;F1;P,1-P
28Ø   NEXT F1
29Ø   END
```

Fig. 15. Computer listing of ROC program.

Hence, to study the 0 dB (or equal signal-to-noise) case, we must select

$$10 \log\left[\frac{\sqrt{C}}{1 - \sqrt{C}}\right] = 0, \tag{3.11}$$

which implies $C = 0.25$. The ROC curves for $|\gamma|^2 = C = 0.25$ and $N = n_d = 4, 8,$ and 16 independent data segments respectively are given in fig. 13. As can be seen in the figure, performance can be improved by increasing n_d, if a sufficient amount of stationary data exists; if not, n_d can only be increased at the expense of degrading the frequency resolution with its inherent difficulties. If n_d is fixed, performance is determined by the underlying coherence or, equivalently, the signal-to-noise ratio (see e.g., fig. 14). For many particular problems, the performance will be desired for different values of n_d and C. Because of the large number of possible choices for these parameters, we will not present an exhaustive series of results. A basic computer program listing is given in fig. 15 to generate ROC curves.

3.6. Confidence bounds for magnitude-squared coherence estimates

In many applications, two received signals are digitally processed to estimate coherence. Results of computing coherence estimate confidence bounds for stationary Gaussian signals are presented. Computationally difficult examples are given for 80 and 95 percent confidence with independent averages of 8, 16, 32, 64, and 128. A more complete discussion can be found in Scannell and Carter (1978).

The MSC is useful in detection and is also of value in estimating the amount of coherent power common between two received signals. Therefore, it would be desirable, having estimated a particular value of MSC, to state with certain confidence that the true coherence falls in a specified interval. A general discussion of confidence intervals is available in work by Cramér (1946). Early attempts to do this for 95 percent confidence were accomplished by Haubrich (1965) who apparently used precomputed CDF curves and used a different method of presentation than the one used here. Related confidence work for the magnitude coherence (MC) is presented by Koopmans (1974). Empirical results for 95 percent confidence are given by Benignus (1969, 1970). The confidence limits given here appear to agree with approximate results in work by Bendat and Piersol (1971), and Enochson and Goodman (1965), and with results of Brillinger (1975) from tabulated densities. Gosselin (1977) compared MSC detectors with other detectors using the notion of ROC curves.

A computer program has been written (see Scannell and Carter 1978) to evaluate the CDF and confidence limits. Recall the CDF is a finite sum of $_2F_1$ hypergeometric functions, each of which is a polynomial, as given in table 1. When C equals zero or unity, CDF values can be computed in closed form.

Let C be the true value of an unknown parameter and let $\hat{C}$ be its estimate. $\hat{C}$ is a random variable (RV) with a known probability density function (pdf), $p(\hat{C}|C)$. [The conditioning on C indicates that the shape of the pdf of $\hat{C}$ depends on the exact (unknown) value of C.]

Suppose we choose $A_L(C)$ and $A_u(C)$ such that

$$\text{Prob}\big(A_L(C) < \hat{C}|C\big) = \int_{A_L(C)}^{\infty} \mathrm{d}\hat{C}\, p(\hat{C}|C) = 0.95 \text{ (say)}, \quad (3.12a)$$

and

$$\text{Prob}\big(\hat{C} < A_u(C)|C\big) = \int_{-\infty}^{A_u(C)} \mathrm{d}\hat{C}\, p(\hat{C}|C) = 0.95 \text{ (say)}. \quad (3.12b)$$

Then the probability that RV $\hat{C}$ lies in the range $(A_L(C), A_u(C))$ is

$$\text{Prob}\big(A_L(C) < \hat{C} < A_u(C)|C\big) = 0.90. \tag{3.13}$$

Now assume that $A_L(C)$ and $A_u(C)$ are monotonically increasing with C, and are continuous. Then there follows

$$\text{Prob}\big(A_u^{-1}(\hat{C}) < C < A_L^{-1}(\hat{C})|C\big) = 0.90. \tag{3.14}$$

Therefore, the confidence interval for C is

$$\big(A_u^{-1}(\hat{C}), A_L^{-1}(\hat{C})\big), \quad \text{with confidence coefficient } 0.90. \tag{3.15}$$

Given a measurement $\hat{C}$, this interval can be computed once the functions $A_u^{-1}(\cdot)$ and $A_L^{-1}(\cdot)$ are known.

Figure 16 presents computer-generated 80 percent and 95 percent confidence limits. The computer program is listed in the conference paper by Scannell and Carter (1978). The five pairs of curves in each figure are for $n_d = 8, 16, 32, 64$, and 128 from outer to inner, respectively. Because an estimate involves a particular value of n_d, only one pair of curves applies. An excellent discussion of the types of statements that can be made with confidence bonds is given by Cramér (1946). Suppose we obtain an estimated MSC of 0.7 from $n_d = 8$ disjoint FFT's: then we draw a horizontal line from 0.7 on fig. 16 for 95 percent confidence limits and see where it intersects the pair of $n_d = 8$ (outer) curves. This occurs at (approximate abscissa values) 0.3 and 0.86. Thus, we state with 95 percent confidence that the true but unknown parameter C falls in the interval $(0.3, 0.86)$. No matter what the true value of C is, we have a 5 percent probability of giving an incorrect statement. That is, if we make many estimates of MSC and keep applying the rule described (whether or not C is random or constant), we will correctly include the true value of C in the interval that we specify 95 percent of the time. Sometimes the method of applying the rule is in doubt; for example in fig. 16 if the estimate comes out to be 0.3 and $n_d = 8$, then a horizontal line does not intersect the upper confidence limit unless we extrapolate it backwards. Thus, we could say that with 95 percent confidence the true MSC is in the region $(-0.1, 0.62)$. Since we know a priori that the true value of C is nonnegative, we could just as easily say (but with no more confidence) that with 95 percent confidence (for $n_d = 8$ and $C = 0.3$) the true MSC falls in the region $(0.0, 0.62)$. Moreover, if both intersections result in negative regions (as for example when $C = 0.001$ and $n_d = 8$), we will have to say that with 80 percent confidence the true MSC lies in $(0.0, 0.0)$. (This is in part due to the fact that the estimator is biased.)

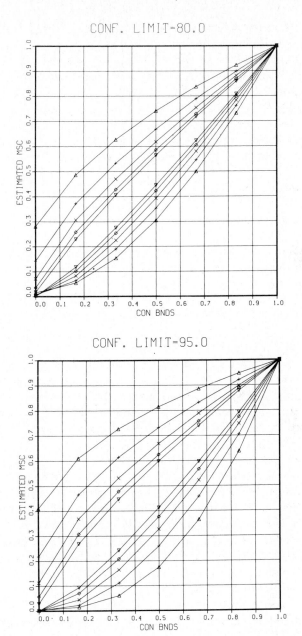

Fig. 16. 80 percent (top) and 95 percent (bottom) MSC estimate confidence bounds for *N* = 8, 16, 32, 64 and 128.

However, if we continue to apply the rule and run the experimental trials we will make correct statements "80 percent" of the time. It is interesting to note that larger values of N do not always result in the upper confidence bound being lower. This also occurs in MC estimate confidence limits (see Koopmans 1974). It is also interesting to note that while increasing n_d is desirable, the confidence bounds for $n_d = 128$ are still very large. For example, even when $n_d = 128$, if $C = 0.3$, the 95 percent confidence intervals are still $(0.2, 0.38)$ and the 80 percent confidence intervals $(0.24, 0.36)$ are not much better.

4. Time delay estimation

4.1. Introduction

A signal emanating from an underwater acoustic source and monitored in the presence of noise at two spatially separated sensors can be mathematically modeled in the direct path as

$$x_1(t) = s(t) + n_1(t), \tag{4.1a}$$

$$x_2(t) = s(t - D) + n_2(t), \tag{4.1b}$$

where $s(t)$, $n_1(t)$ and $n_2(t)$ are real, jointly stationary random processes. Signal $s(t)$ is assumed to be uncorrelated with noise $n_1(t)$ and $n_2(t)$.

There are many applications in which it is of interest to estimate the time delay D. This section reviews the derivation of a maximum likelihood (ML) estimator given by Knapp and Carter (1976). While the model of the physical phenomena presumes stationarity, the techniques to be developed herein are usually employed in slowly varying environments were the characteristics of the signal and noise remain stationary only for finite observation time T. Studies of more complex effects are given by Bjørnø (1981), Tacconi (1977), Griffiths et al. (1973), and Chan (1984). Further, the time delay D may also change slowly requiring time varying or adaptive techniques such as those of Griffiths (1984), Owsley (1984), Picinbono (1984), Meyr (1976), and Lindsey and Meyr (1977). Other investigations of the time varying case are studied by Knapp and Carter (1977), Adams et al. (1980), Carter and Abraham (1980), and Schultheiss and Weinstein (1981).

Another important consideration in estimator design is the available amount of prior knowledge of the signal and noise statistics. In many problems, this information is negligible. For example, in passive detection, unlike the usual communications problems, the source spectrum is unknown or only known approximately.

4.2. Derivation of the ML estimator

The ML estimator is derived as follows. Assume that signals and noises are Gaussian. Denote the Fourier coefficients of $x_i(t)$ by

$$X_i(k) = \frac{1}{T} \int_{-T/2}^{T/2} x_i(t) e^{-jkt\omega_\Delta} \, dt, \tag{4.2a}$$

where

$$\omega_\Delta = \frac{2\pi}{T}. \tag{4.2b}$$

Note that the linear transformation $X_i(k)$ is Gaussian since $x_i(t)$ is Gaussian. Further, as T goes to infinity and k goes to infinity such that $k\omega_\Delta = \omega$ is constant, then

$$\tilde{X}_i(\omega) = \lim_{T \to \infty} T X_i(k) \tag{4.3a}$$

$$= \int_{-\infty}^{\infty} x_i(t) e^{-j\omega t} dt, \tag{4.3b}$$

where $\tilde{X}_i$ is the Fourier transform of $x_i(t)$. A more complete discussion on Fourier transforms and their convergence is given in standard texts. When the observation time is large compared to the correlation time of the signal plus the magnitude of the delay, then

$$E[X_1(k)X_2^*(l)] \cong \begin{cases} \dfrac{1}{T} G_{x_1 x_2}(k\omega_\Delta), & k = l, \\ 0, & k \neq l. \end{cases} \tag{4.4}$$

Now let the vector

$$X(k) = [X_1(k), X_2(k)]', \tag{4.5}$$

where the prime denotes transpose. Define the power spectral density matrix Q such that

$$E[X(k)X^{*\prime}(k)] = E\begin{bmatrix} X_1(k)X_1^*(k) & X_1(k)X_2^*(k) \\ X_2(k)X_1^*(k) & X_2(k)X_2^*(k) \end{bmatrix} \tag{4.6a}$$

$$= \frac{1}{T} \begin{bmatrix} G_{x_1 x_1}(k\omega_\Delta) & G_{x_1 x_2}(k\omega_\Delta) \\ G_{x_1 x_2}^*(k\omega_\Delta) & G_{x_2 x_2}(k\omega_\Delta) \end{bmatrix} \tag{4.6b}$$

$$\triangleq \frac{1}{T} Q(k\omega_\Delta). \tag{4.6c}$$

Recall from section 1 that the magnitude-squared coherence (MSC),

$$C_{x_1x_2}(f) = \frac{\left|G_{x_1x_2}(f)\right|^2}{G_{x_1x_1}(f)G_{x_2x_2}(f)}. \tag{4.7}$$

Using the positive semidefinite properties of the spectral density matrix Q one can readily show that C is bounded by zero and unity. The vectors $X(k)$, $k = -N, -N+1, \ldots, N$ are uncorrelated Gaussian (hence, independent) random variables. More explicitly, the probability density function for $X = [X(-N), X(-N+1), \ldots, X(N)]$, given the power spectral density matrix Q (or the delay, and spectral characteristics of the signal and noises necessary to determine Q), is

$$p(X|Q) = p\left(X|G_{ss}, G_{n_1n_1}, G_{n_2n_2}, G_{n_1n_2}, D\right) \tag{4.8a}$$

$$= c_p \exp\left(-\tfrac{1}{2}J_1\right), \tag{4.8b}$$

where

$$J_1 = \sum_{k=-N}^{N} X^{*\prime}(k)Q^{-1}(k\omega_\Delta)X(k)T, \tag{4.8c}$$

and c_p is a function of $|Q(k\omega_\Delta)|$. With proper substitution we obtain

$$J_1 = \sum_{k=-N}^{N} \tilde{X}^{*\prime}(k\omega_\Delta)Q^{-1}(k\omega_\Delta)\tilde{X}(k\omega_\Delta)\frac{1}{T}. \tag{4.8d}$$

For ML estimation it is desired to choose D to maximize $p(X|Q, D))$.

In general, the parameter D affects both c_p and J_1 in $p(\cdot)$. However, under certain simplifying assumptions, c_p is constant or is only weakly related to the delay. Specifically, suppressing the frequency argument,

$$|Q| = \left(G_{ss} + G_{n_1n_1}\right)\left(G_{ss} + G_{n_2n_2}\right)$$

$$- \left(G_{n_1n_2} + G_{ss}\,e^{-j2\pi fD}\right)\left(G_{n_1n_2}^* + G_{ss}\,e^{+j2\pi fD}\right), \tag{4.9}$$

which is independent of D if $G_{n_1n_2} = 0$ (i.e., the noises are uncorrelated). For large observation times we have

$$J_1 \cong \int_{-\infty}^{\infty} \tilde{X}^{*\prime}(f)Q^{-1}(f)\tilde{X}(f)\,\mathrm{d}f. \tag{4.10}$$

The inverse of the spectral density matrix is given by

$$Q^{-1}(f) = \frac{\begin{bmatrix} G_{x_2x_2}(f) & -G_{x_1x_2}(f) \\ -G_{x_1x_2}^*(f) & G_{x_1x_1}(f) \end{bmatrix}}{G_{x_1x_1}(f)G_{x_2x_2}(f) - |G_{x_1x_2}(f)|^2}. \tag{4.11}$$

When the noises are uncorrelated:

$$G_{x_1x_1}(f) = G_{ss}(f) + G_{n_1n_1}(f), \tag{4.12a}$$

$$G_{x_2x_2}(f) = G_{ss}(f) + G_{n_2n_2}(f), \tag{4.12b}$$

$$G_{x_1x_2}(f) = G_{ss}(f)\,e^{-j2\pi fD}, \tag{4.12c}$$

and it follows that

$$J_1 = \int_{-\infty}^{\infty} \tilde{X}^{*\prime}(f)Q^{-1}(f)\tilde{X}(f)\,\mathrm{d}f = J_2 + J_3, \tag{4.13a}$$

where

$$J_2 = \int_{-\infty}^{\infty} \left[\frac{|\tilde{X}_1(f)|^2}{G_{x_1x_1}(f)} + \frac{|\tilde{X}_2(f)|^2}{G_{x_2x_2}(f)} \right] \cdot \frac{1}{[1 - C_{12}(f)]}\,\mathrm{d}f, \tag{4.13b}$$

$$-J_3 = \int_{-\infty}^{\infty} A(f) + A^*(f)\,\mathrm{d}f, \tag{4.13c}$$

where

$$A(f) = \tilde{X}_1(f)\tilde{X}_2^*(f) \cdot \frac{G_{ss}(f)\,e^{j2\pi fD}}{G_{x_1x_1}(f)G_{x_2x_2}(f)[1 - C_{12}(f)]}. \tag{4.13d}$$

Note that for real signals and noise, $A^*(f) = A(-f)$ and it follows that

$$-J_3 = \int_{-\infty}^{\infty} A(f)\,\mathrm{d}f + \int_{-\infty}^{\infty} A(-f)\,\mathrm{d}f = 2\int_{-\infty}^{\infty} A(f)\,\mathrm{d}f. \tag{4.14}$$

Letting $T\hat{G}_{x_1x_2}(f)$ be defined as $\tilde{X}_1(f)\tilde{X}_2^*(f)$, we have

$$-J_3 = 2T\int_{-\infty}^{\infty} \hat{G}_{x_1x_2}(f) \frac{1}{|G_{x_1x_2}(f)|} \frac{C_{12}(f)}{[1 - C_{12}(f)]} \cdot e^{j2\pi fD}\,\mathrm{d}f. \tag{4.15}$$

Notice that the ML estimator for D will minimize $J_1 = J_2 + J_3$, but the selection of D has no effect on J_2. Thus, D should maximize $-J_3$.

Equivalently, the ML estimator selects as the estimate of delay the value of τ at which

$$R_{y_1y_2}(\tau) = \int_{-\infty}^{\infty} \hat{G}_{x_1x_2}(f) \frac{1}{\left| G_{x_1x_2}(f) \right|} \frac{C_{12}(f)}{[1 - C_{12}(f)]} \, e^{j2\pi f\tau} \mathrm{d}f \quad (4.16)$$

achieves a peak.

These results compare favorably with closely related work by MacDonald and Schultheiss (1969), Hannan and Thomson (1971, 1973, 1981), Hahn and Tretter (1973) and Cleveland and Parzen (1975). A more complete discussion of the derivation and related references can be found in work by Knapp and Carter (1976).

4.3. Interpretation of the ML estimator

One common method of determining the time delay D is to compute the standard cross correlation function

$$R_{x_1x_2}(\tau) = E[x_1(t)x_2(t - \tau)], \quad (4.17)$$

where E denotes expectation. The argument τ that maximizes eq. (4.17) provides an estimate of delay. Because of the finite observation time, however, $R_{x_1x_2}(\tau)$ can only be estimated. For example, for ergodic processes, an estimate of the cross correlation is given by

$$\hat{R}_{x_1x_2}(\tau) = \frac{1}{T - \tau} \int_{\tau}^{T} x_1(t)x_2(t - \tau) \, \mathrm{d}t, \quad (4.18)$$

where T represents the observation time. In order to improve the accuracy of the delay estimate $\hat{D}$, it is desirable to pre-filter $x_1(t)$ and $x_2(t)$ prior to cross correlation. We call this simple but very important process generalized cross correlation for lack of a better description. As shown in fig. 17a, x_i may be filtered through H_i to yield y_i for $i = 1, 2$. The resultant y_i are cross-correlated, that is, multiplied and integrated, for a range of hypothesized time delays or time shifts, τ, until the peak is obtained. The time shift causing the peak is an estimate of the true delay $\hat{D}$. When the filters $H_1(f) = H_2(f) = 1$, for all f, the estimate D is simply the abscissa value at which the standard cross-correlation function peaks. Knapp and Carter (1976) provide for a generalized correlation through the introduction of the filters $H_1(f)$ and $H_2(f)$ which, when properly selected, can significantly enhance the estimation of time delay.

a

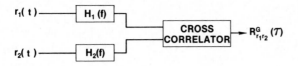

GCC FUNCTION

$$R^G_{r_1 r_2}(T) = \int_{-\infty}^{\infty} W(f)\, G_{r_1 r_2}(f)\, e^{j2\pi f \tau}\, df \quad = \int_{-\infty}^{\infty} W_\phi(f)\, e^{j\phi(f)}\, e^{j2\pi f \tau}\, df$$

WEIGHTING FUNCTION

$$W(f) = H_1(f)\, H_2^*(f), \quad W_\phi(f) = \left| G_{r_1 r_2}(f) \right| W(f)$$

b

METHOD	$W(f) = H_1(f)\, H_2^*(f)$	$W_\phi(f) = W(f)\, \lvert G_{r_1 r_2}(f)\rvert$
SCC	1	$\lvert G_{r_1 r_2}(f)\rvert$
ROTH	$1/G_{r_1 r_1}(f)$	$\lvert G_{r_1 r_2}(f)\rvert / \lvert G_{r_1 r_1}(f)\rvert$
WIENER PROCESSOR	$C_{r_1 r_2}(f)$	$C_{r_1 r_2}(f)\, \lvert G_{r_1 r_2}(f)\rvert$
SCOT	$1/\sqrt{G_{r_1 r_1}(f)\, G_{r_2 r_2}(f)}$	$\sqrt{C_{r_1 r_2}(f)}$
PHAT	$1/\lvert G_{r_1 r_2}(f)\rvert$	1
ML	$\dfrac{C_{r_1 r_2}(f)}{[1 - C_{r_1 r_2}(f)]\lvert G_{r_1 r_2}(f)\rvert}$	$\dfrac{C_{r_1 r_2}(f)}{1 - C_{r_1 r_2}(f)}$

$$C_{r_1 r_2}(f) = \frac{\lvert G_{r_1 r_2}(f)\rvert^2}{G_{r_1 r_1}(f)\, G_{r_2 r_2}(f)}$$

Fig. 17. (a) Generalized cross-correlation (GCC) function. (b) Various GCC functions.

The cross correlation between $x_1(t)$ and $x_2(t)$ is related to the cross-power spectral density function by the well-known Fourier transform relationship

$$R_{x_1 x_2}(\tau) = \int_{-\infty}^{\infty} G_{x_1 x_2}(f)\, e^{j2\pi f \tau} df. \tag{4.19}$$

When $x_1(t)$ and $x_2(t)$ have been filtered with filters having transfer functions H_1 and H_2 respectively, as depicted in fig. 17a, the cross-power spectrum between the filter outputs is given by

$$G_{y_1y_2}(f) = H_1(f)H_2^*(f)G_{x_1x_2}(f),$$ (4.20)

where the asterisk denotes the complex conjugate. Therefore, the generalized cross-correlation or GCC function between $x_1(t)$ and $x_2(t)$ is

$$R_{y_1y_2}(\tau) = \int_{-\infty}^{\infty} W(f)G_{x_1x_2}(f)\,e^{j2\pi f\tau}df,$$ (4.21a)

where the generalized frequency weighting

$$W(f) = H_1(f)H_2^*(f).$$ (4.21b)

In practice, only an estimate of the cross-power spectral density can be obtained from finite observations of the received signals. Consequently, the integral

$$\hat{R}_{y_1y_2}(\tau) = \int_{-\infty}^{\infty} W(f)\hat{G}_{x_1x_2}(f)\,e^{j2\pi f\tau}df$$ (4.22)

is evaluated and used for estimating delay. Indeed, depending on the particular form of $W(f)$ and prior information available, it may also be necessary to estimate the generalized weighting. For example, when the role of the prefilters is to accentuate the signal passed to the correlator at those frequencies at which the coherence or signal-to-noise ratio (SNR) is highest, then $W(f)$ can be expected to be a function of the coherence or signal and noise spectra which must either be known or estimated. Besides the maximum likelihood (ML) weighting there is an entire family of generalized weightings. See fig. 17b for some common GCC weightings.

The ML weighting is

$$W_{ML}(f) = \frac{1}{|G_{x_1x_2}(f)|} \cdot \frac{C_{12}(f)}{[1 - C_{12}(f)]}.$$ (4.23)

When $|G_{x_1x_2}(f)|$ and $C_{12}(f)$ are known, this is exactly the proper weighting. When these terms are unknown, they can be estimated via lag reshaping spectral estimation techniques discussed earlier or using techniques of Marple (1984), or the WOSA method of Carter et al. (1973a), or classical methods of Jenkins and Watts (1968) and Bendat and Piersol (1980). Substituting estimated weighting for true weighting is entirely a heuristic procedure whereby the ML estimator can approximately be achieved in practice. When the noises are uncorrelated but have the same power spectrum we can show

$$W_{ML}(f) = \frac{G_{ss}(f)/G_{nn}^2(f)}{[1 + 2G_{ss}(f)/G_{nn}(f)]}. \tag{4.24}$$

For small SNR we have

$$W_{ML}(f) \cong \frac{G_{ss}(f)}{G_{nn}^2(f)}, \tag{4.25}$$

which is the well-known Eckart filter (see Eckart 1952) used in optimum signal detection at low SNR. Thus the prefilters used for optimum signal detection at low SNR are the same prefilters as used for minimum variance time-delay estimation at low SNR.

4.4. Fundamental performance limits

The Cramér–Rao lower bound (CRLB) is given by

$$\text{var} \geqslant \frac{-1}{E\left[\dfrac{\partial^2 \ln p(X/Q, \tau)}{\partial \tau^2}\right]}\Bigg|_{\tau = D}. \tag{4.26}$$

The only part of the log density which depends on τ, the hypothesized delay, is J_3. More explicitly,

$$E\left[\frac{\partial^2}{\partial \tau^2} \ln p(X/Q, \tau)\right] = \frac{\partial^2}{\partial \tau^2} E(-\tfrac{1}{2}J_3). \tag{4.27}$$

If $G_{x_1 x_2}(f) = |G_{x_1 x_2}(f)| e^{-j2\pi f D}$, then (since the complex cross-spectral estimator is unbiased) it follows that

$$E(-\tfrac{1}{2}J_3) = T\int_{-\infty}^{\infty} e^{j2\pi f(\tau - D)} \frac{C_{12}(f)}{[1 - C_{12}(f)]}\, df. \tag{4.28}$$

Hence, the minimum variance of any time delay estimator is

$$\min \text{var} = \left[T\int_{-\infty}^{\infty} (2\pi f)^2 \frac{C_{12}(f)}{[1 - C_{12}(f)]}\, df\right]^{-1}. \tag{4.29}$$

This is the minimum variance and that which the ML processor achieves asymptotically for sufficiently large T. For constant signal and noise power spectra it follows that

$$\frac{C_{12}(f)}{1 - C_{12}(f)} = \frac{(\text{SNR})^2}{1 + 2(\text{SNR})},$$ (4.30)

so that

$$\text{min var} = \frac{1}{2T \int_0^B (2\pi)^2 f^2 \left[\dfrac{\text{SNR}^2}{1 + 2(\text{SNR})} \right] df},$$ (4.31a)

or simply

$$\sigma_{\text{CRLB}}^2 = \frac{3}{8\pi^2} \frac{[1 + 2(\text{SNR})]}{(\text{SNR})^2} \cdot \frac{1}{B^3 T}.$$ (4.31b)

When the signal and noise have the same flat power spectra, as noted by Scarbrough et al. (1983), for the time delay estimation (TDE) problem, this form of the Cramér–Rao lower bound (CRLB) is commonly used as the performance standard. The CRLB yields a lower bound on the variance of any unbiased time delay estimate as a function of several parameters [e.g., the signal and noise power spectra and the integration (observation) time]. Part of the appeal of the CRLB is that for cases of practical interest, there is a theorem which states that the maximum likelihood (ML) estimate can be made arbitrarily close to the CRLB for sufficiently long integration times, (see Van Trees 1968). However, the theorem does not specify how long the integration time must be. Thus, while the CRLB sets a lower bound on the variance of the time delay estimate, actual performance can be much worse for a given signal-to-noise ratio (SNR) and observation time. This is corroborated by the simulation results of Scarbrough et al. (1981, 1983) and Hassab and Boucher (1979). Several studies have been conducted to find a bound tighter than the CRLB which would predict performance more accurately. Work in this area has been done by Ianniello et al. (1983), Weiss and Weinstein (1982, 1983), and Chow and Schultheiss (1981). Ianniello (1981) has developed a correlator performance estimate (CPE). It has been shown via simulation that, for the cross correlation technique of TDE, the CPE yields a more accurate estimate of performance than the CRLB, especially at low SNR (see Ianniello 1982). The following discussion presents some additional comparisons of the CPE and CRLB and discusses some implications of these comparisons. In particular, the behavior of the CPE and the CRLB is considered as a function of the observation time and SNR, and the implications of this behavior are considered as related to coherent and incoherent signal processing techniques for time delay estimation. In addition, simulation results are presented to support the inferences of the theoretical analysis.

Chow and Schultheiss (1981), Scarbrough (1984), Betz (1984) and Johnson et al. (1983) have studied the low SNR problem. In the work here, consideration will be limited to signal and noise power spectra $G_{ss}(f)$ and $G_{nn}(f)$, respectively, which are flat (constant) over the frequency range $-B$ to $+B$ Hz and zero outside this range. B is then a measure of the source signal bandwidth. Additionally, it will be assumed that the bandwidth-observation time product BT is large (say $BT \geqslant 100$, e.g., for $B = 100$ Hz, $T \geqslant 1$ s).

In general, proper prefiltering prior to cross correlation is required to achieve the ML estimate of time delay. Under the special conditions here, the ML estimate of the time delay can be obtained by computing the cross-correlation function $R(\tau)$ between $x_1(t)$ and $x_2(t)$. The ML time delay estimate is the value of τ which maximizes $R(\tau)$. For other power spectra, a generalized cross correlator with the proper prefilters is required.

For the signals of interest in this section, the minimum variance of the TDE for the CRLB can be expressed as (see Quazi 1981)

$$\sigma_{\text{CRLB}}^2 = \frac{3}{8\pi^2} \cdot \frac{[1 + 2(\text{SNR})]}{(\text{SNR})^2} \cdot \frac{1}{B^3 T}, \tag{4.32}$$

where T is the observation time and $\text{SNR} = G_{ss}(f)/G_{nn}(f) = $ constant for $|f| \leqslant B$. Note that in this definition, SNR is the signal-to-noise ratio at the input of a single receiver.

The CPE was developed by Ianniello (1982) to provide more accurate performance prediction for the cross-correlation technique for TDE in the presence of large estimation errors or anomalous estimates. The CPE assumes that the anomalous estimates will be uniformly distributed across the correlation window, say, from $-T_0$ to $+T_0$ s. The CPE yields the following estimate of the variance of the TDE error:

$$\sigma_{\text{CPE}}^2 = \frac{PT_0^2}{3} + (1 - P)\sigma_{\text{CRLB}}^2, \tag{4.33}$$

where P is the probability of an anomalous estimate. For the signal and

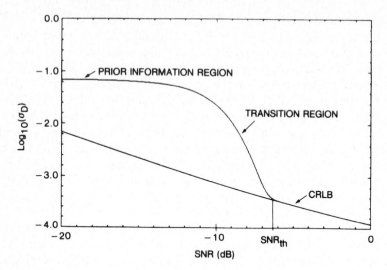

Fig. 18. Plot of CPE and CRLB versus SNR (dB) ($T = 8$ s, $B = 100$ Hz, $T_0 = \frac{1}{8}$ s).

noise spectra considered here, P can be approximated as

$$P \cong 1 - \int_{-\infty}^{\infty} \frac{1}{\sqrt{2\pi}} \exp\left[-\tfrac{1}{2}(x-\alpha)^2\right]\left[\int_{-\infty}^{\beta x} \frac{dy}{\sqrt{2\pi}} \exp\left(-\tfrac{1}{2}y^2\right)\right]^{M-1} dx,$$

$$(4.34a)$$

where

$$\alpha = \frac{\sqrt{2BT}\,(\mathrm{SNR})}{\left[(\mathrm{SNR})^2 + (1+\mathrm{SNR})^2\right]^{1/2}},$$

$$(4.34b)$$

$$\beta = \left[1 + \frac{(\mathrm{SNR})^2}{(1+\mathrm{SNR})^2}\right]^{1/2},$$

$$(4.34c)$$

and $M = 4BT_0$. The probability of anomaly must be evaluated numerically to obtain the probability of anomaly P for a given set of parameters B, T, T_0, and SNR.

The CPE and CRLB are compared in fig. 18 for the case of $B = 100$ Hz, $T = 8$ s, and $T_0 = \frac{1}{8}$ s relative to an assumed sampling frequency of 2048 Hz. Both curves in fig. 18 are plotted as the base 10 logarithm of σ_D, the standard deviation of the time delay error, versus the SNR in dB.

The upper curve is the CPE; the lower curve is the CRLB. The CPE is characterized by three regions: (1) at low SNR there is a region where prior information limits the variance (e.g., the maximum observable delay T_0 is known), (2) at moderate SNR there is a transition region from the prior information limit to the CRLB, and (3) at high SNR the CPE coincides with the CRLB. The SNR at which the CPE begins to deviate from the CRLB is referred to as the threshold SNR (SNR$_{th}$ in fig. 18). We note also that while the CPE is not a bound, recent work has been done to derive a bound tighter than the CRLB. This new bound is close to the CPE, falling just below the CPE in fig. 18. This new bound, the Ziv–Zakai lower bound (ZZLB) is discussed by Ianiello et al. (1983).

4.5. Simulation description and results

A computer simulation was conducted to corroborate the theoretical TDE performance predictions. Earlier simulations have been conducted by Hassab and Boucher (1981). Good agreement among many earlier simulations was shown in a comparison of a large number of simulation results by Kirlin and Bradley (1982). The cross correlation of simulated received sequences was computed using the FFT approach described by Oppenheim and Schafer (1975). After cross correlation the estimate of the time delay was obtained by finding the delay value for which the cross correlation was a maximum. This approach yields the time delay estimate quantized in units of the sampling interval.

The simulation was conducted for two integration times over a range of SNR values. Unlike many earlier simulations the effects of low SNR were studied. Different integration times are obtained by varying the number of cross-power spectra which are averaged before taking the IFT (inverse Fourier transform) to obtain the estimate of the cross-correlation function. 8 and 32 data segments were processed coherently in the simulation to obtain integration times of 2 and 8 s respectively, for the case of 512 point data segments and an assumed sampling frequency of 2048 Hz. A total of 2000 trials was conducted at each SNR to obtain the experimental time delay variances. These results are plotted in fig. 19a along with the corresponding theoretical curves for the CRLB and the CPE. The symbol size of the experimental points is indicative of the 90 percent confidence limits. The theory and results are in very close agreement and support the previous analysis. Of profound importance is that the CRLB is a poor predictor of performance at low SNR. This has significant implications for using techniques to increase coherent

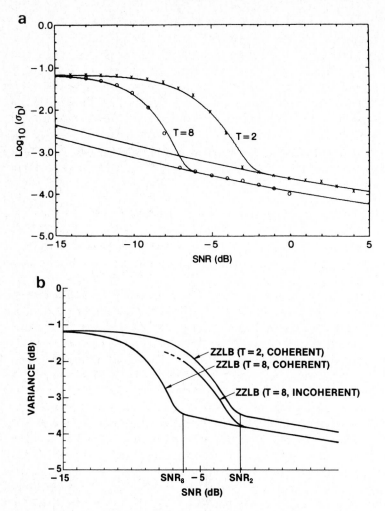

Fig. 19. (a) Comparison of CPE, CRLB, and simulation results ($B = 100\,\text{Hz}$, $T_0 = \frac{1}{8}\,\text{s}$, $T = 2$ and $8\,\text{s}$). (b) Plot of ZZLB and effects of coherent versus incoherent processing.

processing time as discussed by Scarbrough et al. (1983) and by Scarbrough (1984). These results are graphically portrayed in fig. 19b. Note that now we plot the Ziv–Zakai lower bound (ZZLB) as opposed to the CPE. Also we compare a coherent or tracking approach for $T = 8$ with an incoherent approach. The threshold is clearly reduced by coherent processing.

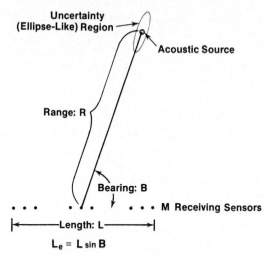

Fig. 20. Array geometry used to estimated source position.

5. Focussed time delay estimates for ranging

5.1. Introduction

The maximum-likelihood (ML) processor is presented for passively estimating range and bearing to an acoustic source. The source signal is observed for a finite-time duration at several sensors in the presence of uncorrelated noise. When the speed of sound in an isovelocity medium and the sensor positions are known, the ML estimator for position constrains the source to sensor delays to be focussed into a point corresponding to a hypothesized source location. The variances of the range error and bearing error are presented for the optimum processor. It is shown that for bearing and range estimation, different sensor configurations are desirable. However, if the area of uncertainty is to be minimized, then the sensors should be divided into equal groups with one-third of the sensors in each group.

An underwater acoustic point source radiating energy to several collinear receiving sensors is shown in fig. 20. The position of the source in two-space can be characterized by range R and bearing B from a given frame of reference. The particular geometry of interest is two-dimensional with an acoustic point source whose range and bearing are to be estimated by a fixed number of receivers. For the purposes of this work, we presume that the receiving hydrophones are collinear. However, regardless of the hydrophone positions, a fixed number of sensors have an inherent uncertainty in estimating source location. This uncertainty region is nominally elliptical, so that by properly defining how range and bearing are measured, the estimation errors can be decoupled (see fig. 20). For a collinear array of sensors, we measure the bearing as the angle between the line array and the major axis of the uncertainty region.

The uncertainty in measuring R and B is characterized by an extremely elongated elliptical uncertainty region. The problem addressed here is how to estimate range R and bearing B to a source when M sensors separated by a maximum of L (meters) have observed T seconds of received data. We will examine the maximum-likelihood (ML) technique for position estimation.

5.2. Mathematical model

For our purposes we assume that each receiving sensor at the tth instant in time corresponds to a signal plus noise. Namely, the ith (of M) sensor outputs is characterized by

$$r_i(t) = s(t + D_i) + n_i(t), \quad i = 1, M, \quad 0 \leqslant t \leqslant T. \tag{5.1}$$

The signal and noises are uncorrelated and the noises are mutually uncorrelated. Without loss of generality, $D_1 = 0$.

For a spatially stationary, that is, nonmoving, source the signal can be viewed as an attenuated and delayed source signal. However, it is felt that the problem of estimating the position of a stationary source is considerably easier than that of a moving source. Thus, the results here serve as a bound on performance; still we will see that it is extremely difficult in the best case to estimate source range.

5.3. Maximum-likelihood performance

The maximum-likelihood (ML) estimate for the time delay vector has been derived for stationary Gaussian processes by Hahn (1975) and Carter (1976). It is not difficult to show that an ML estimate for range and bearing when the sensor (element) positions are known is achieved by a variation of the work by Carter (1976). In particular, by focussing all the time-delay elements at many (hypothesized) range and bearing

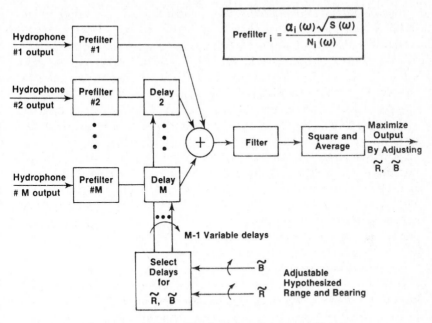

Fig. 21. Maximum-likelihood estimator for range and bearing.

pairs and watching for the peak output of the ML time-delay vector system, the ML position estimate is observed. An ML system realization is shown in fig. 21. This figure is sometimes referred to as a focussed beamformer. More details are given in the paper by Carter (1977a). Another way to look at this problem is that we want to maximize a quantity by adjusting a number of delay parameters subject to the constraint that all the delays must intersect in a single hypothesized position. Such a system and its variance has been examined by Bangs and Schultheiss (1973).

For equal noise spectra, we should utilize a product of pre- and postfilters with magnitude-squared transfer function, given by Bangs and Schultheiss (1973) (see also Knapp and Carter 1976, Hannan and Thomson 1973, MacDonald and Schultheiss 1969, and Carter 1976),

$$|h(\omega)|^2 = \frac{S(\omega)/N^2(\omega)}{1 + M[S(\omega)/N(\omega)]}, \tag{5.2}$$

in order to minimize the variance of delay estimates (where S and N

denote signal and noise-power spectral densities). The variance of the parameter estimate $\hat{\Theta}$, where $\Theta = R$ or $\Theta = B$, is available in the article by Carter (1977a) and is discussed in the following subsection.

Having selected proper prefilters for a specified array geometry, the maximum likelihood estimate of range and bearing is obtained by coherently processing the outputs of the sensing hydrophones (see Carter 1977a). In particular, each hydrophone output is prefiltered to accentuate a high signal-to-noise ratio (SNR), then delayed and summed. The summed signal is fed to a filter, then squared and averaged for the observation time. The output of this network is maximized through the indirect adjustment of the delay parameters. The delay parameters are derived on the basis of two adjustable parameters: hypothesized bearing and hypothesized range. Thus, an operator need only adjust his best estimate of bearing and range. From these two inputs, proper delays are inserted in each hydrophone receiving line. The process of delaying and summing is a focussed beamformer where the delays used cause the beam former to presume that the source wavefront is curved and not planar. The individual sensor-to-sensor delays inserted are directly related to the hypothesized source and sensor locations.

For a particular array type A, Carter (1977a), has shown that, at high output SNR, the minimum variance of bearing and range estimates is given by

$$\sigma_A^2(\hat{B}) \cong \frac{K_B^{(A)}}{TMVL_e^2},$$ (5.3)

and

$$\sigma_A^2(\hat{R}) \cong \frac{K_R^{(A)}R^4}{TMVL_e^4},$$ (5.4)

where $\sigma^2(\hat{B})$ is measured in radians squared, the effective array length $L_e = L \sin B$, and the constants K_R and K_B for four array types are given in table 3. M is the minimum of the number of sensors and the array length divided by the design half-wavelength; also at high output SNR,

$$V = \frac{1}{2\pi C^2} \int_0^\Omega \frac{S(\omega)}{N(\omega)} \omega^2 \, d\omega,$$ (5.5)

where $S(\omega)$ is the signal power spectrum, $N(\omega)$ is the noise power spectrum, C is the speed of sound in the medium, Ω is the highest source

Table 3
Constants for four arrays of interest

Array type	$\sqrt{\dfrac{K_R}{K_B}}$	K_R	K_B
Equispaced line	7.75	360	6
$M/2, 0, M/2$	∞	∞	2
$M/3, M/3, M/3$	6.9	144	3
$M/4, M/2, M/4$	5.7	128	4

(or receiver) frequency, and, as earlier, T is the observation time, R is the range, and B is the bearing.

5.4. Discussion

Doubling the number of sensors M or the observation time T will reduce the standard deviation of either the bearing estimate or range estimate by 1.4. In bearing estimation, we desire to make L_e large and the constant K_B small in order to reduce variance. Note that doubling the array length reduces the variance by four. Thus, array length is a more important factor in bearing estimation than either integration time or the number of hydrophones when operating at high output SNR.

The four different array types studied are an equispaced line array and three arrays with M elements grouped at the two ends and the middle of the array (see table 3). MacDonald and Schultheiss (1969) have shown that, by placing half of the M elements at each end of a line array with none in the middle (in an M over two, zero, M over two grouping), a bound on bearing variance is obtained. This bound, of course, is for a hypotherical array where the elements are collocated, one half at each end. The practical implications of MacDonald and Schultheiss' result are both to provide a bound on how well bearing can be estimated under ideal conditions, and to suggest how to place a limited number of hydrophones over a large aperture. Namely, half of the hydrophones should be positioned at each end of the array, placed at half-wavelength spacing for the design frequency and none should be placed in the middle of the available aperture.

It is noteworthy that the variance of the range estimate depends on the fourth power of the range relative to the effective baseline; thus, the variance of the range estimate is reduced by making the effective array

Table 4
Optimum sensor configurations

For Estimating	Best array configuration		
Bearing	...... $M/2$		...... $M/2$
Range	... $M/4$	...... $M/2$	... $M/4$
Position	.... $M/3$	.... $M/3$	.... $M/3$

length L_e large. This can be done by making the array length L large or by physically steering the array broadside to the source. The variance can also be reduced by decreasing the range to the source. Of course, reducing the range to the source can also increase SNR depending upon propagation conditions.

The constant K_R depends on the array type. For an equispaced line array, K_R is 360. A bound is provided by an array configured with a quarter of the hydrophones at each end, and half in the middle. Thus, we see that the array configuration desired for bearing estimation and the one for range estimation differ. The bearing array should have its elements toward the array ends, while the ranging array should have half of its elements in the central portion. However, an array with a third of its elements at each end and the middle will minimize the uncertainty region (see Carter 1977a). Thus in this sense, an array physically segmented into three equal groups of elements will outperform all other arrays for passively locating an acoustic source. This result is summarized in table 4. The optimum processor coherently combines all M hydrophone outputs. If, however, only the beamformer output from each subarray is used for coherent processing, a nearly optimum technique is believed to result.

Of considerable concern when attempting to predict the performance of a localization technique are values such as SNR, number of sensors, and integration time. It is interesting that these terms, together with constants such as 2π, can all be attributed to the standard deviation of the bearing estimates (measured in radians). Then the relative range error given by the standard deviation of the range estimate divided by the true range is given by a constant times the standard deviation of the bearing estimate times a term that depends linearly on the range to the source relative to the effective array length. In particular,

$$\frac{\sigma_A(\hat{R})}{R} = \sqrt{\frac{K_R}{K_B}} \frac{R}{L_e} \sigma_A(\hat{B}).$$ (5.6)

For example, suppose an equispaced line array had an inherent standard deviation of $\frac{1}{10}$ rad (5.7°) and was to estimate the range to a source ten times as far away as the effective array length. In that case, the relative range error is 7.75, or more than seven hundred percent. Hence, we see that it is extremely difficult to passively estimate the range of a distant source even under ideal conditions with high output SNR.

One of the advantages of expressing relative range errors in this form is that the standard deviation of bearing estimates is a term familiar to sonar engineers and signal processors. Moreover, the ocean medium may inherently limit the practical ability to estimate bearing even though theory predicts that with enough SNR or integration time, the bearing can be measured arbitrarily well. The expression given here clearly points out the need to make the array length large when the source range cannot be reduced. Of interest is that this conclusion is extremely insensitive to the type of array, provided the array has some ranging capability. This can be seen from the similarity of the constants given in table 3.

To summarize, we desire to know how to place a limited number of hydrophones over a baseline of fixed length. The hydrophones should be placed in groups, with the hydrophones in each group placed at half-wavelength spacing for the design frequency. For bearing estimation, half of the M hydrophones should be placed at each end of the array. For range estimation, a quarter of the hydrophones should be placed at each end of the array and half placed in the middle. For simultaneously estimating range and bearing, the hydrophones are placed in three groups, each with M over three hydrophones. If the baseline remained of fixed length and we had more hydrophones to add, we would add the hydrophones at half-wavelength spacing approaching an equispaced line array. On the other hand, if the number of hydrophones was limited but the baseline was not, we would keep the hydrophones at half-wavelength spacing and increase the distance between subarrays.

Acknowledgements

As evident from the contents and the list of references this work is a review of research by this author conducted alone and with the following

568   *G.C. Carter*

colleagues: Dr. Charles Knapp, Dr. Kent Scarbrough, Mr. E. Scannell,
Dr. Peter Cable, Dr. Albert Nuttall, Mr. Roger Tremblay and Dr. Philip
Abraham. The author has been honored by Prof. G. Tacconi, Prof.
L. Bjørnø, Prof. Y.T. Chan, Prof. J. Plant, Dr. H. Urban, Prof. T.
Durrani and Prof. J.L. Lacoume to present this work at several meetings
over the past decade in Italy, France, the United Kingdom, Denmark,
the Federal Republic of Germany and Canada as well as in the United
States.

References

Adams, W.B, J.P. Kuhn and W.P. Whyland, 1980, Correlator compensation requirements
for passive time delay estimation with moving source or receivers, IEEE Trans. Acoust.
Speech & Signal Process. **ASSP-28** (2), 158–168.
Akaike, H., and Y. Yamanouchi, 1963, On the statistical estimation of frequency response
function, Ann. Inst. Statist. Math., **14**, 23–56.
Bangs, W.J., and P.M. Schultheiss, 1973, Space Time Processing for Optimal Parameter
Estimation, in: Signal Processing, eds. J.W.R. Griffiths, P.L. Stocklin and C. Van
Schooneveld (Academic Press, New York) pp. 577–590.
Bendat, J.S., and A.G. Piersol, 1971, Random Data: Analysis and Measurement Procedures
(Wiley, New York).
Bendat, J.S., and A.G. Piersol, 1980, Engineering Applications of Correlation and Spectral
Analysis (Wiley, New York).
Benignus, V.A., 1969, Estimation of coherence spectrum of non-Gaussian time series
populations, IEEE Trans. Audio Electro-Acoust. **AU-17**, 198–201.
Benignus, V.A., 1970, IEEE Trans. Audio Electro-Acoust. **AU-18**, 320.
Betz, J., 1984, Ph.D. Thesis (Northeastern University, Boston, MA).
Bjørnø, L., ed., 1981, Proc. 1980 NATO Advanced Study Institute on Underwater
Acoustics and Signal Processing (Reidel, Boston, MA).
Blackman, R.B., and J.W. Tukey, 1958, The Measurement of Power Spectra (Dover, New
York).
Brillinger, D.R., 1975, Time Series Data Analysis and Theory (Holt, Rinehart and Winston,
New York).
Carter, G.C., 1976, Time Delay Estimation, Ph.D. Dissertation (University of Connecticut,
Storrs, CT).
Carter, G.C., 1977a, Variance bounds for passively locating an acoustic source with a
symmetric line array, J. Acoust. Soc. Am. **62**, 922–926.
Carter, G.C., 1977b, Receiver operating characteristics for a linearly threshold coherence
estimation detector, IEEE Trans. Acoust. Speech & Signal Process. **ASSP-27**, 90–94.
Carter, G.C., 1978, A brief description of the fundamental difficulties of passive ranging,
IEEE Trans. on OE **OE-3** (3), 65–66.
Carter, G.C., 1980, Bias in magnitude-squared coherence estimation due to misalignment,
IEEE Trans. Acoust. Speech & Signal Process. **ASSP-28**, 97–99.
Carter, G.C., 1981, Time delay estimation for passive sonar signal processing, IEEE Trans.
Acoust. Speech & Signal Process. **ASSP-29** (3), 463–470.

Carter, G.C., and P.B. Abraham, 1980, Estimation of source motion from time delay and time compression measurement, J. Acoust. Soc. Am. **67** (3), 830–832.

Carter, G.C., and C.H. Knapp, 1975, Coherence and its estimation via the partitioned modified chirp-z transform, IEEE Trans. Acoust. Speech & Signal Process. **ASSP-23**, 257–264.

Carter, G.C., C.H. Knapp and A.H. Nuttall, 1973a, Estimation of the magnitude-squared coherence function via overlapped fast Fourier transform processing, IEEE Trans. Audio Electro-Acoust. **AU-21**, 337–344.

Carter, G.C., A.H. Nuttall and P.G. Cable, 1973b, The smoothed coherence transform, Proc. IEEE **61**, 1497–1498.

Chan, Y.T., 1984, Time delay estimation in the presence of multipath propagation, Proc. 1984 NATO ASI, Lünburg, FRG, Series C, Vol. 151, ed. H. Urban (Reidel, Dordrecht).

Chow, S.K., and P.M. Schultheiss, 1981, Delay estimation using narrow-band processes, IEEE Trans. Acoust., Speech & Signal Process. **ASSP-29**, 478–484.

Cleveland, W.S., and E. Parzen, 1975, The estimation of coherence, frequency response and envelope delay, Technometrics **17** (2), 167–172.

Cramér, H., 1946, Mathematical Methods of Statistics (Princeton University Press, Princeton, NJ).

Eckart, C., 1952, Optimal rectifier systems for the detection of steady signals, Marine Physical Lab. Rep. SIO Ref. 52-11 (Scripps Institute of Oceanography, University of California).

Enochson, L.D., and N.R. Goodman, 1965, Gaussian approximations to the distribution of sample coherence, Wright Patterson Air Force Techn. Rep. AFFDL-TR-65-57, June 1965 (AD520987).

Gevins, A.S., C.L. Yeager, S.L. Diamond, J.P. Spire, G.M. Zeitlin and A.H. Gevins, 1975, Automated analysis of the electrical activity of the human brain (EEG): A progress report, Proc. IEEE **63**, 1382–1399.

Gosselin, J.J., 1977, Comparative study of two-sensor (magnitude-squared coherence) and single-sensor (square-law) receiver operating characteristics, in Proc. IEEE Int. Conf. on Acoustics, Speech and Signal Processing, Hartford, CT, 1977, IEEE Catalog No. 77CH1197-3, pp. 311–314.

Griffiths, L., 1984, Time varying filtering and spectrum estimation, Proc. 1984 NATO ASI.

Griffiths, J.W.R., P.L. Stocklin and C. Van Schooneveld, eds., 1973, Signal Processing (Academic Press, New York) (including E.B. Lunde, Wavefront stability in the ocean; W.J. Bangs and P.M. Schultheiss, Space–time processing for optimal parameter estimation).

Hahn, W.R., 1975, Optimum signal processing for passive sonar range and bearing estimation, J. Acoust. Soc. Am. **58**, 201–207.

Hahn, W.R., and S.A. Tretter, 1973, Optimum processing for delay-vector estimation in passive signal arrays, IEEE Trans. Inform. Theory **IT-19** (5), 608–614.

Halvorsen, W.G., and J.S. Bendat, 1975, Noise source identification using coherent output power spectra, J. Sound & Vib., **August 1975**, 15–24.

Hannan, E.J., and P.J. Thomson, 1971, The estimation of coherence and group delay, Biometrika **58**, 469–481.

Hannan, E.J., and P.J. Thomson, 1973, Estimating group delay, Biometrika **60**, 241–253.

Hannan, E.J., and P.J. Thomson, Delay estimation, IEEE Trans. Acoust. Speech & Signal Process. **ASSP-29**, 485–490.

Hassab, J.C., and R.E. Boucher, 1979, A quantitative study of optimum and suboptimum filters in the generalized correlator, Proc. IEEE Int. Conf. on Acoustics, Speech and Signal Processing, Washington, DC, 1979, IEEE Catalog No. 79CH1379-7, pp. 124–127.

Hassab, J.C., and R.E. Boucher, 1981, Performance of the generalized cross correlator in the presence of a strong spectral peak in the signal, IEEE Trans. Acoust. Speech & Signal Process. **ASSP-29**, 549–555.

Haubrich, R.A., 1965, Earch noise 5 to 500 millicycles per second, 1. Spectral stationarity, normality and nonlinearity, J. Geophys. Res. **70**, 1415–1427.

Ianniello, J.P., 1982, Time delay estimation via cross-correlation in the presence of large estimation errors, IEEE Trans. Acoust. Speech & Signal Process. **ASSP-30**, 998–1003.

Ianniello, J.P., E. Weinstein and A. Weiss, 1983, Comparison of the Ziv–Zakai lower bound on time delay estimation with correlator performance, in Proc. IEEE Int. Conf. on Acoustics, Speech and Signal Processing, Boston, MA, 1983, IEEE Catalog No. 83CH1841-6, pp. 875–878.

Jenkins, G.M., and D.G. Watts, 1968, Spectral Analysis and Its Applications (Holden-Day, San Francisco).

Johnson, G.W., D.E. Ohlms and M.L. Hampton, 1983, Broadband correlation processing, Proc. IEEE Int. Conf. on Acoustics, Speech and Signal Processing (1983) pp. 583–586.

Kirlin, R.L., and J.N. Bradley, 1982, Delay estimation simulations and a normalized comparison of published results, IEEE Trans. Acoust. Speech & Signal Process. **ASSP-30**, 508–511.

Knapp, C.H., and G.C. Carter, 1976, The generalized correlation method for estimation of time delay, IEEE Trans. Acoust. Speech & Signal Process. **ASSP-24**, 320–327.

Knapp, C.H., and G.C. Carter, 1977. Estimation of time delay in the presence of source or receiver motion, J. Acoust. Soc. Am. **61**, 1545–1549.

Koopmans, L.H., 1974, The Spectral Analysis of Time Series (Academic Press, New York).

Kuhn, J.P., 1978, Detection performance of the smoothed coherence transform (SCOT), Proc. IEEE Int. Conf. on Acoustics, Speech and Signal Processing, p. 678–683.

Lindsey, W.C., and H. Meyr, 1977, Complete statistical description of the phase-error process generated by correlative tracking system, IEEE Trans. Inf. Theory **IT-23** (2), 194–202.

MacDonald, V.H., and P.M. Schultheiss, 1969, Optimum passive bearing estimation in a spatially incoherent noise environment, J. Acoust. Soc. Am. **46**, 37–43.

Marple Jr., S.L., 1984, Spectrum analysis; overview of classical and high resolution spectral estimation, Proc. 1984 NATO Advanced Study Institute, Lünberg, FRG, Series C, Vol. 151, ed. H. Urban (Reidel, Dordrecht).

Meyr, H., 1976, Delay lock tracking of stochastic signals, IEEE Trans. Commun. **COM-24** (3), 331–339.

Nuttall, A.H., 1958, Theory and Application of the Separable Class of Random Processes, Ph.D. dissertation, RLE Report 343 (Massachusetts Institute of Technology, Cambridge).

Nuttall, A.H., 1971, Spectral estimation by means of overlapped FFT processing of windowed data, Report 4169 (Naval Underwater Systems Center, New London, CT).

Nuttall, A.H., and G.C. Carter, 1976, Bias of the estimate of magnitude squared coherence, IEEE Trans. Acoust. Speech & Signal Process. **ASSP-24**, 582, 583.

Nuttall, A.H., and G.C. Carter, 1980, A generalized framework for power spectral estimation, IEEE Trans. Acoust. Speech & Signal Process. **ASSP-28** (3), 334–335.

Nuttall, A.H., and G.C. Carter, 1982, Spectral estimation using combined time and lag weighting, Proc. IEEE **70** (9), 1115–1125.

Oppenheim, A.V., and R.W. Schafer, 1975, Digital Signal Processing (Prentice-Hall,

Englewood Cliffs, NJ).

Owsley, N., 1984, Overview of adaptive array processing techniques, Proc. 1984 NATO ASI, Lünberg, FRG, Series C, Vol. 151, ed. H. Urban (Reidel, Dordrecht).

Picinbono, B., 1984, Adaptive, robust and non-parametric methods in signal detection, Proc. 1984 NATO ASI, Lünberg, FRG, Series C, Vol. 151, ed. H. Urban (Reidel, Dordrecht).

Quazi, A.H., 1981, An overview on the time delay estimate in active and passive systems for target localization, IEEE Trans. Acoust. Speech & Signal Process. **ASSP-29**, 527–533.

Scannell Jr., E.H., and G.C. Carter, 1978, Confidence bounds for magnitude-squared coherence estimates, Proc. IEEE Int. Conf. on Acoustics, Speech and Signal Processing, Tulsa, OK, 1978, IEEE Catalog No. 78CH1285-6, pp. 670–673.

Scarbrough, K., 1984, Ph.D. thesis (Kansas State University, Manhattan, KS).

Scarbrough, K., N. Ahmed and G.C. Carter, 1981, On the simulation of a class of time delay estimation algorithms, IEEE Trans. Acoust. Speech & Signal Process. **ASSP-29**, 534–540.

Scarbrough, K., R.J. Tremblay and G.C. Carter, 1983, Performance predictions for coherent and incoherent processing techniques of time delay estimation, IEEE Trans. Acoust. Speech & Signal Process. **ASSP-31** (5), 1191–1196.

Schultheiss, P.M., and E. Weinstein, 1981, Source tracking using passive array data, IEEE Trans. Acoust. Speech & Signal Processing, **ASSP-29**, 600–607.

Tacconi, G., ed., Aspects of Signal Processing (Reidel, Boston, MA).

Van Trees, H.L., 1968, Detection, Estimation, and Modulation Theory, Part I (Wiley, New York).

Weiss, A., and E. Weinstein, 1982, Composite bound on the attainable mean-square error in passive time-delay estimation from ambiguity prone signals, IEEE Trans. Inf. Theory, **IT-28**, 977–979.

Weiss, A., and E. Weinstein, 1983, Fundamental limitations in passive time delay estimation, Part I: Narrow-band systems, IEEE Trans. Acoust. Speech & Signal Process. **ASSP-31**, 472–485.

COURSE 10

DIGITAL IMAGE PROCESSING AND ANALYSIS

Anastasios N. VENETSANOPOULOS

Department of Electrical Engineering
University of Toronto
Toronto, Ontario, Canada M5S 1A4

J.L. Lacoume, T.S. Durrani and R. Stora, eds.
Les Houches, Session XLV, 1985
Traitement du signal / Signal processing
© Elsevier Science Publishers B.V., 1987

Contents

576 *Contents*

List of symbols

u, v	space variables on the object plane
x, y	space variables on the image plane
W_x, W_y	frequency variables
T	time duration interval
V	velocity
α, c, β, δ	constants
θ	angle
m, n	discrete space variables
k_1, k_2	threshold levels
e	error
$a(i), a(k, l),$	
$b(k, l)$	weights or coefficients
$f(u, v)$	radiant energy on the object plane
$\hat{f}(u, v)$	estimate of the radiant energy
$l(x, y)$	radiant energy on the image plane
$h(x, y, u, v)$	impulse response of a linear image formation system
$h(x, y)$	impulse response of a linear time-invariant system
$F(W_x, W_y)$	Fourier transform of $f(u, v)$
$L(W_x, W_y)$	Fourier transform of $l(x, y)$
$H(W_x, W_y)$	Fourier transform of $h(x, y)$ (frequency response)
$\delta(x, y)$	impulse function in the (x, y) plane
$a(t)$	displacement in the x-direction
$b(t)$	displacement in the y-direction
$J_1(x)$	first order Bessel function
$m(x, y)$	image after nonlinear point-mapping operation
$r(x, y)$	noise of the recording process
$g(x, y)$	image after recording
$g(m, n)$	digital image after recording
$g_L(m, n)$	local average of $g(m, n)$

$s(\cdot)$	nonlinear transformation
$\nabla^2 l = \dfrac{\partial^2 l}{\partial x^2} + \dfrac{\partial^2 l}{\partial y^2}$	Laplacian operator
$L(\cdot)$	operator mapping g into y
σ	standard deviation of noise
$u(\cdot), w(\cdot)$	single valued analytic transformations
$u^{-1}(\cdot)$	inverse transformation to $u(\cdot)$
$D(e^{j\omega_1}, e^{j\omega_2})$	desired frequency response
$d(m, n)$	ideal impulse response
$w(m, n)$	window function
$H(e^{j\omega_1}, e^{j\omega_2})$	Fourier transform of $h(m, n)$
$W(e^{j\omega_1}, e^{j\omega_2})$	Fourier transform of $w(m, n)$
$A, B, C, D,$	
$t(p, q)$	constants of the McClellan transformation
$H(s_1, s_2)$	analog filter transfer function
$I(W_x, W_y)$	inverse filter
$E(\cdot)$	expectation or ensemble average
$t(x, y)$	impulse response of the restoration filter
$R_{fg}(\cdot)$	cross-correlation function between f and g
$R_{gg}(\cdot)$	autocorrelation function of g
$S_{fg}(W_x, W_y)$	cross-spectral density between f and g
$S_{gg}(W_x, W_y)$	spectral density of g
$T(W_x, W_y)$	Wiener filter
$S_{rr}(W_x, W_y)$	spectral density of noise
$S_{ff}(W_x, W_y)$	spectral density of f
$M(W_x, W_y)$	power spectrum equalization filter
$N(W_x, W_y)$	generalized Wiener filter
H	Burg's entropy
H'	Frieden's entropy
$p(f/g)$	a posteriori density function
$i_0(x_0, y_0)$	X-ray intensity distribution on the focal spot
$\mu'(x_s, y_s, z)$	the absorption coefficient of the exposed body at depth z
x_s, y_s	coordinates of sth layer at depth z_s
D_x, D_y	gradient function along x and y
U_x, U_y	edge detector masks
$G(x, y)$	Gaussian function
$F(x)$	cumulative distribution function

1. Introduction

1.1. Introduction to digital image processing

There are three main types of information signals that are generated, transmitted and received and stored. These are voice, data and images.

Voice is continuous and is communicated in analog or digital form. Voice information requires about 900–3000 bits per phoneme or a few thousand bits per word (at 10^4 samples per second). It has been estimated that over an individual lifetime (650 million words over 70 years), the average amount of speech information generated per individual is 2×10^{12} bits.

Data is mainly discrete and its generation and transmission is rapidly growing, over evolving computer communication networks. Data rates at medium speed are 9600 bits/s, but in certain applications exceed 10^6 bits/s. Integrated networks carrying voice, data and images have been proposed in recent years.

Images are continuous or discrete and their transmission over analog and digital channels has been a more recent development. In broadcast television, the NTSC color video signal is sampled at 10.7 MHz (three times the color subcarrier frequency), and each sample is quantized to 8 bits (256 levels), resulting in 2.85×10^6 bits/image and 85.6×10^6 bits/s. An individual, watching two hours of TV per day, will be exposed to 15.7×10^{15} bits of TV information over a lifetime. It is more than true therefore that "an image is worth a thousand words".

The question that now arises is what is an image. Webster defines the word image as "a reproduction or imitation of the form of a person or object." This is however a limited definition, since many images are not natural, such as those depicting mathematical functions. In addition there are some images, which can be represented by a collection of other signals, such as the three two-dimensional (2-D) signals that are required to represent a color image. A more general definition of the word image is that of a vector of multidimensional (m-D) functions. For the pur-

579

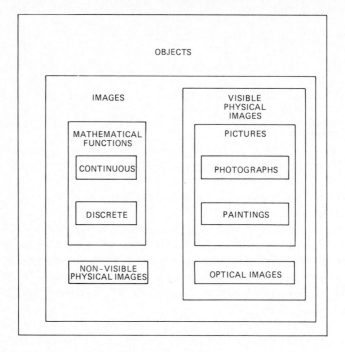

Fig. 1. Types of images.

poses of this course we shall however focus on scalar, 2-D functions $l(x, y)$, where $l(x, y)$ will be a real-valued function, whose value at point (x, y) is called brightness or gray level. If the variables x and y are continuous, the image is analog or continuous. If x and y are sampled, the function becomes discrete and is represented by $l(m, n)$. If in addition $l(x, y)$ is quantized (i.e. it can take only a finite set of values), the image will be called digital. In order to relate the mathematical function $l(x, y)$ with the physical image, retaining mathematical tractability, we usually require that $l(x, y)$ satisfies the following two properties:

(1) Be well-behaved analytically.

(2) Be nonnegative and bounded $(0 \leqslant l(x, y) \leqslant M)$ for all (x, y), within the domain of its definition.

Figure 1 shows the classification of images into different types based on their form or method of generation [1]. Digital image processing will

Table 1
Image processing and related areas

Output	Input	
	Image	Description
Image	Image processing	Computer graphics
Description	Image analysis	Description transformation

refer to that part of digital signal processing, in which the signal is an image.

Image processing is one of four related areas shown in table 1. When the input is an image and the output is also an image, we refer to the process as image processing. When the input is an image and the output is a description, we refer to the process as image analysis. The burst of activity over the past twenty years, with a present publication rate over a thousand papers per year on this subject, makes it impossible to provide a truly comprehensive survey in one course. More extensive surveys exist in some of the more recent texts [1–6]. However, even the sum total of these books does not provide a completely comprehensive or current survey. Therefore the goal of this course will be to expose the reader to the main methods and techniques of digital image processing, and a few selected applications. We shall also indicate the areas which are of current research interest, and relate digital image processing to image analysis. Tables 2a and 2b present an outline of the subjects addressed in this course and their interrelations.

Work on digital image processing was stimulated mainly by space research in the 1960's. Space missions to the moon and Mars sent images to earth that had undergone varied degrees of degradation. To improve their quality or to make meaningful studies on them, some form of digital processing was used. Such processing included sampling, quantization, compression, coding, transmission, enhancement, restoration and storage. Since those early days, digital image processing has expanded rapidly. Some of the most important areas of present application of image processing and analysis are now outlined [4]:

– *Telecommunications.* The main applications here are broadcast television (TV), facsimile, video conferencing and video phone. Images here may be digitized and then compressed, transmitted through space or networks, received, decoded, filtered and displayed. Interpolation techniques are used to achieve higher resolution.

Table 2a
Digital image processing

Objective	Application				
	Point & local operations	Linear shift invariant filters	Nonlinear filters	Adaptive filters	Knowledge-based systems
Image enhancement	Intensity mapping (grey-level expansion and compression, histogram equalization) Local operations (selective averaging, unsharp masking) Eye modeling Pseudocolour Edge sharpening	Low-pass filter High-pass filter High-emphasis filter Finite impulse response (FIR) filter Infinite impulse response (IIR) filter Recursive and non-recursive realizations 3-D filters	Homomorphic filter Median filter (nonrecursive, separable) Order statistics α-trimmed filter max(min) filter MTM filter DWMTM filter L and M filters Generalized mean filter Nonlinear mean filter Nonlinear order statistics filter Morphological filters	Heuristic approaches (local area Gamma control, simultaneous contrast enhancement and dynamic range reduction) Kalman filters	Expert systems
			Nonlinear adaptive filters		
Image restoration	Inverse filter Pseudoinverse filter Wiener filter Power spectrum equalization Generalized Wiener filter Constrained least squares Regularization approaches		Homomorphic filter Maximum entropy Bayesian filter Maximum a posteriori probability	Kalman (scalar and vector, reduced update filters) Edge enhancing Kalman filters	Expert systems

Digital image analysis

Objective	Method — Low-level techniques: Basic techniques	Method — Low-level techniques: Other techniques	Method — High-level techniques
Image detection, classification and recognition	**Feature extraction** Size characteristics area integrated optical density length and width perimeter	Statistical techniques Maximization of the a posteriori probability Minimization of the Bayes risk Sequential detection Parametric techniques Morphological techniques	**Pattern recognition and artificial intelligence techniques** *Level 1:* Formulation of the problem What should we do?
Type 1 problems Known object in noise	Shape characteristics rectangularity circularity moments contours edges curvature	Template matching Matched filters Correlation techniques	*Level 2:* System design What kind of patterns should be searched for? Obtain appropriate patterns for a given application
Type 2 problems Known object with unknown parameters in noise.	polar and boundary representation boundary chain code other shape descriptors Fourier descriptors	Generalized matched filters	*Level 3:* System realization How should knowledge be represented? Production rules Generalization/specialization Decomposition/aggregation Instantiation methods Control structure
Type 3 problems Random object in noise.		Estimator correlator detectors	
Image Estimation	Other characteristics color texture signature	Statistical techniques Maximum likelihood estimation Bayesian estimation	*Level 4:* System implementation What central structure should we use? Data-directed search Hypothesis-directed search Mechanism for ranking hypotheses
Deterministic problems:	**Image segmentation** Global techniques	Matching method Fourier method Method of differentials	
Image sequence analysis General 2-D motion General 3-D motion 3-D estimation from a 2-D image	Adaptive techniques Edge detection techniques Line tracking and gap filling		
Statistical problems:	Region growing techniques	Linear estimation	*Level 5:* System verification Evaluate and modify the system
Type 1 problems Known object in noise	Other segmentation techniques	Nonlinear estimation Whitening approach	
Type 2 problems Known object with unknown parameters in noise	Image matching techniques	The composite hypothesis problem	
Type 3 problems Random object in noise		Power spectral estimation techniques	Expert systems

– *Biomedicine.* The main applications are radiography, echography, nuclear medicine, electron microscopy, scintigraphy, thermography, computer-aided tomography (CAT), nuclear magnetic resonance (NMR), and tomographic filtering. Images may be static or dynamic. They may require real-time or off-line techniques, image compression, transmission, storage and retrieval. Standard TV resolution is not adequate for some applications.

– *Remote sensing.* Weather prediction, remote sensing of earth resources, cartography, space imagery, satellite imagery, aerial imagery, air reconnaissance, astronomy, are some of the applications here.

– *Non-physical images.* Non-physical images are 2-D and m-D functions, such as those representing collections of signals received by distributed sensors. Seismic records, signals generated by sonar and radioastronomical arrays, forward looking infrared (FLIR) and side looking radar (SLR), voice prints, sonograms, range-time and range-rate planes, ambiguity functions, Wigner distributions and multispectral representations are examples of such signals.

– *Other Applications.* Other recent applications include industrial radiographs, non-destructive testing, moving object recognition and tracking, traffic monitoring (of planes, rockets, satellites, cars and fishes), robotics, automatic industrial control and inspection, art processing (restoration, art-work data banks and analysis).

Many of the previously mentioned applications require hard copy or live imagery. Others require real-time or off-line processing. Some may require secure communications, while others may require vast memory storage requirements. Some designers place enormous emphasis on architecture and technology, while others are primarily concerned with mathematical methodology. To clarify the many issues involved in the design of image processing systems, we propose a top-down process described by the following five levels:

Level 1: Formulation of the problem. This level is concerned with the answer to the question "What should be done?". A satisfactory answer to this question requires an understanding of the image formulation process, the image degradation, the intended application and the means and knowledge available. This level may lead to the specification of the imaging system to be designed.

Level 2: System design. This level is concerned with the answer to the question "How should we do it?". Once the objective has been decided upon, a course of action is chosen, which may sometimes lead to an analytical solution, a solution resulting from computer-aided design (CAD), a heuristic or an ad hoc solution. This solution may be constrained, such as when we are searching for the optimal linear filter to optimize some objective function, or unconstrained.

Level 3: System realization. The system described by level 2 is now realized. This level examines available algorithms and architectures, which may be used for the realization of the system designed, and is concerned with various issues, such as speed, modularity and the effects of finite precision arithmetic (input quantization, parameter quantization and roundoff accumulation).

Level 4: System implementation. The architecture described by level 3 is now implemented in software and/or hardware. This level is concerned with languages and technology. Cost, speed, flexibility, transportability and accuracy are some of the main issues.

Level 5: System verification. The system obtained is finally verified by utilizing it in the application intended, and evaluating its performance. Often it is modified at this level to achieve improved performance.

While it is convenient to represent the total design process by these five levels, it is important to note that these levels are interrelated, since a decision made at any one of them affects the others. In addition, not all levels are considered in each image processing problem. This course deals mainly with level 1, 2 and 3 issues.

Since image processing can be accomplished optically or digitally, we now briefly consider these two approaches with respect to their advantages and limitations. Optical image processing depends on lenses, apertures, and phase plates, which are used to process light. These components are characterized by high resolution and parallelism, resulting in very high data rates. Their main limitations are:
- The requirement for display, which restricts the dynamic range of the processor and degrades its processing speed.
- The relatively large size and inflexibility of components, with very tight alignment tolerances.

Digital image processing has the following advantages, due to which it has increased in popularity in recent years:
(1) Images are increasingly acquired in a digital format for processing, and we are witnessing a reduction in cost and an improvement in

efficiency of digital image input/output (I/O) devices, such as scanners, digitizers and displays.

(2) Its great flexibility in applications involving nonlinear and adaptive operations, decision making processes and others.

(3) The recent innovations in algorithms and architectures, which include fast transforms, fast convolution algorithms, distributed arithmetic, wavefront and systolic arrays.

(4) The vast improvement in the price/performance capabilities of digital hardware, mainly due to the micro-electronics revolution during the past ten years.

(5) The popularity of digital data transmission and the availability of powerful techniques for data compression, channel encoding and encryption.

Present limitations of digital image-processing systems lie in their speed, size and frequency range. However, these limitations are gradually removed, due to the progress made in algorithms, architectures and technology.

1.2. Overview of digital image processing and related topics

Image processing is a broad field and is often used as a "catchall phrase". For this reason we now briefly describe the main subjects of this and related areas, and their relationship with other fields of engineering and science.

(1) Image acquisition, formation, sampling, display and recording. These subjects are closest to the areas of physics, electronics and systems science. Image acquisition is essentially a level 4 type subject, with typical characteristic parameters described in table 3. Image formation is concerned with level 1 issues, such as the development of models for image degradation and the study of different types of noise. Image sampling is concerned with the mathematical representation of 2-D and m-D signals, in terms of a discrete set of samples, and can be seen as the extension of the sampling theorem to 2-D and m-D. Quantization is concerned with the errors introduced, due to the quantization of the input image and the effect of those errors on the output. Image display and recording are concerned with the study of display and recording technologies. Most of these subjects will not be presented in this course. However, sections 1.3, 1.4, 1.5 and 1.6 will present some models for image formation, degradation and recording.

Table 3
Typical characteristics of image acquisition, formation, sampling, display and recording

Item	Typical characterizing parameters	Item	Typical characterizing parameters
Image acquisition		**Image sampling and quantization**	
Optics (to include original sensing, digitizer and printer)	Numerical aperture Focal length Spectral transmission Aberrations Distortions		Sampling grid Accuracy Precision Noise characteristics Dynamic range Spot shape
Detector (to include original image, reproductions & digitizer-film and solid state)	Quantum efficiency Dynamic range Spectral sensitivity Noise characteristics Geometric fidelity	**Image display (electronic readout)**	Luminance range Luminance constancy over format Constant ratio (large and small area) Geometric fidelity Spot size/line spacing Color characteristics

Table 3 (cont'd)

Item	Typical characterizing parameters	Item	Typical characterizing parameters
Image formation		**Image sensing and recording**	
Scene	Reflectivity as a function of wavelength Emissivity as a function of wavelength Object or feature size or geometry	Photochemical (film)	Geometric fidelity Spot shape Light source characteristics Modulator accuracy.
Illumination	Level Angle Coherent or non-coherent Spectral distribution Specular or diffuse	Photoelectronic	Geometric fidelity Dynamic range Spectral sensitivity Noise characteristics
Path	Absorption as a function of wavelength Scatter as a function of wavelength Turbulence Bandwidth Other noise sources	Visual system	Sensitivity Acuity Contrast response Pattern recognition Color response
Imaging conditions (to include original image, reproductions and final image or display)	Image motion Angle Distance		

(2) Image enhancement, restoration and filtering. Image enhancement is concerned with methods, which attempt to put the image in a form more suitable for a given purpose. It is related to mathematics, filtering and psychology. There are three fundamental types of enhancement (spatial, spectral and temporal). Spatial methods take advantage of the geometric features of the image, to improve its apparent quality. Spectral methods enhance the image by operating in the spectral domain. Temporal methods seek to take advantage of the frame-to-frame information differences, to enhance other types of information. Section 2 describes some spatial and spectral enhancement techniques.

Image restoration seeks to obtain an image as close to ideal as possible. It is related to filtering and estimation theory. Some of the degradations it attempts to remove are random noise, interference, geometrical distortion, field nonuniformity, contrast loss and blurring. It utilizes the mathematics of random fields to describe images. It employs linear techniques, such as inverse filters, Wiener filters, and Kalman filters, as well as nonlinear techniques, such as homomorphic filters, maximum entropy methods and Bayes estimation. It will be considered in more detail in section 4.

2-D and m-D linear filtering are concerned with the extension of 1-D filtering techniques to two and more dimensions. This subject has received considerable attention, due to its importance in enhancement and restoration [5, 6]. One fundamental difficulty in the extension of 1-D techniques to 2-D and m-D is the lack of a Fundamental Theorem of Algebra for polynomials of two or more variables. 2-D and 3-D techniques will be considered in section 3 of this course. Recently, considerable attention has been given to the development of nonlinear filtering techniques, which are generally superior to linear techniques, in image processing. A brief survey is given in section 2.7. Adaptive filtering is also a topic of considerable importance and many 1-D adaptive filters have been extended to 2-D. A brief survey is given in section 2.8.

(3) Image transmission and encoding. This area is a direct application of communication techniques to images. It deals with source encoding, which attempts to compress the image, by removing redundancies. Some of the techniques here are point-to-point processing, line-to-line processing, and frame-to-frame processing. The first is the easiest, that is processing in the direction of the scan. The most complex is frame-to-frame processing, where the redundancy is greatest, and hence it offers the greatest potential for image compression. Delta modulation, predic-

tive encoding, adaptive delta modulation are all examples of such techniques. In addition to spatial techniques, transform techniques are also used, which are able to compress images without loss of information. Channel encoding attempts to introduce controlled redundancy to the compressed signal, in order to enable its transmission through the channel with noise immunity. These techniques parallel those developed in the area of communications. Finally, a number of encoding methods for secure communications have been used, such as spread spectrum techniques. The design of communication networks to transmit images is another area of current interest. Image transmission and encoding will not be considered in this course.

(4) Image reconstruction. The problem of reconstructing 3-D objects from their projections has received considerable attention in recent years, with CAT, NMR, radioastronomical and seismic applications. Digital signal processing techniques have had a great impact in the solution of this problem. This topic will not be considered in this course.

(5) Image analysis, vision and computer graphics. Image analysis and vision are concerned with the derivation of useful information from images or scenes. They are related to statistics, detection and estimation theory. However, with only a few exceptions, such as spectral estimation and motion estimation, their present status and the complexity of the problems considered, often result in approaches closer to "softer" sciences, such as artificial intelligence, pattern recognition and psychology. Examples of applications in these areas include optical character recognition and automatic diagnosis in biomedicine. The understanding of the human visual process is of fundamental importance for the improvement of image processing systems and it involves many sciences, such as physiology, anatomy, psychology and artificial intelligence.

Within this broad subject area we can identify a number of important subareas.
- *Image detection* is concerned with the decision of the presence or absence of an object or a class of objects in an image.
- *Image estimation* is concerned with the estimation of certain features or parameters contained in an image.
- *Image spectral estimation* deals with the development of 2-D and m-D power spectral density estimates for images.

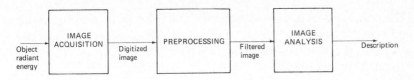

Fig. 2. Digital image processing–analysis system.

– *Image analysis* encompasses all previous areas, but in its present status it is primarily concerned with the representation of image information in a form compatible with present automated, machine-based processing.
– *Computer graphics* attempt to synthesize artificial images. This area is closely related to digital image processing and analysis, but will not be discussed in this course.

Some aspects of the previous areas will be considered in section 5, in an attempt to relate them to conventional image processing. A typical image processing-analysis system is shown in fig. 2.

1.3. Image formation

We now introduce a mathematical model for image formation and degradation. Assuming that the mechanism of image formation is linear, we introduce the following notation (fig. 3):

$f(u, v)$:   the radiant energy on the object plane,
$l(x, y)$:   the radiant energy on the image plane,
$h(x, y, u, v)$:   the impulse response of a linear image formation system.

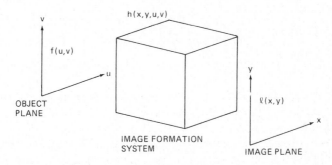

Fig. 3. Image formation.

This model describes the mapping of a 2-D image of the object plane, into the 2-D image of the image plane. Other models also exist, which map a higher dimensional object into the image observed, which is of reduced dimensionality.

Using the assumption of linearity, the object $f(u, v)$ is degraded by the linear operator $h(x, y, u, v)$, to give the image $l(x, y)$ described by [3]

$$l(x, y) = \int \int f(u, v) h(x, y, u, v) \, du \, dv. \tag{1}$$

Note that $h(x, y, u, v)$ is a function of four variables. The axes of the object (u, v) can be chosen for convenience to depend upon certain properties of the object. The axes of the image may be different, since the imaging system can cause an axis translation. The impulse response $h(x, y, u, v)$ is the image formed by a point source of light at the object plane. It is also referred to as "unit pulse response" and "point-spread function". When the point source of light explores the object plane and the image changes, the impulse response is a spatially variant point spread function and eq. (1) represents a spatially variant system. For a spatially (or space) invariant system, eq. (1) is reduced to a convolution equation

$$l(x, y) = \int \int f(u, v) h(x - u, y - v) \, du \, dv = f(x, y) * h(x, y). \tag{2}$$

The Fourier transform of eq. (2) is given by

$$L(W_x, W_y) = F(W_x, W_y) H(W_x, W_y), \tag{3}$$

where $L(\cdot)$, $F(\cdot)$ and $H(\cdot)$ are the Fourier transforms of $l(\cdot)$, $f(\cdot)$ and $h(\cdot)$, respectively.

1.4. Image degradation

Types of degradation that are experienced in imaging systems include point degradation, spatial degradation, temporal degradation, chromatic degradation and a combination of the above. The main goal of modeling is to lead to some form of representation of $h(\cdot)$, the degradation phenomenon. In the ensuing models a number of expressions are provided.

(1) An "ideal imaging system" has an impulse response which is an impulse function, such that

$$l(x, y) = \int \int f(u, v)\delta(x - u, y - v)\, du\, dv = f(x, y). \qquad (4)$$

(2) A system that produces no spatial smearing, but includes a point degradation is modeled by

$$l(x, y) = \int \int f(u, v)h(x, y)\delta[p_1(x, y) - u, p_2(x, y) - v]\, du\, dv$$

$$= f(p_1, p_2)h(x, y). \qquad (5)$$

Such a system produces distortions due to the coordinate change and the point degradation.

(3) If $p_1(x, y) = x$, and $p_2(x, y) = y$ in eq. (5), then

$$l(x, y) = f(x, y)h(x, y). \qquad (6)$$

The system described by eq. (6) allows for multiplicative point degradation effects, which may be caused by lens or tube shading or any other sensor defects.

(4) If the impulse response $h(\cdot)$ is a function of u and v, some form of smearing is noticed, depicted as a loss of resolution. In the mathematical model, this amounts to an integration over the variables u and v. This is the most common form of spatial degradation and practical cases include systems with: object-film plane image motion, defocused images, atmospheric turbulence, clouds, rain, etc. Some of these degradation phenomena are now considered.

(a) Relative motion between scene and camera. It is known that the total exposure at any point of the film can be obtained by integration of the instantaneous exposure during which the shutter is open. Let $a(t)$ and $b(t)$ be the x and y displacements and consider the motion of the object in a time of duration T. The total exposure is given by

$$l(x, y) = \int_{-T/2}^{T/2} f(x - a(t), y - b(t))\, dt.$$

By Fourier transforming both sides of this equation one obtains

$$L(W_x, W_y) = F(W_x, W_y)\int_{-T/2}^{T/2} \exp\{-j2\pi[W_x a(t) + W_y b(t)]\}\, dt.$$

The frequency response of the degradation is therefore given by

$$H(W_x, W_y) = \int_{-T/2}^{T/2} \exp\{-j2\pi[W_x a(t) + W_y b(t)]\}\, dt.$$

In the specific case of uniform motion in the x-direction, with velocity V,

$$a(t) = Vt, \qquad b(t) = 0,$$

$$H(W_x, W_y) = \frac{\sin(\pi W_x VT)}{\pi W_x V} = T \operatorname{sinc}(W_x VT). \tag{7}$$

(b) Defocused images. For defocused images,

$$H(W_x, W_y) = \frac{J_1(a\rho)}{a\rho}, \tag{8}$$

where

$$\rho = \sqrt{W_x^2 + W_y^2},$$

and J_1 is the first-order Bessel function, a is the total displacement during exposure.

(c) Atmospheric turbulence. Atmospheric turbulence may be caused by atmospheric inhomogeneities which cause wave aberration. It can be shown that

$$H(W_x, W_y) = \exp\left[-c\left(W_x^2 + W_y^2\right)^{5/6}\right], \tag{9}$$

where c is a constant which depends on the type of atmospheric turbulence.

1.5. Estimation of the degradation

If the degradation is of unknown nature or if the phenomenon underlying the degradation is too complex for an analytical determination of the point-spread function, then it is necessary to estimate it from the degraded image. This is called a posteriori determination of the point-spread function. Some properties of images that are used when known are:

(1) The image of a sharp point of the object forms the point spread function of the degraded image. In astronomical pictures for instance, an image of a faint star could be used as the point-spread function.

(2) Where the object contains sharp lines, their image can be used to determine the point-spread function. The image of such a line is related to the point-spread function by

$$h_l(y) = \int_{-\infty}^{\infty} h(x, y) \, dx.$$

Taking the Fourier transform and letting $W_x = 0$, we obtain the result

$$H(0, W_y) = H_l(W_y).$$

If the image of a line is Fourier transformed, it will be equal to the frequency response $H(W_x, W_y)$, with $W_x = 0$. If the line is at an angle θ to the x-axis on the object, then its Fourier transform produces the frequency response $H(W_x, W_y)$ along a line of slope $\theta + 90°$ to W_x. With several lines a close approximation of $H(W_x, W_y)$ can be obtained. Edges can also be used for estimating the point-spread function.

(3) The image is divided into N regions of identical size $l_i(x, y)$, $i = 1, 2, \ldots, N$, and $f_i(x, y)$ are the gray levels in the object corresponding to the image elements. If we assume that the extent of the point-spread function is small, compared to the size of an image element, and if we ignore the edge effects, then

$$l_i(x, y) = \int\int h(x - u, y - v) f_i(u, v)\, \mathrm{d}u\, \mathrm{d}v, \quad i = 1, 2, \ldots, N,$$

whose Fourier transform is

$$L_i(W_x, W_y) = F_i(W_x, W_y) H(W_x, W_y), \quad i = 1, 2, \ldots, N,$$

and

$$\prod_{i=1}^{N} L_i(W_x, W_y) = \prod_{i=1}^{N} F_i(W_x, W_y) H^N(W_x, W_y).$$

Solving for the magnitude response and taking the logarithm of both sides we obtain

$$\ln\left| H(W_x, W_y) \right| = \frac{1}{N} \sum_{i=1}^{N} \ln\left| L_i(W_x, W_y) \right| - \frac{1}{N} \sum_{i=1}^{N} \ln\left| F_i(W_x, W_y) \right|$$

The second term on the right-hand side is the average of the logarithms of the Fourier transforms of the object elements.

1.6. Image recording

Since an image is known only after being sensed and recorded by some suitable medium (e.g. retina, film, photo-electronic medium), it is important to provide a model describing the total process of image formation and recording. Such a model is shown in fig. 4. The object radiant energy is transformed by a linear transformation into the image radiant energy. This is further transformed by a nonlinear point-mapping operation $s(\cdot)$ to the light intensity $m(x, y)$. The noise is often assumed to be additive. It may be due to stray illumination, or other reasons. The

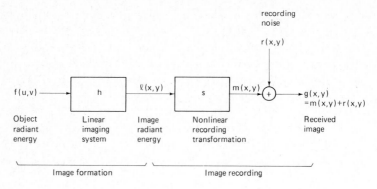

Fig. 4. Image formation and recording.

system output is now given by

$$g(x, y) = s\left(\int \int f(u, v) h(x, y, u, v) \, du \, dv \right) + r(x, y), \qquad (10a)$$

which describes the image formation and recording process. Since $s(\cdot)$ is a nonlinear function, transforming eq. (10a) by an inverse transformation $s^{-1}(\cdot)$ does not yield an additive combination of $f(u, v)$ linearly transformed by $h(\cdot)$, plus noise. Hence the recovery of $f(x, y)$ from $g(x, y)$ is accomplished only approximately. One usual approximation, often considered in the literature, in an attempt to obtain the solution of eq. (10a), is to assume that the nonlinearity $s(\cdot)$ can be neglected, and $h(\cdot)$ represents a space-invariant system. One may then approximate eq. (10a) by

$$g(x, y) = f(x, y) * h(x, y) + r(x, y). \qquad (10b)$$

Equation (10b) indicates that the problem of recovery of $f(x, y)$ from $g(x, y)$ is a problem of deconvolution. This problem is difficult, because it is ill-conditioned. In addition many image point-spread functions are singular, which further complicates the problem.

In practice we often deal with eqs. (10a) and (10b) in discrete form. In such a case eq. (10b) becomes

$$g(m, n) = \sum_{k} \sum_{l} f(k, l) h(m - k, n - l) + r(m, n). \qquad (10c)$$

Equation (10c) suggests that the image recovery process can be formulated as the solution of a set of linear equations corrupted by noise.

The models described by eqs. (10) are widely accepted for image formation and recovery. However, more elaborate models also exist, describing more accurately certain phenomena. Additive noise can have short-tailed, medium-tailed or long-tailed probability density functions. Moreover, photographic recording and other phenomena are more accurately modeled by signal-dependent noise (multiplicative, film-grain and other types).

2. Image enhancement

2.1. Introduction

Enhancement results from the processing of an image, to present to the viewer (or the machine) additional information or insight into some aspect of the pre-enhanced image. Unlike restoration, which has a specific goal and a mathematical basis, little or no attempt is made here to estimate the actual degradation process. Enhancement can be viewed as selective emphasis and/or suppression of information in the image, with the aim to increase its usefulness. There are numerous widely used techniques in image enhancement, such as intensity mapping, local operations, eye modeling, pseudocolor, edge sharpening, linear filtering, nonlinear filtering, adaptive filtering and others. These techniques are now reviewed.

2.2. Intensity mapping

By definition, intensity mapping is a point process. This is an operation on a point-by-point basis to map one gray level into another. Recent developments of digital image displays, with instantaneous point-mapping capabilities, have greatly increased our ability to perform fast point-mapping operations. Examples of intensity mapping operations are given below:

(1) Gray scale transformation. Let the mapping be a transformation from one gray scale z to another scale z', where

$$z' = t(z).$$

Suppose that the allowable range of gray levels is the same on both scales, $z_1 \leqslant z \leqslant z_k$ and $z_1 \leqslant z' \leqslant z_k$. Then, a transformation that has enhancing capabilities can be applied. Consider an underexposed image, whose dynamic range is smaller than $[z_1, z_k]$. The image, which does not occupy its full allowable gray level range, can be made to do so, and the result is an increase in contrast. If the dynamic range of this image is the subinterval $[a, b]$ of $[z_1, z_k]$, then the following transformation

$$z' = \frac{z_k - z_1}{b - a} z + \frac{z_1 b - z_k a}{b - a}$$

will stretch and shift the gray scale to occupy the full range.

(2) Gray scale suppression. If we require to suppress the gray scale outside the subrange $[a, b]$, the following transformation can be used:

$$z' = \begin{cases} \dfrac{z_k - z_1}{b - a}(z - a) + z_1, & a \leqslant z \leqslant b, \\ z_1, & z < a, \\ z_k, & z > b. \end{cases}$$

This is a piecewise linear transformation, which stretches the region $[a, b]$ of the original scale and compresses the subintervals $[z_1, a]$ and $[b, z_k]$ into points. This is desirable only when there is little information in these subintervals. Any selected interval can be stretched at the cost of compressing its complements. The transformation need not be linear. Any desired transformation, such as logarithmic or quadratic, can be used, depending on the requirements.

(3) Histogram equalization. Another form of intensity mapping is known as "histogram modification". The histogram is a graph of relative frequency, with which different gray levels occur in the image, in the gray level range $[z_1, z_k]$. Methods exist allowing us to transform an image's gray level into another, in order to give the image a specified histogram. A special case of interest is "histogram equalization". The need for it arises when comparing two images taken under different lighting conditions. The two images must be referred to the same "base", if meaningful comparisons or measurements are to be made. The base that is used as a standard has a uniformly distributed (equalized) histogram. Note that from Information Theory, a uniform histogram signifies maximum information content in the image.

2.3. Local operations

When the operation is limited to a small neighborhood of a pixel, we refer to it as a local space operation. A number of heuristic operations have been proposed and are used in a variety of areas, where simple processing and high speed is essential. A few are reported here:

An operation proposed in ref. [7] replaces the gray level of pixel x_0, by the average of itself and the values in the neighborhood (consisting of eight adjacent pixels), except those which have gray level differences greater than a fixed threshold k_1. The updated value x_0' is given by

$$x_0' = \frac{1}{N} \sum_{x_i \in S} x_i,$$

where

$$S = \{ x_i : |x_i - x_0| \leqslant k_1 \}, \quad i = 0, 1, \ldots, 8,$$

and N is the number of terms in the sum.

If, according to a second operation in ref. [7], the previous average in the neighborhood of x_0 is evaluated, and if all the signals x_i differ from x_0 by less than a suitable threshold k_2, and x_0 differs from x_0' by more than $k_2 + \delta$ ($\delta > 0$), the value of x_0 is set equal to x_0'. Otherwise x_0 maintains its original value.

Using "unsharp masking", the blurred negative of an image is added to itself. Small features, which were essentially removed in the blurred image are preserved, while large features are compressed in dynamic range.

2.4. Eye modeling and pseudocolor

Eye modeling bases enhancement on models of human vision. Such models have been proposed in the literature [4]. However, a good model describing human visual perception is far from known at present.

Pseudocolor techniques use a mapping of gray levels to different colors to increase the effective viewing dynamic range. For example, mapping of one specific gray level to red, in an aerial image of a field, may highlight the regions where a crop is dry or diseased.

2.5. Edge sharpening

Blurring makes edges and sharp lines on images indistinct. Since blurring is an averaging, or an integration operation, differentiation may be used to remove the blur. Consider a linear derivative operation. For this to be applicable in the removal of blur in a homogeneous image it is required to be isotropic. Isotropy means that the operator has to be rotation invariant. This will ensure that features that run in any direction can be sharpened equally. It is known that the Laplacian operator is such an operator. When applied to the image $l(x, y)$,

$$\nabla^2 l \equiv \frac{\partial^2 l}{\partial x^2} + \frac{\partial^2 l}{\partial y^2}.$$

It is also known that blur can be reduced by subtracting a multiple of the Laplacian from the blurred image, as

$$y = l - \beta \nabla^2 l,$$

where β is a constant and y is the resulting image.

Blurring has been found to weaken the high spatial frequencies more than the low frequencies. Consequently, sharpening can be done by emphasizing the higher spatial frequencies. Differentiation has this high-frequency emphasis effect. Other approaches for edge sharpening have been studied in the literature. Edge detection techniques will be presented in section 5.3.

2.6. Linear filtering

A digital filter is described by an operator $L(\cdot)$, which maps a signal g into a signal y, as

$$y = L(g).$$

When $L(\cdot)$ is a linear operator, i.e. it satisfies both the superposition and proportionality principles, the filter is linear. When in addition the operator is shift-invariant, the filter is linear shift-invariant (LSI).

Linear filtering is another simple approach, which is used for noise removal. Its mathematical properties are described in the course by V. Cappellini. In their simplest form linear filters can be described by simple averages. To show their effect of averaging, consider the image points, whose gray level averages are $m_1, m_2, \ldots, m_N$, and the independent, zero mean noise values $n_1, n_2, \ldots, n_N$, with standard deviation σ. The sample average is given by

$$\frac{1}{N} \left[\sum_{i=1}^{N} m_i + \sum_{i=1}^{N} n_i \right].$$

The second term has zero mean and standard deviation $\sigma/N^{1/2}$. This implies that averaging has reduced the amplitude fluctuations of the noise. Due to their mathematical simplicity and importance linear filters have been studied extensively in the literature, and will be considered in more detail in section 3. Linear filters, however, tend to blur the object boundaries and edges, which are perceptually important in object recognition.

2.7. Nonlinear filtering

Numerous nonlinear filters have been proposed for image enhancement. Homomorphic filters are used in the presence of multiplicative interference or degradation. They reduce multiplicative degradation to additive noise, which is removed by linear filtering. The result is finally exponentiated and approximates the ideal image [1–4]. Median filters are also well known [8–10]. Some others are α-trimmed filters [5], quadratic filters [11], generalized mean filters [12], nonlinear mean filters [13] and nonlinear order statistic filters [14]. Morphological filters are another class of filters recently introduced. Typical nonlinear filters are now described:

(1) Median filters. Median filters are useful in reducing random noise, especially when the noise amplitude probability density has large tails, and periodic patterns. The median filtering process is accomplished by sliding a window over the image. The filtered image is obtained by placing the median of the values in the input window, at the location of the center of that window, at the output image. For relatively uniform areas the median is near the mean, and the filter smooths out noise. As an edge is crossed, one side or the other dominates the window, and the output switches sharply between the values. Thus, the edge is not blurred. The disadvantage of such filters is that in the presence of small signal-to-noise ratios they tend to break up image edges and produce false noise edges.

Median filters of both recursive and non-recursive types have been considered in the literature. Recursive median filters were shown to be more efficient than those of the non-recursive type. A useful special class of median filters are separable median filters. These filters are particularly easy to implement, by performing successive operations over the rows and the columns of the image.

Bovik, Huang and Munson [9] recently introduced a generalization of the median filter. They defined an order statistic (OS) filter, in which the input value at a point is replaced by a linear combination of the ordered values in the neighborhood of the point. The class of OS filters includes as special cases the median filter, the linear filter, the α-trimmed mean filter, and the max (min) filter, which uses an extreme value instead of the median. For a constant signal immersed in additive white noise, an explicit expression was derived for the optimal OS filter coefficients. Both qualitative and quantitative comparisons suggest that OS filters (designed for a constant signal) can perform better than median and linear filters in some applications.

Lee and Kassam [10], introduced another generalization of the median filter, which stems from robust estimation theory. According to different estimators the L filter and the M filter were proposed, in which the filtering procedure uses a running L estimator and an M estimator, respectively. Because the L estimator uses a linear combination of ordered samples for the estimation of location parameters, the use of a running L estimator for filtering resembles the use of an OS filter. Another variation of median filters is the modified trimmed mean (MTM) filter. This filter selects the sample median from a window centered around a point and then averages only those samples inside the window close to the sample mean. MTM filters were shown to provide good overall characteristics. They can preserve edges even better than median filters. The same authors also introduced a double-window modified trimmed mean (DWMTM) filter. In this filter a small and a large window are used to produce each output point. The small window results in the retention of the fine details of the signal and the large window allows adequate additive noise suppression. The DWMTM filter has good performance characteristics. However, it often fails to smooth out the signal dependent components.

(2) Nonlinear mean filters. Recently, a new class of nonlinear filters, which include linear, homomorphic, generalized mean filters, as well as other kinds of filters as special cases, was proposed [13, 14]. These filters are described by the following operation:

$$
y = u^{-1} \left[\frac{\displaystyle\sum_{i=1}^{N} a(i)u(g_i)}{\displaystyle\sum_{i=1}^{N} a(i)} \right], \tag{11}
$$

where the g_i, $i = 1, 2, \ldots, N$, denote the image samples, $u(\cdot)$ is a single valued, analytic function with an inverse function u^{-1}, and $a(i)$ are weights. If these weights are constants, then eq. (11) reduces to the well-known generalized homomorphic filter. Some of these filters of particular interest in image processing are now listed:

$$u(g) = \begin{cases} g, & \text{arithmetic mean} & \bar{g}, \\ 1/g, & \text{harmonic mean} & y_H, \\ \log(g), & \text{geometric mean} & y_G, \\ g^p, p \in Q - [-1, 0, 1], & L_p \text{ mean} & y_{L_p}. \end{cases}$$

If the weights $a(i)$ are not constant, another class of nonlinear means can be obtained, which can be made to exhibit desirable characteristics, by an appropriate choice of the parameters $a(i)$. A useful nonlinear mean of this class is the *contraharmonic mean*

$$y_{CH_p} = \frac{\displaystyle\sum_{i=1}^{N} g_i^{p+1}}{\displaystyle\sum_{i=1}^{N} g_i^{p}}, \tag{12}$$

which can be interpreted as the arithmetic mean with weights given by $a(i) = g_i^p$. The nonlinear mean filters previously described satisfy the following inequalities, which are used in the analysis of their performance.

$$\min[g_i] \leqslant y_{CH_{-p}} \leqslant y_{L_{-p}} \leqslant y_H \leqslant y_G \leqslant \bar{g} \leqslant y_{L_p} \leqslant y_{CH_p} \leqslant \max[g_i]$$

The performance of nonlinear mean filters has been considered in ref. [13]. Their properties and a comparison with some other filters in the presence of different kinds of noise are summarized in table 4.

(3) Nonlinear order statistic filters. A more general class of nonlinear filters, which encompasses homomorphic filters, order statistic filters, median filters and nonlinear mean filters was recently proposed [14]. The structure of these nonlinear filters is shown in fig. 5. It consists of a point-wise nonlinear function, $u(\cdot)$, a network which sorts signals according to their magnitude, multiplication by the coefficients $a(i)$, $i = 1, 2, \ldots, N$, a summation and the nonlinear point operation function $w(\cdot)$. $g_1, g_2, \ldots, g_N$ are image samples contained in a window and y is the output sample. If the sorting input is not activated, then no ordering is performed and for $w = u^{-1}$, the filter reduces to a nonlinear mean filter. If the coefficients $a(i)$ are independent of the signal samples g_i, $i = 1, 2, \ldots, N$, and $w = u^{-1}$, the filter further reduces to a homomor-

Table 4
Overview of the performance of various filters in the presence of different kinds of noise

Kind of noise	Performance of filters*				
	Arithmetic mean	Geometric mean	L_p mean	Contra-harmonic mean	Median
Short-tailed additive noise	+	+	+	+	−
Heavy-tailed additive noise	−	−	−	−	+
Positive spikes	−	−	+	+	+
Negative spikes	−	−	+	+	+
Mixed spikes	−	−	−	−	+
Multiplicative noise	−	+	−	−	−
Film-grain noise	−	−	+	−	−
Edge preservation	−	−	+	+	+

* + : good performance
  − : poor performance

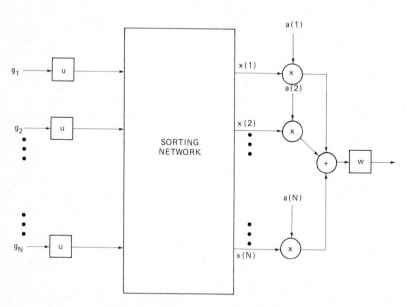

Fig. 5. Nonlinear order statistic filter.

phic filter. On the other hand, if the nonlinearities are removed, the nonlinear order statistic filter reduces to an order statistic filter, whose special cases are median filters and α-trimmed filters.

The characteristics of this general class of nonlinear filters are controlled by the choice of the nonlinearities u and w, and the filter coefficients $a(i)$, $i = 1, 2, \ldots, N$. These are chosen according to the kind of noise which is encountered. The very same structure of fig. 5 can be used as an edge detector by appropriate choice of $a(i)$, $i = 1, 2, \ldots, N$. The resulting edge detectors can be very simple and very efficient. Different such edge detectors were studied and their performance was shown to be better than the performance of some well-known edge detectors [14]. Their results can be further upgraded by choosing the nonlinearities to perform histogram equalization. A combination of the edge detector and the filter can produce an adaptive general nonlinear filter, which was shown to have good performance in image filtering [14]. In addition, nonlinear mean filters followed by DWMTM filters were recently shown to have potential in the simultaneous removal of impulsive and signal dependent noise.

2.8. Adaptive filtering

A number of heuristic approaches have been considered in the literature, utilizing adaptive filtering techniques in image enhancement. Such adaptive filters can be space-variant linear or nonlinear filters or time-variant filters that vary from frame to frame. Some typical space-variant linear filters will be summarized.

Local area gamma control works by measuring the local area variance and adapting the gain. Therefore, details not clearly distinguished are enhanced. Enhancement of noise is prevented by reducing the gain, when the variance is large.

When an image with a large dynamic range is recorded on a medium with a smaller dynamic range, the details of the image in the very high and/or low luminance regions are not well preserved. One approach to such a problem is a simultaneous contrast enhancement and dynamic range reduction [15]. A block diagram of this filter is shown in fig. 6. In the figure $g(m, n)$ denotes the unprocessed digital image and $g_L(m, n)$, which denotes the local mean of $g(m, n)$, is obtained by low-pass filtering. The low-pass filtering operation is a simple local averaging

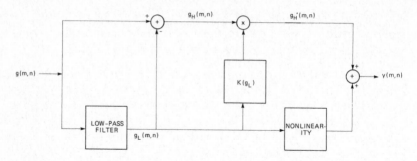

Fig. 6. Contrast enhancement and dynamic range reduction.

given by the Wallis algorithm

$$g_L(m, n) = \frac{1}{(2M + 1)(2N + 1)} \sum_{k=-M+m}^{M+m} \sum_{l=-N+n}^{N+n} g(k, l). \quad (13)$$

The sequence $g_H(m, n)$, which denotes the local contrast, is obtained by subtracting $g_L(m, n)$ from $g(m, n)$. The local contrast is modified by multiplying $g_H(m, n)$ by $K(g_L)$, a scalar, which is a function of $g_L(m, n)$. The specific functional form of $K(g_L)$ depends on the particular application under consideration, and $K(g_L) > 1$ represents a local contrast increase, while $K(g_L) < 1$ represents a local contrast decrease. The local mean is modified by a point nonlinearity, which is chosen so that the overall dynamic range of the resulting image is approximately the same as the dynamic range of the recording medium. The modified local contrast and local mean are then combined to obtain $y(n, m)$, the processed image. The algorithm is applicable to the enhancement of images degraded by varying amounts of smoke, haze, fog, etc. Its major advantage is that it is both conceptually and computationally simple, relative to other techniques used for similar purposes.

Homomorphic filters have also found application in image enhancement. Such filters will be described in section 4.6. Kalman filters have also been used. They are reported in section 4.7.

2.9. An application of image enhancement [13]

An example of image enhancement is shown in fig. 7. The original image is shown in fig. 7a. The same image corrupted by positive impulse noise with probability 30%, is shown in fig. 7b. The output of a 3 × 3 median

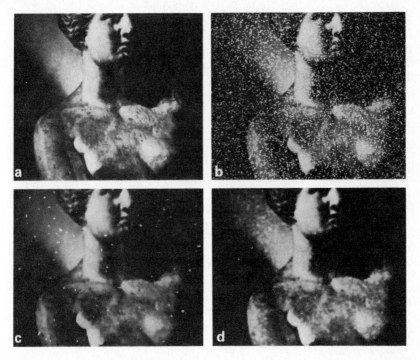

Fig. 7. Comparison of nonlinear filtering techniques: (a) test image, (b) test image corrupted by impulse noise, (c) image filtered by 3×3 median filter, (d) image filtered by a 3×3 CH_2 mean filter.

filter is shown in fig. 7c. The median filter removes some noise, but fails to remove all of it. The output of a 3×3 contraharmonic mean filter is shown in figure 7d. All the spikes have been removed. Linear filtering however (not shown here) degrades the image considerably by smoothing the edges.

3. Multi-dimensional filters

3.1. Introduction

Two- and three-dimensional digital filters have been successfully applied to many areas of digital image processing. Some of the main types of filters and their applications are now summarized:

– *Low-pass 2-D filters.* These result in the smoothing of the image, reducing high spatial frequency noise components.

– *High-pass 2-D filters.* These enhance very low contrast features, when they are superimposed on a very dark or very light background.

– *High-emphasis and band-pass 2-D filters.* These have an effect on the image, which is more subtle than that of the previous two classes. They generally tend to sharpen the edges and enhance small details.

Digital filters have also been used to reduce the amount of information displayed in images, and compress those images for more efficient transmission and storage.

In this section, the focus will be on design, realization and implementation techniques for 2-D and 3-D digital filters. The area of stability testing and stabilization of unstable filters will not be considered here. The reader is referred to refs. [5], [6] and [16] for this topic.

We shall only discuss LSI 2-D digital filters, that relate the input $x(m,n)$, and the output $y(m, n)$ sequences by a linear constant coefficient difference equation [5]

$$\sum_k \sum_l b(k, l) y(m - k, n - l) = \sum_k \sum_l a(k, l) x(m - k, n - l),$$

$$(14)$$

where $a(\cdot)$ and $b(\cdot)$ in eq. (14) are constant coefficients. In the frequency domain, the corresponding transfer function is

$$H(z_1, z_2) = \frac{A(z_1, z_2)}{B(z_1, z_2)} = \frac{\sum_k \sum_l a(k, l) z_1^{-k} z_2^{-l}}{\sum_k \sum_l b(k, l) z_1^{-k} z_2^{-l}}. \qquad (15)$$

The inverse z-transform of $H(z_1, z_2)$ is $h(m, n)$, the impulse response of the filter. If $h(m, n)$ has only a finite number of non-zero values, the filter is called a "finite impulse-response" (FIR) 2-D filter. Otherwise, it is an "infinite impulse-response" (IIR) 2-D filter. If the filter only requires samples of the input image to evaluate the output image, it is called "non-recursive". If on the other hand samples of the input image, together with evaluated output image samples are required, the filter is called "recursive". Digital filter design techniques are subdivided into conventional techniques and CAD techniques. Figures 8 and 9 outline the main features of those two design classes.

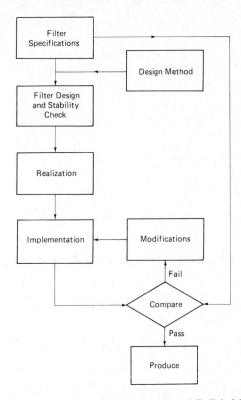

Fig. 8. Conventional design technique for a 2-D digital filter.

3.2. *Two-dimensional FIR digital filter design*

In designing 2-D FIR filters, stability is not an issue. With a bounded input the general convolution

$$
\begin{aligned}
y(m, n) &= \sum_{k=-\infty}^{+\infty} \sum_{l=-\infty}^{+\infty} x(k, l) h(m - k, n - l) \\
&= x(m, n) * h(m, n)
\end{aligned}
\tag{16}
$$

shows that the output will also be bounded, since the impulse response $h(m, n)$ is non-zero for only a finite number of points (m, n). Many

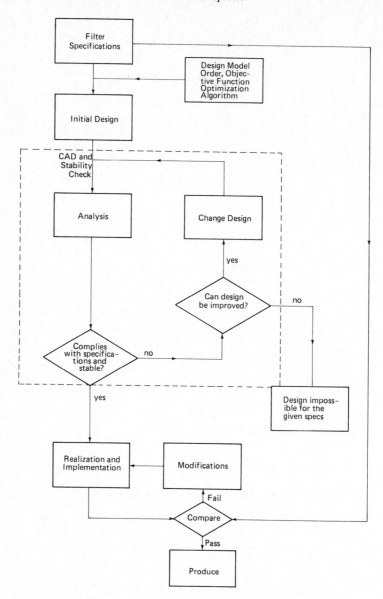

Fig. 9. Computer-aided design technique for a 2-D digital filter.

design techniques for 2-D linear phase FIR filters are generalizations of familiar 1-D methods. The simplest is the class of window techniques.

Window techniques. Given a desired frequency response, $D(e^{j\omega_1}, e^{j\omega_2})$, the inverse 2-D Fourier transform results in the ideal impulse response $d(m, n)$. In general, $d(m, n)$ is of infinite extent, whereas the FIR filter that is to be designed should have an impulse response of finite extent. The ideal impulse response is then multiplied by a finite extent function $w(m, n)$ called a window:

$$h(m, n) = d(m, n)w(m, n). \tag{17}$$

The Fourier transform of $h(m, n)$, denoted by $H(e^{j\omega_1}, e^{j\omega_2})$, is the 2-D

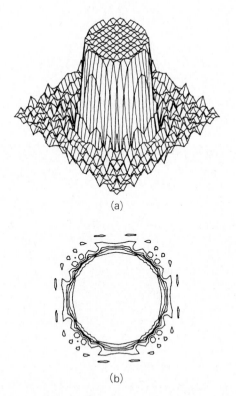

(a)

(b)

Fig. 10. Magnitude response (in dB) of a Kaiser-window circularly-symmetric low-pass FIR filter with an impulse response of size 31×31 samples.

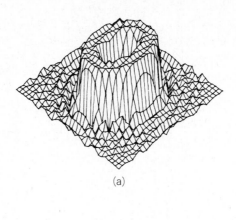

(a)

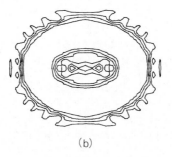

(b)

Fig. 11. Magnitude response (in dB) of a Kaiser-window elliptically-shaped band-pass FIR filter with an impulse response of size 31 × 31 samples.

convolution of $D(e^{j\omega_1}, e^{j\omega_2})$ and the Fourier transform of the window $w(m, n)$, denoted by $W(e^{j\omega_1}, e^{j\omega_2})$, in the frequency domain. It has been shown [5], that if $w(x)$ is a good symmetric 1-D window, then $w(x, y) = w([x^2 + y^2]^{1/2})$ is a good circularly symmetric 2-D window. Discrete 2-D versions of 1-D windows, such as Kaiser's, are used to design the filter. In general, windowing techniques are the easiest to apply and can be used to design filters of any order, with arbitrary magnitude and phase responses. Figures 10 and 11 show two filters designed by this method.

Frequency sampling techniques. Another class of design techniques for FIR filters, that can be extended from 1-D methods, is the class of frequency sampling techniques. The DFT coefficients of the filter are fixed at certain frequencies, typically being one in the desired passband

and zero in the desired stopband. The values in the transition band are left as free variables, in order to improve the approximation by minimizing some error criterion. Hu and Rabiner [17] showed that if the filter being designed is symmetric through the origin and across the 45° diagonals of the (ω_1, ω_2) plane (octagonally symmetric), then the frequency response is real with linear phase terms. Furthermore, the optimization problem reduces to a linear programming problem. They were able to design such filters as large as 25 × 25, but the serious drawback is that the method was computationally very expensive.

Optimization techniques. Various techniques to design 2-D FIR filters, optimal in the minimax or Chebyshev sense, have been studied [5]. Hu and Rabiner [17] extended their frequency sampling technique, by making all the frequency samples variable, but the computation involved was so great that only a 9 × 9 filter was attempted. Kamp and Thiran [18] extended the 1-D Remez exchange algorithm to design FIR circularly symmetric low-pass filters. The resulting filters were more selective than those designed by the windowing or the frequency sampling techniques, but again the computation time was excessive, of the order of 40 minutes of computer time for a 6 × 6 filter. Mersereau also tried a similar algorithm, but concluded that the use of optimization techniques was only worthwhile, in terms of cost, for relatively small impulse responses [5].

Separable filter techniques. There exist methods of designing 2-D FIR filters from 1-D FIR filters, which require relatively little computer time, but are not generally optimal, although they can be quite good. A convenient case is when the desired frequency response is separable:

$$D(e^{j\omega_1}, e^{j\omega_2}) = D_1(e^{j\omega_1}) D_2(e^{j\omega_2}),$$ (18)

where D_1 and D_2 can be designed by any 1-D method. Twogood and Mitra [19] proposed a CAD technique, based on singular value decomposition, to design a separable approximation to the desired 2-D frequency response. In some cases the gain in computational efficiency is great enough to offset the error in approximating a nonseparable 2-D filter by a separable one.

Transformation techniques. McClellan [20] proposed a powerful design method for FIR zero-phase digital filters. A change of variables transforms an optimal 1-D filter into a 2-D filter. The 1-D filter is of high-order, but efficient algorithms are known for the design of 1-D FIR filters. The change of variables is effectively a low-order 2-D filter, allowing a high-order 2-D filter to be designed with the computational effort of a 1-D design. An added advantage is that in many cases, the 2-D filter obtained is also optimal. A disadvantage of this technique is that the class of FIR filters that can be designed is limited. McClellan showed that a 1-D zero-phase FIR filter of odd length can be transformed into a 2-D FIR filter with quadrantal symmetry, by the change of variables:

$$\cos \omega = A \cos \omega_1 + B \cos \omega_2 + C \cos \omega_1 \cos \omega_2 + D. \tag{19}$$

The class of filters that can be approximated in this way includes those filters whose magnitude response is constant over certain regions of the 2-D frequency plane, such as low-pass and band-pass circularly symmetric filters, as well as fan filters. The change of variables (19) results in a contour in the 2-D plane for each fixed ω, with the 2-D frequency response along this contour constant, equal to the 1-D frequency response at ω. The problem reduces to picking A, B, C, D to obtain suitable contours that approximate the desired 2-D frequency response. McClellan was able to design a 31×31 circularly symmetric low-pass filter with 5 seconds of computer time, a dramatic reduction compared to the techniques previously mentioned. Mersereau et al. [21] generalized the McClellan technique by making the change of variables more general:

$$\cos \omega = \sum_{p=0}^{P} \sum_{q=0}^{Q} t(p, q) \cos(p\omega_1) \cos(q\omega_2). \tag{20}$$

Several algorithms were also presented for designing the generalized transformation (20). Thus, the kinds of filters that can be designed in this way are more extensive. A filter designed with a generalized McClellan transformation can also be implemented with the number of multiplications proportional to N, faster than the FFT implementation, for filters up to about 45×45. Later, Mersereau [22] extended the technique still further, so that even the restriction to filters of quadrantal symmetry was relaxed. For example, eq. (19) is generalized to

$$\cos \omega = A + B \cos \omega_1 + C \cos \omega_2 + (D + E)\cos \omega_1 \cos \omega_2$$
$$+ (E - D)\sin \omega_1 \sin \omega_2. \tag{21}$$

Arbitrary zero-phase 2-D FIR filters can be designed with this method.

3.3. Two-dimensional IIR digital filter design

A 2-D IIR digital filter is described by a 2-D difference equation of the form

$$y(m, n) = \sum_{k=0}^{M_a} \sum_{l=0}^{N_a} a(k, l) x(m - k, n - l)$$

$$- \sum_{\substack{k=0 \\ k+l \neq 0}}^{M_b} \sum_{l=0}^{N_b} b(k, l) y(m - k, n - l). \qquad (22)$$

This is the causal or quarter plane form, in which the recursion is in the $(+m, +n)$ direction. IIR filters have the advantage over FIR filters, that they can generally be implemented with fewer multiplications. Their disadvantage, however, is that it is not easy to test or guarantee stability of 2-D IIR filters, while stability is guaranteed for FIR filters. For 1-D IIR filters the denominator of the system function can be factored, and the resulting poles are easily checked for stability. In 2-D, due to the lack of the Fundamental Theorem of Algebra, no factorization exists. Much work has been done on this problem, with the basic stability checks described in refs. [5], [6] and [16]. Discussion of stability of 2-D IIR filters is beyond the scope of this course, but methods for the design of such filters will be reviewed.

Transformation Techniques. In their early paper Shanks and Justice [23] presented a design technique in the frequency domain by transforming a 1-D analog filter. Let $H(s)$ be the transfer function of an analog filter. This can be considered a 2-D analog filter that varies in 1-D only if written as

$$H(s_1, s_2) = H(s_1). \qquad (23)$$

Now, rotate the (s_1, s_2) plane by an angle β, onto the (s_1', s_2') plane, with the transformations

$$s_1 = s_1' \cos \theta + s_2' \sin \theta,$$
$$s_2 = s_2' \cos \theta - s_1' \sin \theta. \qquad (24)$$

Finally, apply the bilinear transformation

$$s_1' = c \frac{1 - z_1}{1 + z_1}, \qquad s_2' = c \frac{1 - z_2}{1 + z_2}, \qquad (25)$$

where c is a constant. By cascading various prototype 1-D filters, rotated by various angles, different 2-D filters can be obtained. Zero phase filters are also possible, which are not possible in the 1-D case, due to the constraint of causality.

Costa and Venetsanopoulos [24] used this method to design good circularly symmetric 2-D IIR filters. They made the procedure more exact, showing how to obtain a desired cutoff frequency in a given direction with a specified maximum error. In addition, they proved that rotated filters are marginally stable, if the angle θ lies in the range $270° < \theta < 360°$. To obtain filters with other angles of rotation, the input and output data can be rotated by appropriate multiples of 90°. The resulting frequency response of rotated filters is distorted to an extent, depending on the angle of rotation. In a later paper, Mneney and the above authors [25] showed how to perturb the parameters of the filters designed, to make them strictly stable under finite precision arithmetic.

More recently, an algorithm was proposed by which the frequency response is approximated in as many directions as the number of rotated filters being cascaded. The cutoff frequency of each rotated filter is adjusted separately and iteratively, according to the frequency response obtained when all the filters in the cascade are considered. This approach has not only the advantage of giving a better circularly-symmetric frequency response, but also noncircularly-symmetric filters can be designed, by specifying different cutoff frequencies in each direction. However, since only low-pass filters are cascaded, the relationships among these cutoff frequencies cannot be arbitrary, because the locus of the cutoff frequency of the 2-D filter is a smooth curve, such as a circle or an ellipse. Using this new algorithm problems of convergence may arise, if too much accuracy is specified in too many directions. The proposed algorithm takes care of this problem, by aborting the iteration as soon as it ceases to converge. An error code is returned, which specifies if convergence was attained. There is a compromise between accuracy in the cutoff frequency and the number of directions in which it is specified. Two filters designed using this algorithm are shown in Figs. 12 and 13.

A different approach for producing stable 2-D IIR filter designs is modifying existing designs that are known to be stable. Pendergrass et al. [26] studied such spectral transformations. They proposed transformations that produce stable first quadrant transfer functions, from prototype stable first quadrant transfer functions. They also guaranteed the transform of real rational functions, into real rational functions, so that the result can be realized with real arithmetic. Finally, the transformations were required to preserve some characteristics of the magnitude response, such as peak ripple, while altering other characteristics, such as

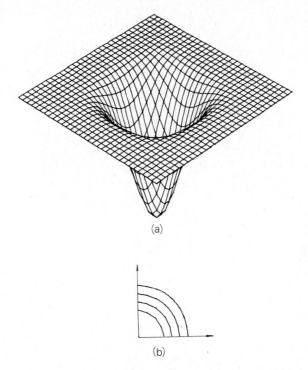

(a)

(b)

Fig. 12. Magnitude response of a 2-D circularly-symmetric high-pass IIR filter with zero-phase response (derived from a fourth-order Gaussian magnitude approximation continuous filter and rotations by multiples of 30°).

cutoff frequencies. They found and catalogued ten stable 2-D all-pass functions, that met these requirements, simplifying the design process for many useful filter types.

Optimization techniques. As in the case with FIR filters, many computer-aided optimization procedures have been proposed to design 2-D IIR digital filters. A major problem is ensuring the stability of the final filter design. Maria and Fahmy [27] constrained the filters to have the form of products of simple first- and/or second-order terms. Stability was tested at each step, with simple inequalities relating the coefficients of the low-order filters.

Later, basing their approach on some new theoretical results on filter symmetries [28], which imply that quadrantally symmetric causal 2-D

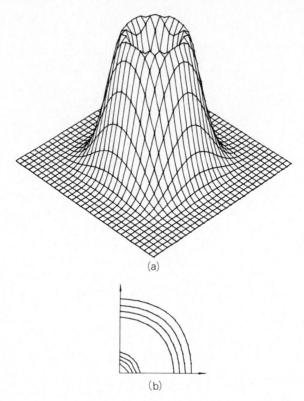

(a)

(b)

Fig. 13. Magnitude response of a 2-D circularly-symmetric band-pass IIR filter with zero-phase response (derived from a fourth order-Gaussian magnitude approximation continuous filter and rotations by multiples of 30°).

filters have a separable denominator, George and Venetsanopoulos [29] designed quarter plane filters possessing quadrantal and octagonal symmetries, with a reduced number of coefficients and significantly reduced optimization cost. This approach was also extended to half-plane filters and multiplierless 2-D IIR filters [30]. These were magnitude response designs only, but the design of all-pass phase equalizers was used to equalize the phase and produce suitable magnitude and phase responses.

While CAD techniques may be computationally expensive, they are characterized by the following attributes:

(1) They result in optimal designs.

(2) The filters designed can be useful as a bench mark for examining suboptimal designs.

(3) The resulting filters are completely general. They can satisfy arbitrary specifications.

(4) These methods can be used to match the requirements of VLSI implementation.

3.4. Two-dimensional filter realization and implementation

In realizing the transfer function of a 2-D LSI filter, there is an infinite variety of structures that analytically will result in the same relationship between the input sequence and the output sequence. Different realizations of the same transfer function have widely diverse performance, in terms of roundoff errors and coefficient sensitivity. A fundamental limitation of 2-D filtering is that the Fundamental Theorem of Algebra, in one variable, does not extend to two or more variables. One implication of the limitation is that we cannot realize 2-D filters in cascade form, to reduce the effect of quantization noise. Another implication is that an arbitrary 2-D transfer function cannot be manipulated into a form required for a particular realization. This lead to the joint study of the design and realization [5].

FIR 2-D digital filters are traditionally realized by convolution or fast-convolution (through fast transforms), with the exception of special kinds, such as those obtained through transformations or those designed to possess special structures.

IIR 2-D digital filters, however, were until recently realized by their difference equation realization, in a direct form, with the exception of special kinds, which had a modular structure imposed on them by the design procedure. Motivated by the desire to give a general filter realization structure, for the sake of modularity, a number of contributions were made [31–33], which resulted in a method for the exact expansion of a general 2-D transfer function in first-order terms, each one of which is a function of one of the two variables only. This method, which was based on matrix decomposition, resulted in a rich set of realizations, which were compared in refs. [32] and [33]. These and other comparisons indicate that it is now possible to realize 2-D FIR and IIR filters in an enormous variety of ways, and to evaluate their cost, speed and noise performance.

The effects of input quantization, coefficient quantization and round-off accumulation were considered in ref. [34] for fixed-point arithmetic, and it was shown how these three kinds of errors affect the filter output. The way in which these errors combine to produce the output error was presented in ref. [35], and it was found that in general they combine in a nonlinear way, except if the registers are of sufficient length, in which case the individual noise variances can be added to obtain the noise variance at the filter output.

Hardware was also designed to perform fast convolution of an FIR 2-D digital filter with an image, based on Fermat number transforms [36]. It was shown that in the filtering of a 256 × 256 image, this method was faster by a factor of four, than that obtained by fast convolution based on the Fast Fourier Transform (FFT). A first-quadrant 2-D, second-order digital filter was recently implemented on a chip [37], using 4-micron NMOS VLSI technology, and theoretical investigations are currently under way to develop the theory for the realization of general 2-D digital filters, implemented via such chips.

In addition, the desire for fast processing led to investigations of different approaches which can be used to achieve real-time image processing [6]. Jaggernauth et al. [38] presented the architecture and the hardware of a 2-D IIR digital filter, which was implemented on the basis of distributed arithmetic. Faster architectures were considered in ref. [39].

3.5. Multi-dimensional filters

Three-dimensional (3-D) image processing has recently emerged as an essential tool in many areas of current interest. This is due to the increasing importance of many applications, where it plays a significant role, and the availability of high-speed computers with large memory. 3-D signals offer significant advantages over 2-D signals, because they preserve spatial information. Depth, surface orientation or edges can be easily detected from 3-D data.

The most important applications of 3-D image processing are the following:

– *Biomedicine.* 3-D CAT, which has the ability to preserve structures, finds applications in craniofacial surgical planning, the study of the central nervous system, stereostatic neurosurgery, stereostatic biopsy, irradiation, radiation therapy, reconstructive therapy, and the study of moving parts.

– *Time-varying 2-D signals.* In video it is customary to model the time varying 2-D signals as 3-D signals, to preserve spatial and temporal correlations.

– *Robotics.* Robotics is one of the most rapidly growing areas of current technology. There is an increasing effort to use robots in more sophisticated applications. For such applications the availability of a 3-D visual system is necessary.

– *Geophysics.* 3-D seismic data processing was shown to have many advantages over 2-D processing. 3-D migrated data describe more accurately the geology of the area under study, than the corresponding 2-D data.

One of the main applications of 3-D filtering is noise reduction. Morgenthaler [40], assuming different resolution in the z-axis from that in the x- and y-axes, developed an experimental model to study and compare the effects of 2-D and 3-D filtering on simulated 3-D data. In most of the operations he used, 3-D techniques achieved greater reduction in the standard deviation of the noise, but increased blurring effects at the edges, compared to those of 2-D techniques. Hurt and Rosenfeld [41], using series of autoradiographs with equal resolution on all three axes, found that 3-D techniques are more effective for noise removal and edge preservation than 2-D techniques.

The design of efficient 3-D filters is only now starting to emerge. The McClellan transformation technique was extended to three dimensions [42], and 3-D FIR spherically symmetric filters, as well as fan filters were designed. 3-D IIR filters were also designed by cascading 2-D rotated filters. The method of rotated filters was also recently extended to obtain IIR 3-D spherically symmetric designs. Hirano, Sakane and Mulk [43] developed 3-D filters for TV-signal applications, based on cascade and/or parallel combinations of 1-D component filters. Pitas and Venetsanopoulos [44] introduced symmetries to the design of m-D filters, and were able to design good 3-D IIR spherically symmetric and fan filters, by optimization techniques. The filters designed are efficient and have easy stability checking. Finally, Venetsanopoulos and Mertzios [45, 46] extended the realization techniques based on matrix decomposition to m-D, and presented modular structures for m-D filters, based on Walsh and other matrices.

4. Image restoration

4.1. Introduction

Restoration is the process by which we infer the original distribution of object radiant energy from the image received. Restoration techniques can be classified into deterministic and stochastic. The first class is appropriate for a deterministic image formation model, where noise can be neglected. Deterministic techniques attempt to recover the object radiant energies by "inversion". Stochastic techniques are appropriate in the case of a stochastic image formation process. Since the object radiant energy is usually transformed by the imaging system and corrupted by noise, it is not possible to restore it completely. In this class of techniques an optimum restoration system is defined as one which optimizes some measure of the distance between the actual and the estimate radiant energies.

Restoration techniques can also be distinguished into those which utilize a priori knowledge, and those which are based on a posteriori knowledge, exclusively. A priori knowledge includes a variety of parameters, such as geometrical equations of motion or relative position between the object and image film plane during exposure. Algorithms have been developed for geometrical coordinate, as well as other transformations, which allow the correction of such degradations. Other forms of a priori knowledge might include knowledge of image models or statistics. By a posteriori knowledge we refer to the knowledge that can be acquired by observing the image. Obvious examples might include point spread function determination from edges or points in the image, that are known to exist in the object. Other examples of a posteriori knowledge include obtaining estimates of the noise variance and power spectrum from relatively smooth regions of the image. Some of the methods used for the estimation of the degradation were described in section 1.5.

There are two difficulties in designing restoration algorithms. The one relates to the metric of optimization, which should ideally reflect the psychovisual effects of noise on the observer. The second difficulty relates to the fact that statistical properties of the image and the noise should be known by some techniques to a degree that is not justified in practical situations.

We now describe some of the main restoration techniques.

4.2. Inverse filters

When the effect of image recording and additive noise can be ignored, eq. (10b) can be Fourier transformed to result in eq. (3). Since we would like to remove the degradation phenomenon, which is $H(W_x, W_y)$ in the frequency domain, the obvious solution is to perform inverse filtering, described by

$$I(W_x, W_y) = \frac{1}{H(W_x, W_y)}, \tag{26}$$

where W_x, W_y are spatial frequencies in the x- and y-axes, respectively. For linear motion blur the inverse restoration filter is

$$I(W_x, W_y) = c\frac{w}{\sin(aw)} \tag{27}$$

where

a is the total displacement,

$w = W_x\cos\theta + W_y\sin\theta$,

θ is an angle defining the direction of motion relative to the image,

c is an arbitrary constant.

It is clear that the denominator is zero when $aw = n\pi$, $n = 1, 2, \dots$. This is a reason, in addition to the simplifying assumptions made, that inverse filters do not perform well in situations characterized by a small signal-to-noise ratio or a significant degree of blur. A modification of the inverse filter is the "pseudo-inverse" filter, which has a small additive constant introduced in the denominator of the inverse filter, to alleviate the effect of denominator zeros.

4.3. Wiener filters

This method attempts to determine an estimate of the object radiant energy $\hat{f}(u, v)$, which minimizes some measure of the difference between $\hat{f}(u, v)$ and $f(u, v)$. A common measure for such minimization is the mean square error criterion. We now assume the image formation eq. (10a) and evaluate the mean square error between f and $\hat{f}$:

$$e^2 = E\left[(f - \hat{f})^2\right], \tag{28}$$

where $E[\cdot]$ denotes expectation. It is easy to show that this is minimized by the mean of the a posteriori density or the "conditional mean"

$$\hat{f} = E[f/g]. \tag{29}$$

This optimum estimate, in the least-squares sense, is in general a nonlinear function of the signal, and requires knowledge of the a posteriori density of f conditioned on g, which is rarely available. To simplify the solution, we often make a linearization assumption, and seek to obtain the optimum linear estimate $\hat{f}_L$, which minimizes eq. (28). It is noted that, when we assume eq. (10b) and f and r are Gaussian, the optimum nonlinear estimate coincides with the linear least-squares estimate. However, the Gaussian assumption is generally not valid for image statistics.

We can derive the optimum linear estimate, in the least-squares sense, for eq. (10c). Suppose $g(x, y)$ is the image and $t(x, y)$ is the restoration filter, which restores the image to $\hat{f}(x, y)$. For a space invariant system

$$\hat{f}(x, y) = \int \int t(x - x', y - y') g(x', y') \, \mathrm{d}x' \, \mathrm{d}y'. \tag{30}$$

It is easily shown that the $t(\cdot)$ satisfying eq. (31) minimizes the mean square error e^2:

$$E\big[\big[f(x, y) - \hat{f}(x, y)\big] g(u, v)\big] = 0. \tag{31}$$

For homogeneous random variables, which implies that f and g are wide-sense stationary, the following relation results between the auto-correlation function of g and the cross-correlation function between f and g:

$$R_{fg}(x - u, y - v)$$

$$= \int \int t(x - x', y - y') R_{gg}(x' - u, y' - v) \, \mathrm{d}x' \, \mathrm{d}y'.$$

Let $x - u = a$, $y - v = b$, $x - x' = l$, $u - y' = r$, then

$$R_{fg}(a, b) = \int \int t(l - a, r - b) R_{gg}(l, r) \, \mathrm{d}l \, \mathrm{d}r.$$

In the frequency domain this results in the restoration filter

$$T(W_x, W_y) = \frac{S_{fg}(W_x, W_y)}{S_{gg}(W_x, W_y)}, \tag{32}$$

where $S_{fg}(W_x, W_y)$ denotes the cross-spectral density between the degraded and undegraded images and $S_{gg}(W_x, W_y)$ the spectral density of the degraded image.

Expression (32) requires statistical knowledge of both the degraded and the undegraded images.

$$S_{fg}(W_x, W_y) = H^*(W_x, W_y) S_{ff}(W_x, W_y),$$

where $S_{ff}(W_x, W_y)$ is the spectral density of the object radiant energy and the asterisk denotes complex conjugation. In the presence of uncorrelated additive noise of spectral density $S_{rr}(W_x, W_y)$, we obtain

$$S_{gg}(W_x, W_y) = |H(W_x, W_y)|^2 S_{ff}(W_x, W_y) + S_{rr}(W_x, W_y). \qquad (33)$$

Substituting in eq. (32) we obtain the Wiener-filter frequency response

$$T(W_x, W_y) = \frac{H^*(W_x, W_y)}{|H(W_x, W_y)|^2 + \dfrac{S_{rr}(W_x, W_y)}{S_{ff}(W_x, W_y)}}. \qquad (34)$$

In the absence of noise this filter, called the Wiener filter, reduces to the inverse filter. In the derivation of the Wiener filter, it was assumed that the undegraded image and the noise belong to homogeneous random fields and that their power spectra are known. Generally, the Wiener filter performs better than the inverse filter in low signal-to-noise ratio situations. This is because when the signal power is small, the gain of the Wiener filter also reduces, controlling the ill-conditioned behavior of the noise effect. Notice, however, that the a priori knowledge required by the Wiener filter may not be available in some situations.

4.4. Power spectrum equalization

This method determines $\hat{f}(u, v)$, which is a linear estimate of $g(x, y)$, in such a way that the power spectrum of the estimate is equal to the power spectrum of the original image. Thus, the method is a power spectrum equalization method. The restoration filter is then given by [3]

$$M(W_x, W_y) = \left[\frac{1}{|H(W_x, W_y)|^2 + \dfrac{S_{rr}(W_x, W_y)}{S_{ff}(W_x, W_y)}} \right]^{1/2}. \qquad (35)$$

The filter $M(W_x, W_y)$ reduces to the inverse filter for zero noise. For low signal amplitudes, the filter gain goes to zero. Between these two extremes the filter gain is greater than that of $T(W_x, W_y)$, but is less than that of $I(W_x, W_y)$. This filter usually results in images of better visual quality than those of the inverse filter or the Wiener filter. Notice that if we multiply both the numerator and denominator by $[S_{ff}(W_x, W_y)]^{1/2}$ the denominator of the filter represents the spectral density of the output signal, which can be easily estimated from the image received. S_{ff} can also be estimated from similar unblurred images.

4.5. Generalized Wiener filters [1]

The generalized Wiener filters are defined by

$$N(W_x, W_y)$$

$$= \left[\frac{1}{H(W_x, W_y)} \right]^\alpha \left[\frac{H^*(W_x, W_y)}{\left| H(W_x, W_y) \right|^2 + \beta \dfrac{S_{rr}(W_x, W_y)}{S_{ff}(W_x, W_y)}} \right]^{1-\alpha},$$

$$(36)$$

where α and β are positive real constants. Notice that for $\alpha = 1$ or $\beta = 0$, eq. (36) becomes the same as the inverse filter of eq. (26). For $\alpha = 0$ and $\beta = 1$, it becomes the same as the Wiener filter of eq. (34). For $\alpha = 1/2$ and $\beta = 1$, it results in the filter of the power spectrum equalization method of eq. (35), and for $\alpha = 1/2$, it results in a filter which is the geometric mean between the inverse filter and the Wiener filter. This last filter has been shown to give more pleasing results to the eye. Other values of the parameters α and β are possible.

The previous linear techniques are also referred to as "direct" techniques for image restoration. Indirect techniques contain constrained linear restoration techniques, adaptive and nonlinear restoration techniques. "Constrained least-squares" restoration results in the following filter

$$C(W_x, W_y) = \frac{H^*(W_x, W_y)}{\left| H(W_x, W_y) \right|^2 + \delta},$$

$$(37)$$

where δ is a parameter that must satisfy some equation [3]. This parameter can be determined by successive iterations. In addition, "regularization" techniques have been recently introduced in an attempt to solve the image restoration problem [47, 48].

4.6. Nonlinear image restoration

The nature of eqs. (10a) and (29) suggest that nonlinear techniques should be used for image restoration. Such techniques were developed only recently, due to their large computational demands. Usually, nonlinear methods are superior in the sense of accuracy, but not speed. Justification of the continued use of linear filters lies in their simplicity, economy and speed.

One nonlinear method that has been investigated is homomorphic filtering. This amounts to converting the blurred intensities to density values, by taking the logarithm, restoring by linear techniques, and then exponentiating the result. This technique works well for signal-dependent noise of a multiplicative nature. In ref. [14] it was shown that only in the special case of multiplicative noise, can homomorphic filtering transform it to signal-independent noise. In more general noise cases, nonlinear order statistic filters [14] can be used to transform signal-dependent noise into approximately signal-independent noise.

Maximum entropy techniques amount to the application of a positive formalism, which follows from a fundamental physical principle. This formalism results in a regularization of the ill-conditioned problem of image inversion [48]. In Burg's maximum entropy approach, described by Wernecke and D'Addario [49], the signal $f(x, y)$ is regarded as the square of another variable $a(x, y)$. This guarantees that $f(x, y)$ remains nonnegative. The first two moments of $a(x, y)$ are presumed fixed. The solution of the maximum entropy approach shows that biased representation for the probability density of $a(x, y)$ is normal, and this leads to a criterion, which results in the maximization of the quantity

$$H = - \int \int \ln f(x, y) \, \mathrm{d}x \, \mathrm{d}y. \tag{38}$$

According to Frieden's approach [50], a statistical model of the object is assumed, which differs from Burg's choice. This leads to the calculation of the image estimate, by maximizing

$$H' = - \sum_n \sum_m f(x_n, y_m) \ln f(x_n, y_m), \tag{39}$$

where (x_n, y_m) are the centers of cells into which the image is subdivided.

The two approaches differ by the factor $f(x, y)$. This factor gives Frieden's approach a mathematical form that is closer to that of classical entropy, than is Burg's algorithm. However, the price paid for this is the lack of a closed form solution for the image estimate. Both approaches exert a smoothing influence on the estimate $\hat{f}(x, y)$. If any two values $f(x_1, y_1)$ and $f(x_2, y_2)$ are made to differentially change to a pair of more equal values (a smoothing tendency), both H and H' increase. Therefore, a maximum H and H' by either criterion, fosters a "maximally smooth" estimate. The absolute maximum in fact, results from the perfectly smoothed estimate $\hat{f}(x, y)$. The input data, however, act as constraints, which force fluctuations in $\hat{f}(x, y)$ and hence a departure from the absolute maximum.

While the conditional mean is the result of Bayesian estimation for a quadratic cost function, the median of the a posteriori density is the optimum solution for an absolute value criterion and the maximum a posteriori estimate maximizes a uniform cost function [51]. Some of the Bayesian nonlinear restoration techniques were used in image restoration [52]. However, their application requires additional statistical knowledge. For example, the determination of the maximum a posteriori estimate requires knowledge of the a posteriori density function $p(f/g)$, plus the probability density functions of f and the noise of the system. In practical situations these functions are usually unavailable and have to be assumed. This implies that the advantage of these techniques, which lies in their ability to operate in low signal-to-noise ratio conditions, may be reduced due to lack of accuracy in the statistical knowledge required.

4.7. Adaptive image restoration

An image can be modeled by a nonhomogeneous random field. Its statistics change over its various regions. This implies that space-variant filters, which sense the statistical changes and adapt accordingly, are desirable. However, such filters are complex and require considerable computation. Thus, early recursive estimation techniques were based on the assumption that the image statistics were stationary, greatly simplifying the image modeling procedure. This assumption resulted in filters that smooth the image, smoothing edges and reducing contrast. As a result the subjective image quality was degraded. Ingle et al. [53] achieved improved results, by segmenting the image into blocks, and filtering each block independently of the others, while allowing some adaptation to the local image statistics. However, their segmentation is somewhat arbitrary and smoothing of the edges still occurs.

To overcome the computational problems of large global states, the reduced update concept was introduced by Woods [54]. This led to the reduced update Kalman filter, which resulted in good performance for homogeneous images, but led to blurred edges in natural images, which are nonhomogeneous.

To face the crucial problem of non-stationarity of images, a number of approaches were taken, characterized by Kalman filters, which adjust their operation when important image features are changed (in the case of texture) or detected (in the case of edges). Biemond and Gerbrands

[55] apply edge information extracted from the noisy image to control the input of a 1-D raw-wise scan-ordered Kalman filter, such that the filter responds more quickly to the presence of an edge. A recent paper [56] followed the same approach in the 2-D context, with a view towards improving the edge response of an arbitrary first-order 2-D Kalman filter. The first-order restriction allows relatively fast and possibly on-line implementation. This method is more appropriate for images where the image edges are perpendicular to the direction of processing.

Most of the adaptive algorithms discussed are linear; however, nonlinear adaptive algorithms are also being investigated, as they are expected to better meet image processing requirements. Although some preliminary nonlinear adaptive filters have been reported [14], the search for efficient and meaningful nonlinear adaptive filters remains wide open. Each of the previous techniques claims optimality, with respect to differing criteria, and has advantages and limitations in terms of cost, volume of data handling, extent to which it is affected by sensor nonlinearity and others. It is also noted that a degraded image suffers a fundamental loss of information, compared to the image which would have resulted in the absence of degradation. This fundamental loss cannot be recovered by any of the previous techniques. Since the state of the art does not allow an accurate understanding and mathematical modeling of the human visual process, it is not yet possible to define precisely the meaning of optimum restoration.

4.8. Application of image restoration to radiology

4.8.1. Introduction
Conventional radiographs are 2-D displays of 3-D objects. The X-ray image on the film plane consists of images from the individual layers of the exposed object. The main features of the radiologic process are the following:
(1) the X-ray source, the focal spot, emits radiation which penetrates the objects and therefore information from all the layers is displayed;
(2) the size of the focal spot is not infinitesimal and thus contributes to the *penumbra effect*;
(3) the whole process is nonlinear following an exponential law.
The problem of X-ray image enhancement for a layer located at a specific depth of the exposed object was addressed in refs. [57–59]. A linear approximation of a radiologic model was introduced, which took

advantage of the penumbra effect. In this example, the application of a Wiener filter to extract depth information from a single radiograph is described.

4.8.2. Formulation of the problem

The radiologic process is characterized, in general, by the following quantities:

(1) the focal spot X-ray intensity distribution;
(2) the absorption coefficient function of the exposed body, and
(3) the received image on the film plane.

The focal spot has finite dimensions and is described by a spatial intensity distribution. This is usually modeled by a Gaussian or a twin-peaked Gaussian distribution. Let $i_0(x_0, y_0; x, y)$ denote the X-ray intensity distribution of the focal spot, where x_0, y_0 are the coordinates on the focal spot plane; and let $\mu'(x_s, y_s, z)$ be the absorption coefficient of the exposed body at depth z, measured from the focal spot, with x_s, y_s being the coordinates on the sth layer. Finally, let $l'(x, y)$ be the X-ray intensity of the received image at the point (x, y) on the film plane. Then, the equation that governs the system is given by [57]

$$l'(x, y) = \int_D \int i_0(x_0, y_0; x, y) \exp\left[-\int_L \mu'(x_s, y_s, z)\, \mathrm{d}l \right] \mathrm{d}x_0\, \mathrm{d}y_0$$

(40)

where D is the focal spot area, L is the straight line connecting the points (x_0, y_0) and (x, y), and $\mathrm{d}l$ is the differential length along this line at the point (x_s, y_s).

The system described by eq. (40) is nonlinear and space-varying. In ref. [60] the analysis of the radiologic process was carried out based upon the following assumptions:

(1) The focal spot plane and the film plane are assumed to be parallel. This implies that

$$i_0(x_0, y_0; x, y) = i_0(x_0, y_0).$$

(2) The variations of μ' around its mean value $\bar{\mu}$ are assumed to be small. $\mu'(\cdot)$ is expressed as

$$\mu'(x_s, y_s, z) = \bar{\mu} - \mu(x_s, y_s, z),$$

where $\mu(\cdot)$ is the variation around the mean.

(3) The dimensions of the film are much smaller than the distance between the focal spot and the film plane.

The first of these assumptions renders the system spatially invariant, while (2) and (3) permit the linearization of the exponential factor of eq. (40), thus yielding

$$l'(x, y) = e^{-\bar{\mu}d}[\bar{l} + l(x, y)], \tag{41}$$

where

$$\bar{l} = \int_D \int i_0(x_0, y_0) \, dx_0 \, dy_0 \tag{42}$$

is a constant independent of $\mu(\cdot)$, and

$$l(x, y) = \int_L \int_D \int i_0(x_0, y_0)\mu(x_s, y_s, z) \, dx_0 \, dy_0 \, dz. \tag{43}$$

The term described by eq. (42) is the mean value of the image, $\bar{l}$, and can be subtracted before processing. Equation (43) describes the variation of $l'(\cdot)$ around $\bar{l}$.

Since the image $l(x, y)$ is the superposition of the images $l_s(x, y)$ of the various layers, it is clear that

$$l(x, y) = \sum_{s=1}^{N} l_s(x, y), \tag{44}$$

where N is the number of layers. The problem of digital restoration of radiographs can be formulated as follows. Given a single X-ray image $l(x, y)$, design a Wiener filter such that the output image approximates the image of the layer located at depth z_s in the minimum mean-square error sense.

4.8.3. System design
When the focal spot has a Gaussian distribution and the exposed object is not correlated (described by a white-noise process), the transfer function of the filter for depth z_s can be obtained by [60]

$$T_s(f) = Af \exp\left[-2\pi^2\sigma^2\left(\frac{d - z_s}{z_s}\right)^2 f^2\right], \tag{45}$$

where $f = \sqrt{W_x^2 + W_y^2}$, σ is the standard deviation of the Gaussian distribution, d is the distance between the focal spot and the film plane and A is a constant. In contrast, the inverse filter is high-pass with transfer function

$$I_s(f) = B \exp\left[2\pi^2\sigma^2\left(\frac{d - z_s}{z_s}\right)^2 f^2\right], \tag{46}$$

where B is a constant. This filter restores the image $I_s(x, y)$ from its blur, but ignores the contribution of the other layers and has the disadvantage of enhancing high-frequency noise.

Similarly, if the exposed object is modeled as a second-order Markov process, possessing circular symmetry and power spectral density given by

$$S_p(f) = \frac{6\pi\beta^3}{\left(4\pi^2 f^2 + \beta^2\right)^{5/2}}$$

and the focal spot is assumed to be a point source, then, the calculation of the transfer function of the optimal filter yields

$$T_s(f) = \frac{Af\left(8\pi^2 f^2 + 2\beta^2 + \pi f\sqrt{\beta^2 + 4\pi^2 f^2}\right)}{\left[4\pi^2\left(\dfrac{d}{z_s}\right)^2 f^2 + \beta^2\right]^{5/2}}, \tag{47}$$

where β is the parameter of the Markov process.

4.8.4. System implementation and verification

To demonstrate the performance of the optimal filter a number of simulations were performed. Fig. 14a depicts a simulated phantom consisting of an "X" and a "+" of approximately equal size, located at respective depths $z_1 = 300$ mm and $z_2 = 700$ mm, measured from the focal spot. The focal spot-to-film distance was 1 m. The X-ray intensity distribution was Gaussian with standard deviation $\sigma = 2$:

$$i_0(x_0, y_0) = \frac{1}{2\pi\sigma^2} \exp\left(-\frac{x_0^2 + y_0^2}{2\sigma^2}\right).$$

A Hamming window was applied to the image of fig. 14a, to reduce the effect of the edges. Figure 14b is the result after processing the windowed image with the filter of eq. (47). Another example of an inverse tomographic filter is shown in fig. 15. A radiograph of a mock chest (phantom) was obtained. In order to have sufficient resolution, only small areas of 50×50 mm^2 in the radiograph were digitized. Figure 15a shows one example obtained with a focal spot of nominal size 2 mm, and a radiological magnification of 1.5. The regions to be processed were carefully chosen to contain two lesions, one on the side of the chest closest to the focal spot (small white area in the fourth quadrant) and

Fig. 14. (a) Original image, (b) output of the Wiener filter.

one on the side of the chest closest to the film (large white area in the second quadrant). One result is shown in fig. 15b, where the lesion in the second quadrant was clarified. In general, the reconstruction of one layer of the phantom produces a sharper image of the lesion in that layer, and also an apparent reduction of its size. The same filter degrades the lesion of the other layer.

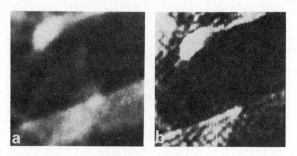

Fig. 15. (a) Original image of lesions in a lung, (b) image digitized and then top lesion filtered by inverse tomographic filter.

5. Image analysis and vision

5.1. Introduction to image analysis [61-63]

Image analysis is the process of deriving a useful description from a given image or a set of images. It is an area closely related to image processing, and utilizes many image processing operations, as well as additional techniques, drawn from detection theory, estimation theory, artificial intelligence and pattern recognition. By artificial intelligence we mean the art of creating machines that perform tasks considered to require intelligence when performed by humans. A brief description of image analysis will be presented in this section.

Image analysis usually consists of a process of simplification. A complex object, the image, is coverted into a simpler form (a description), by some sequence of operations. This simplification is done in order to achieve one or more of a number of possible objectives:

(1) *To acquire a more efficient representation.* This objective is related to image compression and artificial intelligence. An image can be represented by a simpler image, such as a line drawing, and convey the essential formation contained in the original image. While the usual representation of an image is that of an array, where each pixel's value gives the radiation received, illumination, reflectivity, etc., other representations exist, such as those that represent regions by edges, chain codes and graphs [62]. Hierarchical representations have also evolved, as well as object-centered and viewer-centered representations. The fundamental problem here is how to choose the representation, which is more appropriate for a given application.

(2) *To emphasize pertinent features for detection, classification, recognition and estimation.* The second objective is related to detection, classification, pattern recognition and/or estimation theories. Detection and pattern recognition attempt to detect, recognize, identify and/or classify 2-D and 3-D objects, from 2-D or 3-D images. Estimation applications are concerned with the estimation of useful parameters of objects, available images. Such parameters are range, velocity and surface orientation. Detection and estimation problems are closely related, since in some cases an object should be detected before its parameters can be estimated. In other cases the estimation of the characteristics of some regions of an image may lead to the detection of an object.

(3) *To improve the efficiency of operations.* The third objective is related to the observation that certain operations can be performed more efficiently by operating on transformed images. As an example, given an image of a 3-D object, one may obtain a simplified representation of the object contained in the image, and then rotate it, to obtain a different view of the object.

In addition to the conventional image processing techniques, already discussed, there exist many additional techniques, which are widely used in image analysis. Such techniques are summarized in sections 5.2 and 5.3. Section 5.4 outlines knowledge-based image processing. Section 5.5 presents an overview of the image detection, classification and recognition areas, while section 5.6 summarizes image estimation, and outlines some of the most important applications. Section 5.7 describes a knowledge-based image processing system.

5.2. Basic image analysis techniques

5.2.1. Feature extraction
The extraction of features, such as edges and curves, from an image is useful for a variety of purposes. It can also be the first step towards object detection, classification, recognition and/or parameter estimation. Some important characteristics, which are often extracted, are the following [2, 62, 63]:
(a) Size characteristics.

These characteristics are related to the size of the object. Some of the most commonly used are:
– *Area*. This is defined as the number of pixels inside (and including) the boundary of an object multiplied by the area of a single pixel.

- *Integrated optical density.* This is the sum of the gray levels of all pixels inside an object. It is equal to the area multiplied by the mean interior gray level.
- *Length and width.* These correspond to the longest and shortest object dimensions, and can be obtained by locating the major axis of the object and measuring the length and width relative to it.
- *Perimeter.* This is the length of the circumference of the object. It is usually measured after a smoothing of the object boundary has taken place.

(b) Shape characteristics.

These characteristics are related to the shape of the object. Some of the main ones are:

- *Rectangularity.* This can be defined as the ratio of the object's area to the area of the minimum enclosing rectangle.
- *Circularity.* This is usually defined as the ratio of the perimeter squared to the area. This has a minimum ratio of 4π in the case of a circle.
- *Moments.* These are often used in an attempt to summarize image shape information. As an example, moments evaluated around the center of gravity of an object may often contain valuable information about its shape, which may enable the distinction of this object from others.
- *Contours.* Contours of equal gray level are often outlined to describe the shape of paths of objects.
- *Edges.* This is a special class of contours, where the contours are very sharp.
- *Curvature.* This is a special shape characteristic that can be extracted once the edges are known.
- *Polar boundary representation.* A point may be chosen and the length of the vector starting from that point and ending at the contours of the object may be plotted as a function of the vector angle. The resulting plot may characterize the object.
- *Boundary Chain Code* (BCC). This describes a sequence of boundary points, by an arbitrary boundary point which is specified by its coordinates (x, y), and continues by specifying subsequent boundary points. These points are represented by three bits each, since there are eight neighboring points to any boundary point in a 2-D image, and boundaries are continuous.
- *Other shape descriptors.* Such as the differential chain code (DCC), which is the derivative of BCC. Another is the line segment code

(LSC). A third is the medial axis transform (MAT). A fourth is the parametric boundary representation (PBR). Additional descriptors exist and many more can be proposed in specific applications.

- *Fourier descriptors.* Transforms of various representations may contain valuable information characterizing the object. For example the magnitude of the Fourier transform of the polar representation of a polygon is invariant to object rotation, and has been used to characterize the object [4].

(c) Other characteristics.

Other characteristics not related to the size or shape of an object may also be used. Such characteristics are:

- *Color.* Color may be an important distinguishing factor of an object. This may also be used as a basis for image segmentation. Subsequent object identification may be based on color properties.
- *Texture.* Textural properties of image regions are often used for classification or segmentation of the image. Classical texture descriptors have been derived from autocorrelation functions or power spectra. Structural approaches have also been proposed, which may be used for texture discrimination [64]. Texture analysis based on explicit extraction of primitives has also been explored [62].
- *Signatures.* The spatial representation of an object or its transform may contain appropriate characteristics (signatures), which may be used to identify the object. These characteristics may be sought and extracted.

A number of techniques are available, which can be used for the extraction of features, such as those previously described. Some additional low-level techniques used in image analysis are summarized.

5.2.2. Image segmentation

Segmentation is a technique that partitions an image into disjoint regions. This can be done in a number of ways. Notice also that image segmentation is usually done as a first step towards object identification. However, this is a difficult process as geometrical considerations, shadows, noise, etc. often mask the object segment on the 2-D image, and tend to complicate the process.

(a) Global techniques.

Here the value of a threshold or thresholds, used to distinguish an image characteristic (gray level, color, texture, etc.) is kept constant throughout the image. This is used to segment the image into different regions. When pixel characteristics are used for segmentation, pixels can be classified independently, which allows for a parallel and fast implementation, or sequentially, resulting in algorithms which are not as fast, and fuzzy [62].

(b) Adaptive techniques.

Here the threshold or thresholds are considered, which may be slowly varying functions of position in the image. This threshold or thresholds are varied and used to segment the image into different regions.

(c) Edge detection techniques.

In some applications edges provide sufficient information for region segmentation. In others, line tracking and gap filling may have to be used.

(d) Region growing techniques.

These techniques start by dividing the image into many tiny regions. A set of parameters is then assigned to each region, whose values characterize the region (gray level, color, texture, etc.). Boundaries between regions are then examined and a value is assigned to their strength. Strong boundaries are left unchanged, while weak boundaries are dissolved and the adjacent regions are merged. The process continues until no weak boundaries are left. These techniques are generally more costly than the techniques previously mentioned, but work well when there is no a priori knowledge available.

(e) Other segmentation techniques [1–3]

Many other image segmentation techniques exist, such as spectral segmentation techniques, shape segmentation and relaxation techniques.

5.2.3. Other image analysis techniques

Many other techniques have been used for image analysis. Image matching techniques match two or more different images obtained by different sensors in an attempt to achieve a number of objectives. Some of the objectives are:

(1) to determine the points at which they differ,
(2) to determine their distances from the camera,
(3) to obtain another image, which is preferable to the given image.

The low-level techniques described are examples of possible approaches. Knowledge of the nature of the 2-D object depicted in a 2-D image provides additional information, which can enhance our capabilities to perform successful operations. In the case of 2-D images of 3-D objects, knowledge of these 3-D objects, coupled with the geometrical considerations, and an understanding of the image formation process, can result in improved techniques.

5.3. Edge detection

Edge detection is an important area of feature extraction. Much research has been done on this subject [1–4, 61, 63]. Edge detection can be defined as a process that locates a boundary between two adjacent regions of an image, which differ in some characteristics. In this section we use the gray level as the critical characteristic. Table 5 shows some applications of edge detection.

5.3.1. First derivative or gradient operators

Edges occur where intensity changes rapidly. Therefore, edges can be detected by thresholding the image gradient. Many variations of this class of edge operators have been developed.

The earlier gradient methods find the two orthogonal components of the gradient separately. Let the two orthogonal gradient functions of the image be D_x and D_y. Then these gradients can be approximated by

$$D_x = f(x + n, y) - f(x, y),$$

$$D_y = f(x, y + n) - f(x, y),$$

where n is a small integer, usually unity. The gradient magnitude D and directional angle θ can be calculated from

$$D = \sqrt{D_x^2 + D_y^2} \tag{48}$$

$$\theta = \arctan \frac{D_y}{D_x}. \tag{49}$$

The quantities D_x and D_y, can also be expressed in terms of two masks U_x and U_y, respectively. Some of these operators are now defined.

(a) The Prewitt operator (smoothed gradient). This operator is defined in terms of the masks

$$U_x = \begin{bmatrix} -1 & 0 & 1 \\ -1 & 0 & 1 \\ -1 & 0 & 1 \end{bmatrix}, \qquad U_y = \begin{bmatrix} 1 & 1 & 1 \\ 0 & 0 & 0 \\ -1 & -1 & -1 \end{bmatrix}, \tag{50}$$

which imply that the quantities D_x and D_y can be calculated from the

Table 5
Applications on edge detection

Domain	Objects to be located	Examples of tasks
General applications	General images	Location of object Dimension measurements Area measurement Motion measurements Data compression

Robotics	Outdoor scenes
	Indoor scenes
	Mechanical parts
		Recognition and description of objects
		  in scenes
		Detection and identification of product
		  components for automated assembly lines
		Non-contact measurements
		Detection of faults in production

Medical	CT scan images Nuclear medicine images Cells Chromosomes	Measurement of nuclear medicine images for diagnosis Improved images of cells and chromosomes
Satellite and aerial	Terrain Runways Buildings Meteorological images	Improved images Geographic mappings Target location Meteorological analysis
Military	Aerial images Military vehicles Satellite images	Classification of vehicles Scene matching for missile guidance and target acquisition Surveillance and tactical analysis.
Astronomy	Astronomical images	Improved images Counting stars
Neurophysiology and Psychophysics	Visual images	Modelling of the human vision systems
Physics	Particle tracks	Identification of tracks Discovery of new particles
Visual arts	Photographs Paintings Visual images	Measurement of art works Conversion of photographs to line drawings

following difference equations:

$$D_x = f(x - 1, y + 1) + f(x, y + 1) + f(x + 1, y + 1)$$

$$- f(x - 1, y - 1) - f(x, y - 1) - f(x + 1, y - 1),$$

$$D_y = f(x - 1, y - 1) + f(x - 1, y) + f(x - 1, y + 1)$$

$$- f(x + 1, y - 1) - f(x + 1, y) - f(x + 1, y + 1).$$

$$(51)$$

(b) The Sobel operator. This operator is defined in terms of the masks

$$U_x = \begin{bmatrix} -1 & 0 & 1 \\ -2 & 0 & 2 \\ -1 & 0 & 1 \end{bmatrix}, \qquad U_y = \begin{bmatrix} 1 & 2 & 1 \\ 0 & 0 & 0 \\ -1 & -2 & -1 \end{bmatrix}, \qquad (52)$$

which imply two different equations analogous to eq. (51).

After the quantity D is evaluated by one of these operators, it is compared with a threshold to determine if it is part of an edge. Notice that the fact that these operators effect the input image in a linear fashion implies that the effect can be expressed as a small convolution of the image with a mask.

5.3.2. *Template matching operators*

Another class of operators are the template matching operators, which can be implemented without calculating squares, square roots and inverse tangents. Examples of such operators are the Kirsch operator, the Compass Sobel operator, the Compass Prewitt operator, as well as others.

In the case of the Kirsch operator the image is operated upon by eight masks. The maximum values of the outputs of each of the eight masks are taken as gradient magnitudes, and the directions are obtained as those directions which result in the maximum gradient magnitudes. The edge points can be found by thresholding the gradient magnitudes. The choice of the gradient threshold is dependent on the noise levels and background variations. The threshold level should be set higher, when the noise levels are high or the background variations are significant. The eight Kirsch masks are the following:

$$U_1 = \begin{bmatrix} 5 & 5 & 5 \\ -3 & 0 & -3 \\ -3 & -3 & -3 \end{bmatrix}, \quad U_2 = \begin{bmatrix} -3 & 5 & 5 \\ -3 & 0 & 5 \\ -3 & -3 & -3 \end{bmatrix},$$

$$U_3 = \begin{bmatrix} -3 & -3 & 5 \\ -3 & 0 & 5 \\ -3 & -3 & 5 \end{bmatrix}, \quad U_4 = \begin{bmatrix} -3 & -3 & -3 \\ -3 & 0 & 5 \\ -3 & 5 & 5 \end{bmatrix},$$

$$U_5 = \begin{bmatrix} -3 & -3 & -3 \\ -3 & 0 & -3 \\ 5 & 5 & 5 \end{bmatrix}, \quad U_6 = \begin{bmatrix} -3 & -3 & -3 \\ 5 & 0 & -3 \\ 5 & 5 & -3 \end{bmatrix},$$

$$U_7 = \begin{bmatrix} 5 & -3 & -3 \\ 5 & 0 & -3 \\ 5 & -3 & -3 \end{bmatrix}, \quad U_8 = \begin{bmatrix} 5 & 5 & -3 \\ 5 & 0 & -3 \\ -3 & -3 & -3 \end{bmatrix}. \tag{53}$$

The main properties of the previous two classes of operators are summarized as follows:

(1) When the threshold level is high, the edges tend to be thin and the effects of noise and background variations are reduced, but the fine details of the edges are lost.

(2) When the threshold is low, the details of edges are retained, but the effects of noise and background variations are strong, and the edges are thick.

(3) These methods require more than one convolution, and some of them require the computation of squares, square roots and inverse tangents.

5.3.3. Second derivative operators

The edges of an image can also be viewed as the local maxima of the image gradients. Therefore the zero-crossings of the second derivatives will give the edge points. The Laplacian operator approximates the sum of the two partial second derivatives of the image function, with respect to two orthogonal directions (section 2.5). A simple approximation to the Laplacian is given by the mask

$$U_L = \begin{bmatrix} 0 & 1 & 0 \\ 1 & -4 & 1 \\ 0 & 1 & 0 \end{bmatrix}. \tag{54}$$

The edge can be found either by thresholding the Laplacian operated image directly or by thresholding the slope of the zero-crossings. Direct thresholding is simpler, but it may cause thick edges. Thresholding the slope of the zero-crossing is more complex, but it results in thin edges. The advantages of the Laplacian operator can be summarized as follows:
(1) It is directionally independent and can pick up edges from all possible directions by a simple convolution with the image.
(2) It has the potential of producing thin edges.
However, the Laplacian operator has the following two drawbacks:
(1) Directional information is not directly available.
(2) It enhances noise more than the first derivative operators. The noise problem was solved by Kanade, who expanded the Laplacian to a larger area, to average out the noise. His mask is given by

$$
U_K = \begin{bmatrix}
0 & 0 & 0 & -1 & -1 & -1 & 0 & 0 & 0 \\
0 & 0 & 0 & -1 & -1 & -1 & 0 & 0 & 0 \\
0 & 0 & 0 & -1 & -1 & -1 & 0 & 0 & 0 \\
-1 & -1 & -1 & 4 & 4 & 4 & -1 & -1 & -1 \\
-1 & -1 & -1 & 4 & 4 & 4 & -1 & -1 & -1 \\
-1 & -1 & -1 & 4 & 4 & 4 & -1 & -1 & -1 \\
0 & 0 & 0 & -1 & -1 & -1 & 0 & 0 & 0 \\
0 & 0 & 0 & -1 & -1 & -1 & 0 & 0 & 0 \\
0 & 0 & 0 & -1 & -1 & -1 & 0 & 0 & 0
\end{bmatrix}. \tag{55}
$$

5.3.4. The Marr–Hildreth method
Marr and Hildreth applied a smoothing filter to the image before the application of the Laplacian. The smoothing filter was chosen to fulfill the following two requirements:
(1) The rate of intensity changes should be reduced by the filter to limit the image frequency band, so as to avoid aliasing. Due to this, the filter should be band-limited.
(2) The filter should only involve neighboring points to allow good spatial resolution.
However, because of the scaling property of the Fourier transform, these two requirements are contradictory. They are, in fact, related by the space-bandwidth product. The Gaussian function optimizes this trade off. The image is filtered by the Gaussian and then the Laplacian is applied. This can be written as

$$L[G(x, y) * f(x, y)],$$

where L is the Laplacian operator, $G(x, y)$ is a rotationally symmetric

Table 6
Properties of popular edge detectors

Properties	Edge detector				
	Sobel	Prewitt	Kirsh	Laplace	Marr
Number of convolutions	2	2	8	1	1
Size of mask	3×3	3×3	3×3	3×3	variable
Relative number of multiplications (excluding finding the squares)	2	2	8	1	variable
Requires calculations of squares, square roots, and inverse tangent	yes	yes	no	no	no
Relative number required memory	2	2	2	1	1
Threshold level affects edge thickness	yes	yes	yes	no	no
Edge details	low	low	low	high	high
Sensitivity to noise	medium	medium	medium	high	variable
Application on most kinds of edges	yes	yes	yes	yes	yes
1st derivative available	yes	yes	yes	no	no
2nd derivative available	no	no	no	yes	yes
Edge direction available	yes	yes	yes	no	no

Gaussian function and $f(x, y)$ is the image function. Because of linearity,

$$L[G(x, y) * f(x, y)] = L[G(x, y)] * f(x, y) \qquad (56)$$

Therefore, $L[G]$ can be used as the new operator. The directional gradient should be taken in the orientation perpendicular to the edge, in order to give the maximum slope of the zero-crossing. This slope gives valuable information on how sharp the edge is. Marr and Hildreth further demonstrated that the Laplacian can be used to determine second derivatives of the image, without having to worry about the orientation of the edges. The loci of the zero-crossings can be detected economically by searching for the zero-crossings of the convolution of eq. (56).

The disadvantages of the Marr–Hildreth operator are:

(1) A wide Gaussian mask does not give enough details of the edge, while a narrow Gaussian mask results in too many insignificant details.

(2) The operator also rounds off sharp corners.

A summary of the properties of the previous edge detectors can be found in table 6. Many other operators have been reported in the literature such as the Hueckel, Mero–Vassy and Chow operators, as well as global, dynamic and relaxation techniques. Two recent approaches are now summarized.

5.3.5. Point operations in edge detection

A nonlinear point operator $f(x)$, followed by an edge detector was considered in ref. [65]. By choosing $f(x)$ in an appropriate manner the performance of the edge detector can be improved significantly. The point operation functions considered were the following:

(1) $f_1(x) = A \log(x + 1)$, $\qquad\qquad\qquad\qquad\qquad\qquad$ (57)

where A is a constant in order to obtain a suitable output range. This constant A can be chosen as

$$A = \frac{R}{\log(R + 1)},$$

where R is the dynamic range of the gray levels.

(2) $f_2(x) = A \sum_{n=1}^{N-1} \left[arctan\left(\frac{x - B_n}{C} \right) + \frac{\pi}{2} \right],$ $\qquad\qquad$ (58)

where A is a constant for obtaining a suitable output range, N is the number of regions in the image, B_n is a constant chosen between the gray levels of region n and region $n + 1$, and C is also a constant. The smaller C is, the sharper the output $f(i, j)$ will be, but the more details will be lost.

(3) $f_3(x) + RF(x)$, $\qquad\qquad\qquad\qquad\qquad\qquad\qquad$ (59)

where R is the dynamic range of the gray levels and $F(x)$ is the cumulative distribution function (CDF) of the image under question (histogram equilization). The analysis of the previous point operators on edge detection shows that [65]:

(1) The point operator $f_1(\cdot)$ suppresses noise in homogeneous regions with higher mean gray levels. The histogram equilization $f_3(\cdot)$ performs badly for multimodal images. The reason is that it tends to enhance the noise in the homogeneous regions and reduce the edges.

(2) The point operator $f_2(\cdot)$ performs well for multimodal images, enhancing edge information and suppressing noise in homogeneous regions.

The point-operator method was applied on a simulated image, consisting of two regions corrupted with uniformly distributed additive white noise (range 40). One mean gray level was 50, the other was 100. Figure

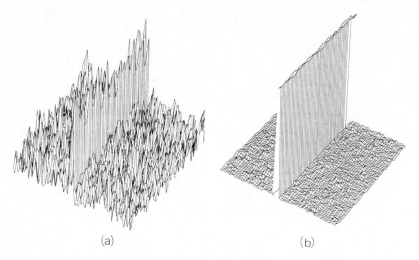

(a)   (b)

Fig. 16. (a) The resulting image using the direct Sobel operator on a simulated image with uniform noise (range 40); (b) the resulting image using $f_2(\cdot)$, followed by the Sobel operator on a simulated image with uniform noise (range 40).

16a shows the resulting edge using the direct Sobel operator (without a point operator). Figure 16b shows the resulting edge using $f_2(\cdot)$, followed by a Sobel operator. It is clear that the edge was improved considerably, through the application of the point operator $f_2(\cdot)$.

The scheme has been also applied on some images to check the analytical results obtained. Figure 17a shows a simple image corrupted with uniformly distributed additive white noise (range 50). The point operator function was chosen as

$$f_2(x) = 60\left[\arctan\left(\frac{x - 70}{20}\right) + \frac{\pi}{2}\right].$$

The result of the application of this operator and the subsequent application of the Sobel operator is shown in fig. 17b.

5.3.6. Edge detectors based on nonlinear filters

Linear filtering techniques were also used in edge detection [66]. Such an edge detector is shown in fig. 18a. However, nonlinear filtering is known to be more successful than linear filtering in image processing, because it removes certain kinds of noise better and preserves edge information. Applications of such techniques [13, 14] in edge detection result in some novel classes of edge detectors with good noise characteristics. The

Fig. 17. (a) The image with uniform noise (range 50); (b) the resulting image using $f_2(\cdot)$, followed by the Sobel operator on the image of (a).

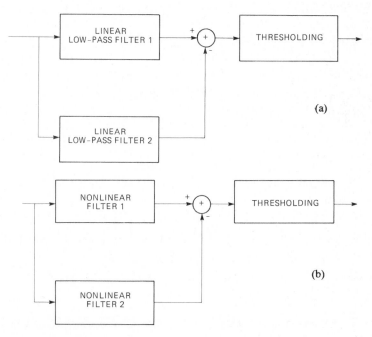

Fig. 18. (a) Structure of an edge detector based on linear filters, (b) structure of an edge detector based on nonlinear filters.

structure of such edge detectors is shown in fig. 18b. The nonlinear filters evaluate a nonlinear function of the luminance. Their difference is a measure of the dispersion of the luminance, within the filter extent. If it is greater than a given threshold, the center of the filter extent is declared an edge point. The use of different nonlinear filters, results in different classes of edge detectors, with different characteristics. Some of these edge detectors were found to have performance comparable to that of well-known techniques coupled with a reduced computational complexity [67].

5.4. Knowledge-based image analysis

Knowledge can be of many different types. It may be "hard" knowledge or "textbook" knowledge. It may be "heuristic", that is "rules of good guessing", that can be developed by a human observer through long experience. The rules are generally accompanied by estimates concerning the weight that each rule should carry in the analysis. The efficient utilization of both "hard" knowledge, which is usually done by numerical processing, and "soft" knowledge, which is usually done by symbolic processing, is a major objective of image analysis.

Low-level techniques mostly require traditional engineering and scientific approaches. They study the problem, its physical and logical constraints and result in a solution which may lead into a program or hardware that attempts to solve the problem. High-level techniques are mostly based on a different approach. In this approach human experts are studied to determine what heuristics they use in solving problems. These rules are then encoded as input to a program that attempts to emulate their behavior.

The development of high-level systems can also be done in the five levels described in section 1.1.

Level 1: Formulation of the problem. This level attempts to answer the question "What should be done?". Places the problem in the context of a specific application, in a way amenable to treatment by symbolic processing.

Level 2: System design. This level attempts to answer the question "What kind of patterns should be searched for?". This is often done through an image segmentation algorithm, that isolates the individual parts in an image, decides which properties of those parts best distinguish the object parts, and tries to identify how to best measure them.

Level 3: System realization. This level considers the realization problem of how knowledge relating to these patterns should be represented. It also studies the control structure for the patterns searched and fixes the various adjustable parameters.

Level 4: System implementation. This level is concerned with the development of the software and/or hardware required. The software is often based on specialized languages, such as LISP. LISP is an elegant computer language, which reflecting its mathematical roots gains a lot of power through recursion. It allows self-modifying code, and it also allows highly flexible information structures that are amenable to representing concepts more complex than numbers.

Level 5: System Verification. This level evaluates the system, calculates the rates of the various misclassification errors and often modifies the system to improve its performance.

Levels 2 and 3 will be considered further in this section.

(1) Choice of patterns. The description of appropriate patterns for a given application, such as object recognition, can be initially done in a narrative form. Although such descriptions may be simple, they are important as a first step towards good knowledge representation. As an example, in a recent system which provides knowledge-based automated interpretation of seismic data [68], the patterns chosen were geological features, such as horizons, anticline traps, salt domes and reefs, unconformities and faults.

The descriptions given to those patterns lead to the following observations [68]:

- The patterns are well-defined and they are quite general and versatile. However, this sometimes creates ambiguities in their search.
- More elementary patterns are parts of more complicated ones, in a semantic sense.
- Some patterns are special kinds of others.
- Some patterns are similar to others.
- Spatial relations are always required (e.g. neighborhood relations, parallelism relations).

(2) Knowledge representation. Early knowledge-based systems were based on production rules of the form "if ... then ... else" [69]. This was a successful framework for some simple knowledge-based systems. However, this approach has various shortcomings, such as the requirement for the specification of the order of application of the rules, and the large number of rules required in complex situations. Other knowledge-based systems use knowledge organization along the axes of generalization/specialization, decomposition/aggregation, instantiation [70, 71].

We shall describe briefly those notions here [72, 73]. The idea to pack knowledge in a modular form is due to Minsky [72]. Each knowledge package is called a "frame" or "class". A specific pattern is described in a class by its components and relations that it has to satisfy. The class "horizon" for example in [68], has as components its points, its length, its strength, its signature and its orientation. The components are called "slots" of the class. Often, certain constraints on the slots have to be satisfied. A generalization/specialization axis is defined by the IS–A link between the classes [71]. For example, a horizontal straight horizon IS–A straight horizon, and a straight horizon IS–A horizon. The classes inherit the properties of their IS–A parents. The decomposition/aggregation axis is defined by a PART-OF link between classes [71]. For example the line segments are PART-OF a horizon. Two horizons are PART-OF an anticline trap, etc. Another knowledge organization axis defined by SIMILARITY links was also proposed [71]. These links suggest other classes to be tried, when the match fails between a given class and data. Finally, another basic link used is the INSTANCE-OF relationship. This link connects a particular entity (called token) to a class. This process is called instantiation. All these constraints of the class must be satisfied for the instantiation to be successful. Having briefly described knowledge representation, we shall now outline how it can be used in pattern searching.

(3) Control structure for pattern search. Knowledge representation models various patterns and facts, but does not reveal how they can be used for the search of such patterns. The mechanism for pattern searching is called control structure. Hypothesize and test is a possible control mechanism used in a number of systems. A hypothesis is formed when we try to create a new instance of a class. The hypothesis tries to verify itself. This means that we try to fill all slots necessary, we test if appropriate slot constraints are valid, or if the appropriate IS–A relations are valid. If the hypothesis verifies itself, then it is inserted in the class (instantiated). If the hypothesis fails, then the insertion in the class

is denied. In [68] two mechanisms were used for hypothesis forming. These correspond to a data directed search and a hypothesis directed search. The data directed search traverses the PART-OF hierarchy in a bottom-up way. Activation of a hypothesis, in this search mode, activates other hypotheses along the PART-OF hierarchy. It also activates the IS–A parents of the hypothesis. The hypothesis directed search, is a top-down traversal of the IS–A and PART-OF hierarchies. The activation of a hypothesis in this mode activates hypotheses for its slots. The hypothesis is instantiated only when all its slots are filled (and the corresponding hypotheses are satisfied).

In addition to the control structures described, a mechanism is usually used for ranking hypotheses [74]. This approach measures the certainty of each hypothesis during the process of hypothesis verification. Hypotheses with low certainty (below a threshold) may be rejected. The hypothesis with high certainty forms the so called "focus" of the system. The certainty of a hypothesis h is a number $R(h)$, $(0 \leqslant R(h) \leqslant 1)$, which receives contributions from three sources:

(1) Contributions from more general hypotheses along the IS–A hierarchy.

(2) Contributions from its components along the PART-OF hierarchy.

(3) Contributions from the successful matching of its internal constraints.

The degree of influence of these contributions is defined by the appropriate choice of factors called "compatibilities". These are "self-compatibility", "IS–A compatibility" and "PART-OF compatibility". Compatibilities are chosen to have values between -1 and 1. The values -1, 0 and 1, corresponding to strongly incompatible, independent, and strongly compatible hypotheses, respectively. The certainty of a hypothesis is described by a cost function $R(h)$, which has been successfully used in many applications [68].

Knowledge-based image processing can be used interactively with an observer or may lead to complete automation. Systems which are able to emulate the performance of a human expert or a team of experts are called "expert systems". An expert system must be able to make decisions, generate hypotheses and make hypotheses based on incomplete or uncertain knowledge, by using a collection of facts and heuristics about a particular application. The typical organization of an expert system consists of a number of components. Such components include a massive database relevant to its area of expertise; a set of rules as to how the database is to be searched; and a control structure that can apply these

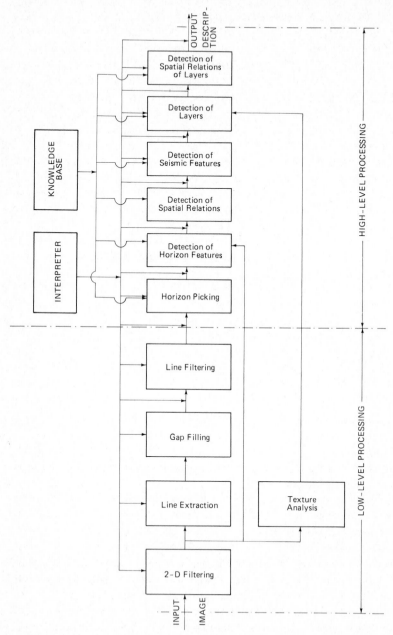

Fig. 19. Structure of an expert system for the automated geophysical interpretation of seismic data (AGIS).

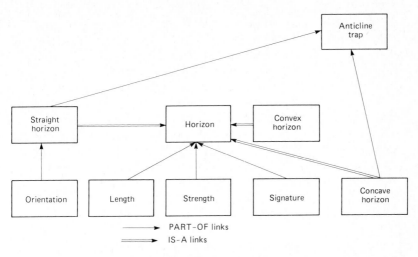

Fig. 20. Control structure of AGIS.

rules to the base of knowledge. The block diagram of such a system presently under development is shown in fig. 19 [68]. Its control structure is shown in fig. 20.

Two of the most important areas of application of image analysis are now described. However, the effect of knowledge-based systems is expected to also apply to other areas of digital image processing, such as enhancement and restoration.

5.5. *Image detection, classification and recognition*

Image detection usually refers to the extension of detection techniques, developed in communications and statistics to problems of image analysis. The work object here refers to an m-D function defined in space. Using an organization scheme similar to that used in ref. [51], we may categorize the problems into three types:

Type 1 problems refer to the detection of a known object in the presence of noise. This is usually treated by template matching [1], and it is analogous to the generalization of the matched filter concept in m-D. Examples of Type 1 problems include automated inspection of integrated circuit chips, and machine recognition of printed characters.

Type 2 problems refer to the detection of a known object with unknown parameters in the presence of noise. The unknown parameters may be the size, orientation, location, but the approach is still analogous to Type 1 problems, and can be treated in a statistical way. Examples of Type 2 problems include the detection of known objects in noisy IR images, as well as radar and sonar images.

Type 3 problems refer to the detection of a random object in the presence of noise. The object may be random by nature, such as clouds and geological formations, or it may be deterministic, but unknown, and available to the observer after it has been affected by random transformations. In this case the matched filter generalizes to an estimator-correlator detector structure for the case of white Gaussian observation noise. This approach is easily extended to the colored-noise case, through the use of the whitening filter [3]. Examples of Type 3 problems are the detection of the lung in X-rays, and the location of ocean eddies in satellite altimetry data [75], in which there is unknown amplitude, radius and velocity.

In addition to the binary detection problem, which is concerned with the answer to the question "is the object there?", it is necessary to also consider the *m*-ary detection problem, which tries to distinguish among *m* possible alternatives. This process is called classification.

The mathematical theory on which object detection is mainly based is statistical decision theory. Approaches have been developed based on the maximization of the a posteriori probability, minimization of the Bayes risk, sequential detection, parametric and non parametric techniques. Non-statistical approaches can also be developed based on regularization theory [47, 48]. When 3-D objects are detected on the basis of 2-D images, the problem is particularly difficult, due to the fact that only one side of the body is visible. Therefore there is an ambiguity on the object recorded, and this leads to an ill-defined problem. This problem can be regularized, by imposing additional constraints with physical significance.

Image recognition deals with a similar problem of attempting to recognize an object imbedded in a noisy image. However the approach is different. It is mainly based on syntactical [4] rather than statistical methods, drawn from artificial intelligence and pattern recognition, and is related to high-level vision. Some of the most challenging problems in this area remain unsolved. Here one may register the image of a scene obtained from different sensors, or by the same sensor at different times. Multisensor characteristics of the image scene can be used to recognize the object. Utilizing images obtained from different locations, it is possible to compute the 3-D positions of a scene point. Comparing images obtained at different times, one may detect changes and predict future changes. The usual approaches are based on computation of cross-correlation and other mismatch measures. 3-D techniques are however needed, which are position, orientation, size and perspective invariant. Naturally many problems can be treated by approaches that can be seen as the combination of statistical and syntactical methods.

5.6. Image estimation

Image estimation usually refers to the estimation of a parameter or parameters of an object or objects in an image or a set of images. Some of the estimation techniques are deterministic and deal with the area of image sequence analysis, which is also known as dynamic scene analysis. This area is concerned with the processing of a sequence of images. Other techniques are statistical and are dealing with the problem of parameter estimation of objects, from images corrupted by noise. Finally, other techniques are syntactical, and are closely related to pattern recognition and artificial intelligence.

A detailed review of these operations of image sequence analysis can be found in work by Nagel [76]. The following is a partial list of applications:

- *Industrial*: Monitoring of industrial automated processes by visual sensors, dynamic control of robot movement from visual information.
- *Military*: Tracking of multiple targets from video data, measuring missile dynamics from video data, target parameter estimation.
- *Commercial*: Using motion estimation for a more efficient signal, resulting in bandwidth compression of video-conferencing and picture phone signals.
- *Medical*: Studying cell motion by micro-cinematography, heart motion from X-ray image sequences, transport phenomena in the circulatory and metabolic systems, tracking of biological objects from a 3-D stack of tomographic data.
- *Other*: Cloud tracking to determine wind velocities, highway traffic monitoring.

Among the deterministic image estimation techniques, 2-D and 3-D motion estimation techniques have received considerable attention [77]. Earlier work in the area of image sequence analysis was almost exclusively concerned with the 2-D case. Here, 2-D translation motion was the first to be studied. There are three major techniques in estimating 2-D translation:

(1) The matching method.
(2) The Fourier method.
(3) The method of differentials.

A problem of more general interest is that of the estimation of general 2-D motion. The 3-D motion estimation problem is more complex, and attempts to determine the relative position of the points of a 3-D object which is moving in space, from a series of perspective projections of the rigid 3-D configuration.

Stochastic image estimation techniques have been developed in a manner analogous to that of object detection. Such techniques may be categorized into three types:

Type 1 problems refer to the estimation of the characteristics of a known object in the presence of noise.

Type 2 problems refer to the estimation of the characteristics of an object, with some unknown parameters, in the presence of noise.

Type 3 problems refer to the estimation of the characteristics of a random object imbedded in noise.

The mathematical theory on which parameter estimation is based is statistical estimation theory [78]. Maximum likelihood estimation assumes that the parameters are fixed and unknown. A given image is considered and the parameter is taken as that value which makes the occurrence of the observed image most likely. Bayesian estimation on the other hand assumes that the parameters are random variables, but that they have a known assumed probability density function before the image was taken. After the image is measured, Bayes theorem is used to update the a priori probability density function.

Syntactical techniques have also been developed, which are analogous to those mentioned in section 5.4, where the objective is now that of parameter estimation.

As can be seen from the previous sections, image analysis involves many different approaches that incorporate different kinds of information about the class of images being analyzed. This area of research is very young, and there is still no general theory of how to best utilize available knowledge. Most successful applications of image analysis are relatively simple and application-specific.

5.7. A knowledge-based image processing system [79, 80]

5.7.1. Formulation of the problem

D'Arcy Thomson was the first to propose a method for analyzing the growth of the head. According to this method, characteristic points, or landmarks were related to characteristic anatomical features. These points were relatively few (thirty six), and were defined anatomically.

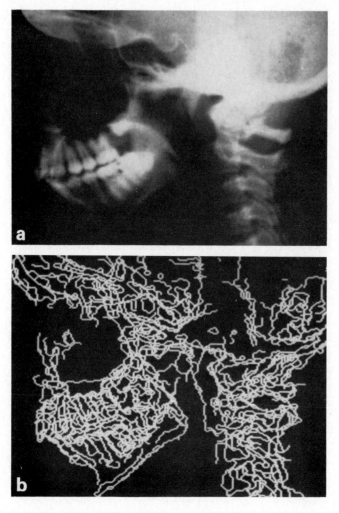

Fig. 21. (a) Original cephalogram, (b) result of global search on (a); c, d overleaf.

The constellation of these points and its evolution in time helps to determine the appropriate orthodontic treatment.

In practice the image used by orthodontists is a lateral head X-ray or cephalogram, such as the one shown in fig. 21a, where the characteristic anatomical features appear either as lines or as edges on a dark background. The orthodontist traces the relevant lines and edges on the X-ray

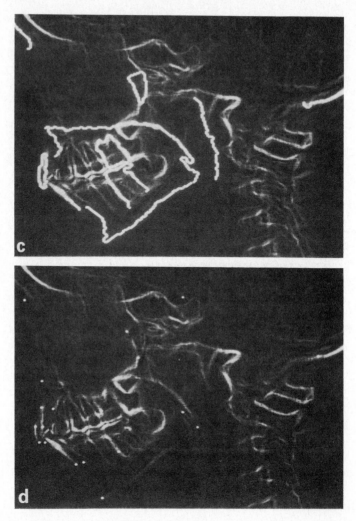

Fig. 21. (Cont'd) (c) lines extracted by knowledge-based system, (d) location of landmarks found by knowledge-based system.

and subsequently locates the characteristic points. Up to now landmarking was always done manually. This was a long and tedious process, in which the results could vary from one judge to the next, in proportions comparable to the variation in position of the landmarks on two different cephalograms. Therefore a system performing automatic recognition of the landmarks was required.

Before starting the landmarking process two types of criteria were examined:

(1) whether the image has been correctly exposed;
(2) whether it was free from extraneous structures and improper positioning effects (anomalies).

It was assumed that the logarithmic histograms of the transparency to X-rays of all the patients' heads were similar, and that their histograms were bell-shaped. The logarithmic histogram of the X-ray is equal to the transform of the logarithmic histogram of the actual patient's head, by the S-shaped characteristic of the film. Therefore the logarithmic histogram of the pixels on a digitized X-ray has two humps, one corresponding to white saturated regions with few details, and the other to black saturated regions. If the X-ray is correctly exposed, the two humps are of approximately equal size. If the first hump is larger, the picture is too dark, therefore overexposed, whereas if the second is larger, the picture is too light, thus underexposed.

There are three main kinds of anomalies. Filled cavities, missing teeth, and the two profiles resulting from a patient whose head is not upright when X-rayed. Cavities, when filled with a metal alloy, are nearly opaque to X-rays. They therefore appear as bright spots, that have a tendency to "bloom" and obscure neighboring lines. The absence of a tooth is also hard to detect. Teeth do not have very sharp edges, and it is difficult to distinguish between the shape of a tooth and that of a hole. Finally, the possibility of the presence of two shifted profiles was not considered in this initial algorithm for the sake of simplicity [79, 80].

5.7.2. System design and realization (low-level)
In this application, the processing consisted of a median filter, which was used to remove the noise. This was followed by a Mero–Vassy operator, which was found to be the best among a number of operators attempted. Subsequently a line tracker was used, and the result of this global search was a line drawing, which contained significant, as well as useless, lines of the image.

5.7.3. System design and realization (high-level)
Since our images represent biological shapes, they could not be described in terms of shifted and rotated patterns, that could be easily recognized by an algorithm. This task was further complicated by the characteristics of the background in the image, which is composed of white noise and irrelevant edges and lines. These are highly correlated with the signal

which corresponds to the relevant lines, and are of approximately the same intensity. The image was also corrupted with quantization noise and anomalies, such as filled cavities. This implies that the algorithm must contain some a priori knowledge about the relevant features. These features appear as both edges and lines on the original image.

We choose to describe these features by categories of knowledge, as follows:

(1) Position: The approximate position of the line is given to the algorithm.

(2) Number and characteristics of constituent segments: A line is modeled as being composed of straight segments, separated by abrupt changes of direction. Each segment has a number of characteristics such as its starting and ending points, its approximate length, its general orientation, and the characteristics of the noise around it.

5.7.4. System implementation and verification

Median filtering, followed by a Mero–Vassy edge detector and a line tracking operator were applied on fig. 20a to obtain the result of the global search, shown in fig. 20b. A knowledge-based line-tracker, guided by a reference map was then used. The map was initialized by a planning step, that gave the gross proportions of the lines of the patient. They were computed with simple and fast operations like thresholding of projections and cuts. The map was progressively updated as more lines were found. The reference map provided the approximate position of the lines. It was thus possible to limit the search to a small rectangular region, where the desired line was found. When a point was found on the line, both halves were tracked (in a bidirectional way). During the tracking of the line, the knowledge described in section 5.7.3 was used, in the search for the next pixel, as well as for a global check conducted after the extraction of the line. The result of the processing, superimposed on the pre-processed image, is shown in fig. 20c. The landmarks were finally extracted and are shown in fig. 20d.

6. Future trends in digital image processing and analysis

The need for imaging systems and processing techniques increases as we move towards an information society, which will require more image information, storage, retrieval, processing, compression, communication,

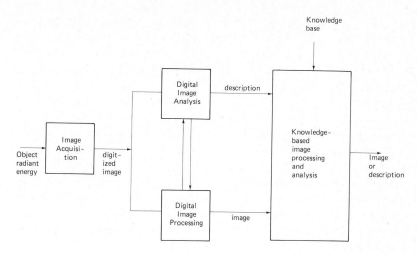

Fig. 22. Proposed image processing/analysis system.

security and analysis. We therefore attempt to chart the future and describe some of the current trends and future directions. Some of the areas where we expect to see a rapid evolution are the following:

(1) A recent trend in digital image processing and analysis has been the infusion of various techniques from the fields of artificial intelligence and pattern recognition, in order to make it possible to automatically extract various kinds of information from an image. This trend is expected to continue in the future and knowledge-based image processing systems are expected to expand. The image processing/analysis system of the future may look like the one depicted in fig. 22. The goal is to ultimately achieve full integration of numerical and symbolic processing and achieve knowledge-based signal/image processing systems.

(2) Improvements are also expected in the person-to-machine interface, which will have an impact on the usage of image processing systems and computers. At first, interpretation will be done interactively. Fifth generation computers may further simplify this process. Automatic interpretation capabilities are expected to arise later.

(3) The area of robotics will expand. Robotic systems will become more flexible and will be able to utilize efficient new algorithms, coupled with recent hardware and architectures. Robots will have an impact on automatic manufacturing and employment.

(4) The present emphasis on the development of expert systems will continue. During the next decade we will see a proliferation of systems with well-defined areas of expertise. Such systems will have a mastery of the knowledge about particular classes of image processing and analysis applications.

(5) The areas of image sequence analysis, 3-D imaging and color imaging will be better understood and utilized. New techniques will be developed for enhancement, restoration and analysis of such images.

(6) Systems will be designed to better satisfy the requirements of human vision, rather than simple mathematical criteria. More meaningful criteria will result in improved system performance. Regularization theory [48] has been recently proposed as a new approach that can be utilized to study ill-defined problems. Coupled with constraints which are physically meaningful, it may result in improved enhancement and restoration techniques.

(7) New mathematical tools may also arise, that are better suited to image processing and analysis. Mathematical morphology was recently introduced as a new tool for image processing and analysis [81, 82]. In this theory each signal is viewed as a set in a Euclidean space. A set of operations based on erosion and dilation can be used to perform nonlinear filtering, image compression and object recognition. The challenge will be to develop new tools that utilize available knowledge and result in a reduced implementation cost.

(8) New results will emerge resulting in a better 3-D representation of objects. Efficient implementations, which are suited to different applications will also be required.

(9) Linear and nonlinear filtering is expected to expand. Not only will known filters be applied to a variety of new applications [83], but new filters will also arise with good properties, reasonable cost and analytical tractability.

(10) New space varying, adaptive and reconfigurable systems, of linear and nonlinear types, will emerge [84]. This will become possible due to improvements in technology and architectures.

(11) There will be an increasing effort placed in advanced VLSI architectures, to implement complex hardware structures. Wavefront and systolic arrays will play a major role. Other architectures will be introduced. We can expect a constantly decreasing cost of memories. In addition, new mathematical tools will be developed, better suited to modular implementations [85, 86].

(12) Better charge coupled devices (CCD) and charge injection devices (CID) will be developed. Both of these classes of devices have advantages in insensitivity to magnetic fields, broad spectral response, immunity to optical overload damage, greater reliability and low cost.

(13) Emphasis will be placed on the development of real-time image processing systems [38, 87] to be used in a variety of applications. Massively parallel image processing systems will be developed.

(14) In image transmission from satellites and space probes the trend will be towards on-board processing, with associated bandwidth compression, based on modern compression techniques.

(15) The stress in future image processing systems will be on the full integration of hardware, software and sensors. The emphasis will be towards total system optimization, as opposed to block optimization.

(16) New sensors will be developed, incorporating front end processing, which will be all digital. High resolution displays will also be developed at a reduced cost.

(17) The methods and results of research in conventional image processing and analysis will also benefit other areas, such as geophysics and array processing, where some of these techniques have been recently applied [88, 89, 90]. Others will have to be adapted to the requirements of new applications. Ultrasound techniques will evolve and efficient image restoration will be applied to improve their resolution.

(18) More powerful computers are expected, which will result in an increase in distributed processing. In addition, we can expect an increase in the number of communication networks. These networks will be fully integrated and will carry voice, data and images efficiently.

References

[1] K.R. Castleman, Digital Image Processing (Prentice Hall, Englewood Cliffs, NJ, 1979).
[2] W.K. Pratt, Digital Image Processing (Wiley, New York, 1978).
[3] M.P. Ekstrom, Digital Image Processing Techniques (Academic Press, New York, 1984).
[4] V. Cappellini, Elaborizione Numerica Delle Immagini (Boringhieri, Torino, 1985).
[5] D.F. Dudgeon and R.M. Mersereau, Multidimensional Digital Signal Processing (Prentice Hall, Englewood Cliffs, NJ, 1984).
[6] S.G. Tzafestas, ed., Progress in Multidimensional Systems Theory (Marcel Dekker, New York, 1986).
[7] V. Cappellini, Some non-linear digital filters for fast noisy-image processing, Proc. IEEE Mediterrenean Electrotechnical Conf. (MELECON), Athens (1983) pp. C2.12–C2.13.

[8] S.G. Tyan, Median filtering: deterministic properties, in: Two-Dimensional Digital Signal Processing, Vol. II, ed. T.S. Huang (Springer, Berlin, 1981).

[9] A.C. Bovik, T.S. Huang and D.C. Munson, "A generalization of median filtering using combinations of order statistics", IEEE Trans. Acoust. Speech & Signal Process. ASSP-31 (December 1983) 1342–1350.

[10] Y.H. Lee and S.A. Kassam, "Generalized median filtering and related nonlinear filtering techniques", IEEE Trans. Acoust. Speech & Signal Process. ASSP-33, No. 3 (June 1985) 672–683.

[11] H.H. Chiang, C.L. Nikias and A.N. Venetsanopoulos, Efficient implementation of quadratic digital filters, IEEE Trans. Acoust. Speech & Signal Process. ASSP-34, No. 6 (Dec. 1986) 1511–1528.

[12] A. Kundu, S.K. Mitra and P.P. Vaidyanathan, Application of two-dimensional generalized mean filtering for removal of impulse noise from images, IEEE Trans. Acoust. Speech & Signal Process. ASSP-32, No-3 (June 1984) 600–610.

[13] I. Pitas and A.N. Venetsanopoulos, Nonlinear mean filters in image processing, IEEE Trans. Acoust. Speech & Signal Process. ASSP-34 (June 1986) 573–584.

[14] I. Pitas and A.N. Venetsanopoulos, Nonlinear order statistic filters for image filtering and edge detection, Signal Process. 10 (June 1986) 395–413.

[15] T. Peli and J.S. Lim, "Adaptive Filtering for Image Enhancement", Proc. IEEE Int. Conf. on Acoustics, Speech and Signal Processing 1981 pp. 1117–1120.

[16] S. Basu and A. Fettweis, Simple proofs of stability results on multi-dimensional polynomials, Proc. of the 7th Eur. Conf. on Circuit Theory and Design, Prague, (September 1985) pp. 447–450.

[17] J.V. Hu and L.R. Rabiner, "Design techniques for two-dimensional digital filters, IEEE Trans. Audio Electroacoust. AU-20, No. 3 (1975) 208–218.

[18] Y. Kamp and J.P. Thiran, Chebyshev approximation for two-dimensional nonrecursive digital filters, IEEE Trans. Circuits & Syst. CAS-22, No. 3 (1975) 208–218.

[19] R.E. Twogood and S.K. Mitra, Computer-aided design of separable two-dimensional filters, IEEE Trans. Acoustics, Speech & Signal Process. ASSP-25, No. 2 (1977) 165–169.

[20] J.H. McClellan, The design of two-dimensional digital filters by transformations, Proc. 7th Annual Princeton Conf. on Information Sciences and Systems (1973) pp. 247–251.

[21] R.M. Mersereau, W.F.G. Mecklenbräuker and T.F. Quatieri Jr, McClellan transformations for two-dimensional digital filtering: I-design, IEEE Trans. Circuits & Syst. CAS-23, No. 7 (1976) 405–414.

[22] R.M. Mersereau, The design of arbitrary 2-D zero-phase FIR filters using transformations, IEEE Trans. Circuits & Syst. CAS-27, No. 2 (1980) 142–144.

[23] J.L. Shanks and J.H. Justice, Stability and synthesis of two-dimensional recursive filters, IEEE Trans. Audio Electroacoust. AU-20, No. 2 (1972) 115–128.

[24] J.M. Costa and A.N. Venetsanopoulos, Design of circularly symmetric two-dimensional recursive filters, IEEE Trans. Acoust. Speech & Signal Process. ASSP-22, No. 6 (1974) 432–443.

[25] S.H. Mneney, A.N. Venetsanopoulos and J.M. Costa, The effects of quantization errors on rotated filters, IEEE Trans. Circuits & Syst. CAS-28, No. 10 (October 1981) 995–1003.

[26] N.A. Pendergrass, S.K. Mitra and E.I. Jury, Spectral transformations for two-dimensional digital filters, IEEE Trans. Circuits & Syst. CAS-23, No. 1 (1976) 26–35.

[27] G.A. Maria and M.M. Fahmy, An I_p design technique for two-dimensional digital recursive filters, IEEE Trans. Acoust. Speech & Signal Process. ASSP-22, No. 1 (1974) 15–21.

[28] P. Karivatharajan and M.N. Swamy, Quadrantal symmetry associated with two-dimensional digital functions, IEEE Trans. Circuits & Syst. CAS-25 (1978) 340–345.

[29] B.P. George and A.N. Venetsanopoulos, Design of two-dimensional digital filters on the basis of quadrantal and octagonal symmetries, Circuits, Systems & Signal Process. 3, No. 1 (January 1984) 59–78.

[30] S.J. Varoufakis and A.N. Venetsanopoulos, A special class of multiplierless two-dimensional recursive digital filters, J. Franklin Inst. 318, No. 2 (August 1984) 105–121.

[31] A.N. Venetsanopoulos and B.G. Mertzios, A decomposition theorem and its implications to the design and realization of two-dimensional filters, IEEE Trans. Acoustics Speech & Signal Process. ASSP-33, No. 6 (Dec. 1985) 1562–1575.

[32] C.L. Nikias, A.P. Chrysafis and A.N. Venetsanopoulos, The LU decomposition theorem and its implications to the realization of two-dimensional digital filters, IEEE Trans. Acoust. Speech & Signal Process. ASSP-33, No. 3 (June 1985) 694–711.

[33] I. Pitas and A.N. Venetsanopoulos, Two-dimensional realization of digital filters by transform decomposition and its analysis, IEEE Trans. Circuits & Syst. CAS-32, No. 10 (October 1985) 1029–1040.

[34] A.N. Venetsanopoulos, B.G. Mertzios and S.H. Mneney, Effect of finite precision in two-dimensional recursive digital filters, Int. J. Electron. 58, No. 1 (1985) 159–174.

[35] B.G. Mertzios and A.N. Venetsanopoulos, Combined error at the output of two-dimensional recursive digital filters, IEEE Trans. Circuits & Syst. CAS-31, No. 10 (October 1984) 888–891.

[36] R.H. VanderKraats and A.N. Venetsanopoulos, Hardware for two-dimensional digital filtering using Fermat number transforms, IEEE Trans. Acoust. Speech & Signal Process. ASSP-30, No. 2 (April 1982) 115–162.

[37] S. Lee and A.N. Venetsanopoulos, A two-dimensional digital filter chip set for modular two-dimensional filter implementation, Proc. IEEE Int. Conf. on Acoustics, Speech and Signal Processing, Tampa, FL (March 1985) pp. 26.8.1–26.8.4.

[38] H. Jaggernauth, A. Loui and A.N. Venetsanopoulos, Real-time image processing by distributed arithmetic implementation of two-dimensional digital filters, IEEE Trans. Acoust. Speech & Signal Process. ASSP-33, No. 6 (Dec. 1985) 1546–1555.

[39] K.M. Ty and A.N. Venetsanopoulos, Two-dimensional digital filters with minimum cycle time, Proc. IEEE Int. Conf. on Acoustics, Speech and Signal Processing, Tampa, FL (March 1985) pp. 40.5.1–40.5.4.

[40] D.J. Morgenthaler, Three-Dimensional Digital Image Processing, Ph.D. Thesis (University of Maryland, 1984).

[41] S.L. Hurt and A. Rosenfeld, Noise reduction in three-dimensional digital images, Pattern Recognition 17, No. 4 (1984) 407–421.

[42] M.E. Zervakis and A.N. Venetsanopoulos, Three-dimensional digital filters using transformations, in: Applied Digital Filtering, Adaptive and Nonadaptive, ed. M.H. Hamza (Acta Press, 1985) pp. 148–151.

[43] K. Hirano, M. Sakane and M.Z. Mulk, Design of three-dimensional recursive digital filters, IEEE Trans. Circuits & Syst. CAS-31, No. 6 (June 1984) 550–561.

[44] I. Pitas and A.N. Venetsanopoulos, The use of symmetries in the design of multidimensional digital filters, IEEE Trans. Circuits & Syst., in press.

[45] A.N. Venetsanopoulos and B.G. Mertzios, Decomposition of multidimensional filters, IEEE Trans. Acoust. Speech & Signal Process. CAS-30, No. 12 (December 1983) 915–917.

[46] B.G. Mertzios and A.N. Venetsanopoulos, Modular realization of multidimensional filters, Signal Process. 7, No. 4 (December 1984) 351–369.

[47] T. Poggio and V. Torre, Ill-Posed problems and regularization analysis in early vision, Artificial Intelligence Lab. Memo, No. 773 (MIT, Cambridge, MA, April 1984).

[48] N.B. Karayiannis and A.N. Venetsanopoulos, Image restoration: a regularization approach, in: Proc. Third Int. Conf. on Image Processing and its Applications, ed. A. Constantinidis (IEEE, New York, 1986) pp. 1–5.

[49] S.J. Wernecke and L.R. D'Addario, Maximum entropy image reconstruction, IEEE Trans. Comput. C-26 (April 1977) 351–364.

[50] B.R. Frieden, Restoring with maximum likelihood and maximum entropy, J. Opt. Soc. Amer. 62 (1972) 511–518.

[51] H.L. Van Trees, Detection, Estimation and Modulation Theory (Wiley, New York, 1968).

[52] V. Cappellini, E. Del Re, P. Francolini and G. Tainti, Advanced MAP restoration techniques with application to digital image processing, Proc. Int. Conf. Image Analysis and Processing (October 1980).

[53] V.K. Ingle, A. Radpour, J.W. Woods and H. Kaufman, Recursive estimation with non-homogeneous image models, Proc. IEEE Conf. on Pattern Recognition and Image Processing, Chicago (May 1978) pp. 105–108.

[54] J.W. Woods, Two-dimensional Kalman filtering, in Two-Dimensional Digital Image Processing, Vol. I ed. T.S. Huang (Springer, Berlin, 1981).

[55] J. Biemond and J.J. Gerbrands, An edge-preserving recursive noise-smoothing algorithm for image data, Tech. Rep. No. 17-79-06 (Technical University, Delft, The Netherlands, 1979).

[56] H.J. Tork and A.N. Venetsanopoulos, Edge-improved Kalman filtering for image restoration, Proc. of the First Image Symposium (CESTA), Biarritz, France (May 1984) pp. 22.B.3.1–22.B.3.6.).

[57] J.M. Costa, A.N. Venetsanopoulos and M. Treffler, Digital tomographic filtering of radiographs, IEEE Trans. Medical Imaging MI-2, No. 2 (June 1983) 76–88.

[58] J.M. Costa, A.N. Venetsanopoulos and M. Treffler, Design and Implementation of Digital Tomographic Filters, IEEE Trans. Medical Imaging MI-2, No. 2 (June 1983) 89–100.

[59] J.M. Costa, A.N. Venetsanopoulos and M. Treffler, Evaluation of digital tomograhic filters, IEEE Trans. Medical Imaging MI-4 (1) (March 1985) 1–13.

[60] G. Mitsiadis and A.N. Venetsanopoulos, Wiener filters for restoration of radiographs, Can. Electr. Eng. J. 12 (1987), in press.

[61] D.H. Ballard and C.M. Brown, Computer Vision (Prentice Hall, New York, 1982).

[62] A. Rosenfeld, Image analysis: problems, progress, and prospects, Pattern Recognition 17 (1) (1984) 3–12.

[63] M.D. Levine, Vision in Man and Machine (McGraw-Hill, New York, 1985).

[64] R.M. Haralick, Statistical and structural approaches to texture, Proc. IEEE 67, No. 5 (1979).

[65] X.Z. Sun and A.N. Venetsanopoulos, Nonlinear point operators for edge detection, Can. Electr. Eng. J. 11, No. 3 (July 1986) 90–103.

[66] W. Geuen, An advanced edge detection method based on a two stage procedure, in: Signal Processing II: Theories and Applications, ed. H.W. Schuessler (North-Holland, Amsterdam, 1983).

[67] I. Pitas and A.N. Venetsanopoulos, Edge detectors based on nonlinear filters, IEEE Trans. Pattern Analysis and Machine Intelligence PAMI-8, No. 4 (July 1986) 538–550.

[68] I. Pitas and A.N. Venetsanopoulos, Knowledge-based image processing for geophysical interpretation, in: Signal Processing III: Theories and Applications, eds. I.T. Young et al. (Elsevier, Amsterdam, 1986) pp. 1211–1214.

[69] A. Barr, P.R. Cohen and E.A. Feigenbaum, The Handbook of Artificial Intelligence, Vol. I, II, III (Heuristech Press, Stanford, CA, 1981).

[70] Special issue on knowledge representation, Computer (October 1983).

[71] H. Levesque and J. Mylopoulos, A procedural semantics for semantic networks, in: Representation and Understanding: Studies in Cognitive Science (Academic Press, New York, 1979).

[72] M. Minsky, A framework for representing knowledge, in: Psychology of Computer Vision (McGraw-Hill, New York, 1975).

[73] M. McGalla and N. Cercone, Approaches to knowledge representation, Computer 16 (October 1983) 12–18.

[74] J.K. Tsotsos, Representational axes and temporal cooperative processes, Vision, Brain and Cooperative Computation, in press.

[75] T.M. Watson and J.W. Woods, Automatic detection of ocean ring signals, RPI Image Processing Laboratory Technical Report, No. IPL-TR-81-018 (November 1981).

[76] H. Nagel, Image sequence analysis: what can we learn from applications, in: Image Sequence Analysis (Springer, Heidelberg, 1981) pp. 19–228.

[77] T.S. Huang, ed., Image Sequence Processing and Dynamic Scene Analysis (Springer, New York, 1983).

[78] A.P. Sage and J. Melsa, Estimation Theory with Application to Communications and Control (McGraw-Hill, New York, 1971).

[79] A.D. Levy-Mandel, A.N. Venetsanopoulos and J.K. Tsotsos, Knowledge-based image processing with application to the landmarking of cephalograms, Proc. of the 10th Symp. on Signal Processing and Applications (GRETSI), Nice, France (May 20–24, 1985).

[80] A.D. Levy-Mandel, J.K. Tsotsos, and A.N. Venetsanopoulos, Knowledge-based landmarking of cephalograms, Comput. Biomed. Res. 19, No. 3 (June 1986) 282–309.

[81] J. Serra, Image Analysis and Mathematical Morphology (Academic Press, New York, 1982).

[82] J.F. Bronskill and A.N. Venetsanopoulos, The Pecstrum, Proc. 3rd ASSP Workshop on Spectrum Estimation and Modeling, Boston, MA, eds. C.L. Nikias and J.G. Proakis (IEEE, New York, 1986) pp. 89–92.

[83] T.V. Ho and A.N. Venetsanopoulos, Optimal homomorphic filtering applied to the tomographic restoration of radiographs, Proc. of ELECTRONICOM 85, Toronto, Canada (IEEE, New York, 1985) pp. 182–185.

[84] A.N. Venetsanopoulos and B.G. Mertzios, Digital adaptive and reconfigurable filters, Proc. of the IV Polish–English Seminar on Real-Time Process Control, Jablonna, Poland (May 31–June 32, 1983) pp. 30–40.

[85] B.G. Mertzios and A.N. Venetsanopoulos, Implementation of two-dimensional digital filters via two-dimensional filter chips, Joint Special Issue on VLSI Analog and Digital Signal Processing, IEEE Trans. Circuits & Syst. CAS-33, No. 2 (Feb. 1986) 239–249,

and IEEE J. Solid-State Circuits SC-21, No. 1 (Feb. 1986) 129–139.

[86] S.Y. Kung, On supercomputing with systolic/wavefront array processors, Proc. IEEE 72, No. 7 (July 1984).

[87] A.N. Venetsanopoulos and V. Cappellini, Real-time image processing, in: Progress in Multidimensional Systems Techniques and Applications, ed. S.G. Tzafestas (Marcel Dekker, New York, 1986).

[88] I. Pitas and A.N. Venetsanopoulos, Bayesian Estimation of Medium Properties in Wavefield Inversion Techniques, Proc. IEEE Int. Conf. on Acoustics, Speech and Signal Processing, Tampa, FL (March 1985) pp. 22.3.1–22.3.4.

[89] I. Pitas and A.N. Venetsanopoulos, Detection of wave reflectors and wave sources in the earth subsurface, IEEE Trans. Acoust. Speech & Signal Process. ASSP-34, No. 5 (Oct. 1986) 1245–1257.

[90] A.N. Venetsanopoulos and C.L. Nikias, Design and realization of multidimensional digital filters via matrix decomposition approaches, Advances in Geophysical Data Processing, Vol 2, ed. M. Simaan (JAI Press, 1985) pp. 263–305.

COURSE 11

IMAGE RECONSTRUCTION

M. KAVEH

*Department of Electrical Engineering, University of Minnesota,
Minneapolis, MN 55455, USA*

J.L. Lacoume, T.S. Durrani and R. Stora, eds.
Les Houches, Session XLV, 1985
Traitement du signal / Signal processing
© *Elsevier Science Publishers B.V., 1987*

Contents

1. Introduction

Image reconstruction is the process by which quantitative two- or three-dimensional characterization of a medium is obtained from a collection of measurements of a lower dimension. Therefore, the subject matter is exclusively multi-dimensional with no one-dimensional counterpart in signal processing. The most extensive use of image reconstruction is computer-aided tomography (CAT) in medicine [1]. However, similar reconstruction techniques have been used in applications as diverse as radio astronomy [2], electron microscopy [3] and seismic signal processing [4]. In CAT, the most common and best developed imaging process is with X-rays, with more recent developments centering around the uses of ultrasound [5], positron emission [6] and nuclear magnetic resonance (NMR) [7].

Three ingredients: source of excitation, medium interaction model, and measurements external to the medium under consideration, henceforth called the object, form the basis of all the tomographic imaging techniques and therefore determine the appropriate physico-mathematical framework for the establishment of an appropriate reconstruction algorithm. Thus, in its most general form, image reconstruction (for our purposes synonimous with CAT) can be considered as an inverse problem of the source–medium interaction model.

Two interaction models stand out. In the more common applications (say, X-rays), the excitation wavelength is much smaller than the fine detail in the object and ray-optics assumptions apply. We denote reconstruction based on this model as straight-path tomography. In the second model, scatterers within the object are on the same order as the excitation wavelength, and diffraction effects must be taken into account. This model gives rise to diffraction tomography which has had much of its development in ultrasonic, microwave and seismic imaging applications. By and large, straight-path tomography is the simplest and most-developed of the two. We will show later, however, that in certain cases both

672

reconstructions can be formulated in a common algorithmic framework. In all our discussions we treat two-dimensional reconstruction, as this is the most common application. Three-dimensional reconstruction, conceptually, follows the same lines of development.

In section 2 we present some representative physical and measurement models that generate the data-bases and mathematical models used in image reconstruction. In section 3 the fundamental relationships of Fourier-plane sampling are developed for straight-path and diffraction tomography. This is followed by a discussion of Fourier-domain and spatial (backprojection or backpropagation) methods of reconstruction in section 4. We then give a brief overview, in section 5, of the limited angle issue in practical tomography. A brief exposition of the algebraic reconstruction technique (ART) for straight-path tomography follows in section 6. We close the presentation with a few numerical examples (section 7). The derivations in this overview are by necessity brief. The reader is encouraged to consult the large body of literature that is available on this general topic in the publications of the seemingly disparate disciplines that use image reconstruction.

Before continuing, it is helpful to give some of the notations and conventions that are used in the remainder of this course. Vectors and matrices are indicated by boldface-type symbols, with their elements shown inside [ ]. The symbol ~ appearing on a function denotes the Fourier transform of that function. The measurement geometry in this chapter requires the object that is under investigation to be rotated by many angles. The angularly dependent functions will then be denoted by a subscript which is the angle of rotation.

2. Tomographic models

In this section we present two physical models and three measurement geometries that give rise to data and mathematical models used in image reconstruction. The first physical model is one on which straight-path tomography is based. For this model we consider two measurement geometries based on the mode of excitation: parallel-beam and fan-beam. The second model gives the assumptions and formulation of transmission diffraction tomography. Here, only plane wave insonification is considered. In all instances we assume without a loss in generality that appropriate sets of data are collected by keeping the source and receiver stationary and rotating the objects.

2.1. Models for straight-path tomography

Let $f(r)$ denote the two-dimensional distribution of the parameter of interest in the object under study. We will generally assume that the object has a finite support D, i.e. $f(r) = 0$ outside of D. We call this function the object function. As indicated in the Introduction, when the object is rotated by an angle θ in a fixed measurement coordinate system, the object is represented by $f_\theta(r)$. Clearly,

$$f_\theta(r) = f(\Theta r), \tag{1}$$

where

$$r = [x, y]^T \quad \text{and} \quad \Theta = \begin{bmatrix} \cos\theta & -\sin\theta \\ \sin\theta & \cos\theta \end{bmatrix},$$

and the superscript T denotes the transpose.

Assuming ray-optics, the straight-path model presents the measurement at any point in space, external to the object, as a line integral along the path of the excitation ray, of $f_\theta(r)$. This model is somewhat restrictive even in the geometrical-optics regime, in that it neglects refraction effects (which is, of course, unknown). It does, however, represent many practical measurement situations such as in X-ray tomography, and is also the simplest, best-understood model for which reconstruction algorithms have been developed.

Figures 1 and 2 show two straight-path tomography measurement geometries. In each case the object is irradiated and a resulting one-

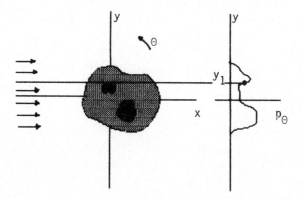

Fig. 1. Parallel-beam straight path geometry.

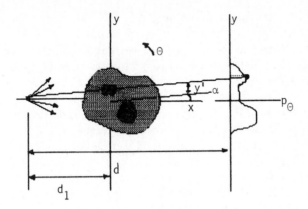

Fig. 2. Fan-beam straight-path geometry.

dimensional function, $p_\theta(y)$, known as a projection, is measured for each rotation angle θ. It is worth noting here that the parallel-beam geometry leads to simple Fourier space reconstruction algorithms. Fan-beam geometries, however, are most often used in practice for ease and speed of measurement.

In practice, a finite number of collimated beams is used for radiation, the object is rotated through a finite number of (usually equispaced) angles and $p_\theta(y)$ is sampled by a finite number of detectors. Thus the practical reconstruction problem is discrete. We will, however, formulate the problem in continuous space and later specialize the solutions to the case of discrete reconstruction.

According to our earlier explanation of the straight-path model, $p_\theta(y)$ is related to the object function by

$$p_\theta(y) = \int_{-\infty}^{\infty} f_\theta(x, y)\,dx, \qquad \text{parallel beam} \tag{2}$$

and

$$p_{\theta'(y)}[y'(y)] = \int_{-\infty}^{\infty} f_\theta\left(x, \frac{d_1 + x}{d}y\right)dx, \qquad \text{fan-beam} \tag{3}$$

where

$$\theta' = \theta - \tan^{-1}\left(\frac{y}{d}\right), \quad \text{and} \quad y' = \frac{d_1 y}{\sqrt{d^2 + y^2}}.$$

Note that eq. (3) relates a single point at y' on a parallel projection at an angle θ', to the value of the fan-beam projection at y for a rotation angle

θ. Thus, in principle, a complete set of parallel projections can be obtained from a continuous 360° continuum of fan-beam projection. Then, parallel beam algorithms can be used for image reconstruction. The drawback of this approach in practice is that the finite number of rotation angles and y values do not in general generate equisampled parallel projections. In such a circumstance, one may use an interpolation scheme to generate approximations to the sampled parallel projections as given in ref. [8]. With proper design of the data collection system, however, intermediate parallel projections can be generated on-the-fly [9]. We therefore concentrate on the algorithms for the parallel projection problem.

An example of the above model is demonstrated in X-ray tomography. Here, a single ray at y is characterized by $N_\theta(y)$ photons input to the object, and $M_\theta(y)$ photons detected outside the object along the ray. The object is to a good approximation represented by a spatially varying linear attenuation coefficient $f(x, y)$. Then, for the parallel-beam case,

$$p_\theta(y) = \ln\left(\frac{M_\theta(y)}{N_\theta(y)}\right).$$

In the next subsection we establish the model for diffraction tomography.

2.2. Diffraction tomography

As mentioned earlier, diffraction tomography differs fundamentally from its straight-path counterpart, in that the actual wave propagation through the medium must be considered. Thus, projections in the form of line integrals are no longer valid. As such, image reconstruction in the diffraction regime is posed as an inverse problem of an appropriate wave equation. The most common wave equation used is the Helmholtz equation with a spatially varying coefficient. We first establish the model for transmission tomography employing plane wave excitation. Details of the derivations can be found in refs. [5] and [10–15].

Figure 3 depicts the most common measurement geometry in two-dimensional transmission diffraction tomography. The source of excitation is assumed to be a plane wave and the array of receiver sensors measures the wave which consists of the primary and scattered components. The data-base for inversion is obtained from coherent measurements of the scattered waves for a number of angular orientations of the object.

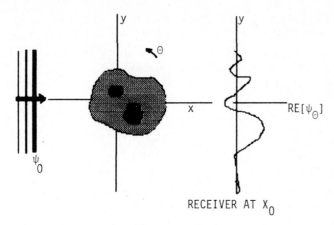

RECEIVER AT X_0

Fig. 3. Measurement geometry for diffraction tomography using plane wave excitation.

The key to the current state of the art in diffraction tomography is the simplification or approximation of the wave equation to make the desired parameters of the inhomogeneity accessible to recovery. The approximations are usually based on the assumption that the wavenumber within the inhomogeneity differs only slightly from that of the known homogeneous background. In solving the direct problem of the wave equation, this assumption has been the basis for several perturbation solutions. A similar approach is taken in solving the inverse problem.

We begin by examining the scalar two-dimensional Helmholtz equation of interest with a space varying coefficient. Accordingly, the complex amplitude of the scalar field, $\psi(r)$, satisfies

$$\left(\nabla^2 + k^2(r)\right)\psi(r) = 0, \tag{4}$$

where ∇^2 is the Laplacian operator, $k(r)$ is the spatially varying wavenumber which in general is complex, with its real part often reflecting variations in the velocity of propagation and its imaginary part signifying loss in the medium under study. In transmission diffraction tomography, k^2 is expressed as

$$k^2(r) = k_0^2[1 + f(r)]$$

where k_0 is the known real constant wavenumber of the homogeneous medium, such as water in ultrasonic tomography, surrounding the object under study, and $f(r)$ signifies a small perturbation, $|f(r)| \ll 1$, to the homogeneous medium due to the presence of the object. The aim of solving the inverse problem is, once again, to approximately reconstruct (image) the object function $f(r)$ based on measurements of $\psi_\theta(x_0, y)$ outside the object.

We now express eq. (4) in the form of an inhomogeneous, constant-coefficient differential equation with a source term containing $f(r)$. This can be accomplished by reformulating the equation in terms of a general transformation on $\psi(r)$. Thus, let $\psi_0(r) = e^{jk_0x}$ be the complex amplitude of the plane incident wave propagating in the x-direction. Define a scalar differentiable function of $\alpha(r) = \psi(r)/\psi_0(r)$ as $U(r) = g(\alpha(r))$. Then we have

$$\psi_0(r)\nabla^2\alpha(r) = \nabla^2\psi(r) - 2jk_0\hat{x}^T\nabla\psi(r) - k_0^2\psi(r), \qquad (5)$$

where ∇ is the gradient operator and $\hat{x}$ is the unit vector along the x-axis. Let g' and g'' denote the first and second derivatives of $g(\alpha(r))$ with respect to $\alpha(r)$. Equation (5) can be expressed as

$$\nabla^2U(r) + 2jk_0\hat{x}^T\nabla U(r) = -k_0^2 f(r)\alpha(r)g' + \left(\frac{\nabla U(r)}{g'}\right)^2 g''$$

$$\equiv -h(r). \qquad (6)$$

Equation (6) is one of the desired forms of the wave equation in terms of the transformed function $U(r)$. The left-hand side of the equation is a linear constant coefficient operation on $U(r)$. The right-hand side of eq. (6) explicitly contains $f(r)$ as a source term as well as an additional term containing $(\nabla U(r))^2 \equiv (\nabla U(r))^T(\nabla U(r))$. The operator on the left-hand side of eq. (6) is invariant under the transformation $g(\cdot)$. Therefore, the effectiveness of the inversion process, i.e. the approximate recovery of $f(r)$, is determined by the nature of the terms other than $f(r)$ on the right-hand side, i.e., the fidelity with which $h(r)$ represents $f(r)$. Here, we only discuss the two most popular transformations.

An examination of the form of $h(r)$ in eq. (6) points to a natural choice for g, i.e. one for which $\alpha(r)g'$ is a constant, for example, unity. Thus,

$$\frac{\partial g(\alpha(r))}{\partial(\alpha(r))}\,\alpha(r) = 1,$$

resulting in

$$g(\alpha(r)) = \ln(\alpha(r)) \qquad (7)$$

where $\ln(\cdot)$ denotes the real or complex natural logarithm depending on the nature of its argument. This transformation is interesting, in that it is a natural choice from the general form of $h(r)$. It is also, coincidentally, exactly the same transformation employed in the method of smooth perturbation [16] in the solution of the direct problem of the wave equation, better known as the Rytov approximation. Consequently, we denote $g(\alpha(r)) = \ln(\alpha(r))$ the Rytov transformation, and the resulting source function, $h(r)$, the Rytov source. This function is now given by

$$h(r) = k_0^2 f(r) + (\nabla U(r))^2. \tag{8}$$

Thus, $h(r)$ is comprised of the object function, $f(r)$, and an additive error term that depends on the variations of a logarithmically smoothed function of the normalized propagating wave. If the variations of this wave over distances of the order of a wavelength are small, $h(r)$ will be a good representation of $f(r)$.

An especially simple transformation that may be used in conjunction with eq. (6) is a linear transformation given by $U(r) = \alpha(r) - 1$. The resulting source function is

$$h(r) = k_0^2 f(r) + k_0^2 f(r) U(r). \tag{9}$$

Since eq. (6), using the above $U(r)$, coupled with neglecting the second term on the right-hand side of eq. (9) is used in solving the forward problem of the wave equation under the Born approximation [16], we denote the above function the Born transformation. It is clear from eq. (9) that the difference between $h(r)$ and $f(r)$ in this case is proportional to the object function itself. So, the requirement that $h(r)$ should be a close approximation to $f(r)$ necessitates that

$$\left| \frac{\psi(r) - \psi_0(r)}{\psi_0(r)} \right| \ll 1;$$

a very severe limitation indeed. The relative merits of the Born and Rytov transformations in diffraction tomography are detailed in refs. [12–14] and [17].

3. Inversion in the Fourier domain

In this section we present the fundamental Fourier domain relationships between the object function and the measurements for different angles in parallel-beam straight-path and plane wave diffraction models. First, we establish two properties of the Fourier transform of a two-dimensional function $f(r)$.

Let $\Lambda^T \equiv [\mu, \nu]$ be the two-dimensional spatial-frequency vector with Cartesian coordinates μ and ν. The first property of interest relates the

Fourier transforms of a function and its rotated version:

$$\tilde{f}_\theta(\Lambda) = \int_{-\infty}^{\infty} \int_{-\infty}^{\infty} f_\theta(r) \exp(-j\Lambda^T r) \, dx \, dy$$

$$= \int_{-\infty}^{\infty} \int_{-\infty}^{\infty} f(\Theta r) \exp(-j\Lambda^T r) \, dx \, dy$$

$$= \int_{-\infty}^{\infty} \int_{-\infty}^{\infty} f(r) \exp\left[-j(\Theta\Lambda)^T r\right] dx \, dy$$

$$= \tilde{f}(\Theta\Lambda). \tag{10}$$

That is, the Fourier transform of a rotated function is the Fourier transform of the unrotated function, rotated by the same angle.

The second property of the Fourier transform that is needed, relates the partial transform, with respect to say the y-coordinate, of a two-dimensional function $g(x, y)$ to its two-dimensional transform $\tilde{g}(\mu, \nu')$. Let $G(x, \nu)$ denote the partial transform defined by

$$G(x, \nu) = \int_{-\infty}^{\infty} g(x, y) e^{-j\nu y} \, dy. \tag{11}$$

Substituting in eq. (11) for $g(x, y)$ in terms of $\tilde{g}(\mu, \nu')$, one obtains

$$G(x, \nu) = \frac{1}{4\pi^2} \int_{-\infty}^{\infty} \int_{-\infty}^{\infty} \tilde{g}(\mu, \nu') e^{j\mu x} \int_{-\infty}^{\infty} e^{j(\nu' - \nu)y} \, dy \, d\nu' \, d\mu$$

$$= \frac{1}{2\pi} \int_{-\infty}^{\infty} \int_{-\infty}^{\infty} \tilde{g}(\mu, \nu') e^{j\mu x} \delta(\nu' - \nu) \, d\nu' \, d\mu$$

$$= \frac{1}{2\pi} \int_{-\infty}^{\infty} \tilde{g}(\mu, \nu) e^{j\mu x} \, d\mu. \tag{12}$$

We now proceed to present the straight-path and diffraction projection-slice theorems.

3.1. Straight-path projection-slice

We are interested in relating the one-dimensional Fourier transform of the projection at angle θ, i.e. $p_\theta(y)$, to $\tilde{f}(\mu, \nu)$. Taking the Fourier transform of eq. (1) and using the property given in eq. (11), we have

$$\tilde{p}_\theta(\nu) = \frac{1}{2\pi} \int_{-\infty}^{\infty} \int_{-\infty}^{\infty} \tilde{f}_\theta(\mu, \nu) e^{j\mu x} \, d\mu \, dx$$

$$= \int_{-\infty}^{\infty} \delta(\mu) \tilde{f}_\theta(\mu, \nu) \, d\mu = \tilde{f}_\theta(0, \nu). \tag{13}$$

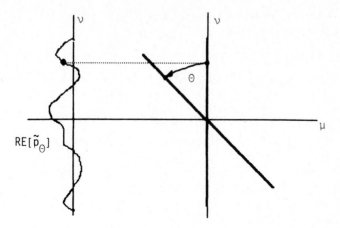

Fig. 4. Straight path projection-slice.

Applying the rotation property (10) to eq. (13) we obtain the Fourier-slice relation as

$$\tilde{p}_\theta(\nu) = \tilde{f}\big(\boldsymbol{\Theta}[0, \nu]^{\mathrm{T}}\big) = \tilde{f}(-\nu \sin\theta, \nu \cos\theta). \tag{14}$$

Equation (14) states that the Fourier transform of the projection at an angle θ gives a radial slice of $\tilde{f}(\Lambda)$ at an angle of θ from the ν-axis, as demonstrated in fig. 4. Therefore, with a sufficiently dense set of θ, one should be able to approximately reconstruct $\tilde{f}(\Lambda)$, and consequently $f(\boldsymbol{r})$. $p_\theta(y)$ is a two-dimensional function (of θ and y) known as the Radon transform of $f(\boldsymbol{r})$. Equation (14) implies that given $p_\theta(y)$, $|y| < \infty$, $0 \leqslant \theta \leqslant \pi$, one can exactly recover $f(\boldsymbol{r})$. This remarkable fact was first proved by Radon [18]. Practical image reconstruction to be addressed in the next section develops approximations of $f(\boldsymbol{r})$ when $p_\theta(y)$ is sampled in both θ and y.

3.2. Diffraction projection-slice

Following the straight-path terminology, let us call the transformed measured data, $U_\theta(\boldsymbol{r})$, the diffraction projection at angle θ. Equation (6) for the object rotated by θ is then given by

$$\nabla^2 U_\theta(\boldsymbol{r}) + 2\mathrm{j}k_0\hat{\boldsymbol{x}}^{\mathrm{T}} \nabla U_\theta(\boldsymbol{r}) = -h_\theta(\boldsymbol{r}). \tag{15}$$

Examining the expression for $h(\boldsymbol{r})$ it becomes obvious that, in general,

$h_\theta(r) \neq h(\Theta r)$. However, under the assumption that $h(r)$ is a good approximation to $f(r)$, we have the relation

$$h_\theta(r) \simeq f_\theta(r) = f(\Theta r).$$

Substituting the approximation $f_\theta(r)$ in the right-hand side of eq. (15), Fourier transforming the result and using relation (10), we obtain

$$\left(\mu^2 + \nu^2 + 2k_0\mu\right)\tilde{U}_\theta(\Lambda) \simeq k_0^2\tilde{f}(\Theta\Lambda). \tag{16}$$

Using relation (12) in conjunction with eq. (16) results in

$$W_\theta(x_0, \nu) \simeq \frac{k_0^2}{2\pi} \int_{-\infty}^{\infty} \frac{\tilde{f}(\Theta\Lambda)e^{j\mu x_0}}{(\mu - \mu_1)(\mu - \mu_2)} \, d\mu, \tag{17}$$

where $W_\Theta(x_0, \nu)$ is the one-dimensional Fourier transform of $U_\Theta(x_0, y)$ along the y-coordinate. The integral in eq. (17) can be evaluated by contour integration in the complex plane. The result is [5, 14]

$$\tilde{f}(\Theta\Lambda_1) \simeq \frac{2j}{k_0^2}\sqrt{k_0^2 - \nu^2} \exp(-j\mu_1 x_0)W_\theta(x_0, \nu), \quad |\nu| \leq k_0, \tag{18}$$

where

$$\mu_1 = -k_0 + \sqrt{k_0^2 - \nu^2}, \quad \text{and} \quad \Lambda_1^T = [\mu_1, \nu].$$

Equation (18) is the diffraction counterpart of the Fourier-slice relation in eq. (14). It gives the relationship between the one-dimensional Fourier transform of the diffraction projection and $\tilde{f}(\Lambda)$. $\Theta\Lambda_1$, $|\nu| \leq k_0$, describes a semi-circular contour in the $[\mu, \nu]$ plane, over which $\tilde{f}(\cdot)$ is obtained from $W_\theta(x_0, \nu)$. Figure 5 shows the contour for a given angle Θ and the process by which values of $W_\theta(x_0, \nu)$ are mapped onto this contour to approximately generate a semi-circular "slice" of the function $\tilde{f}(\Lambda)$. The reconstruction is again accomplished by obtaining a dense coverage of the frequency-plane with these diffraction Fourier-slices by varying θ. We denote the collection of these contours as well as the radial ones in straight-path tomography, the reconstruction contours (RCON's).

We close this section by commenting on the relationship between straight-path tomography and diffraction tomography. The straight-path tomography formulation can be obtained from the diffraction model by considering the limit as the source wavelength approaches zero. Under this assumption, $k_0 \to \infty$ and the RCON in fig. 5, over the object frequency range, is well-approximated by a straight-line tangent to the circle at the origin. This straight line is precisely the slice shown in fig. 4.

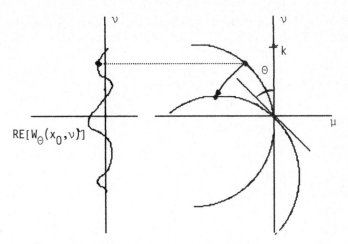

Fig. 5. Diffraction projection-slice.

4. Reconstruction

In the previous section we set the basis for approximately recovering $f(r)$ from a set projections for different angular orientations of the object. The data-base was generated on rotated radial lines or semi-circles denoted as RCON's in the spatial frequency domain. In performing Fourier reconstruction, computational concerns naturally lead one to perform interpolation of the data onto a uniform grid and use the fast Fourier transform (FFT) to obtain the object function. We will develop a general formalism for this type of reconstruction that is equally applicable to a variety of RCON's. We first introduce the general formulation of the problem. This is followed by two types of interpolation techniques: spatial domain and Fourier domain.

We assume that the measured (computed) information about the two-dimensional function $f(r)$ is given in the Fourier domain in terms of an observed function $O(\alpha, \beta)$, where (α, β) are related to the Cartesian coordinates (μ', λ') via a known transformation matrix T. That is

$$\tilde{f}(\mu', \lambda') = O(\alpha, \beta), \tag{19}$$

with

$$\begin{bmatrix} \mu' \\ \lambda' \end{bmatrix} = T(\alpha, \beta) = \begin{bmatrix} T_1(\alpha, \beta) \\ T_2(\alpha, \beta) \end{bmatrix}, \tag{20}$$

where T_1 and T_2 are assumed to be continuous. The loci defined by T are the RCON's. In the context of the parallel-beam straight-path and diffraction tomography problems we have from the previous section:

(a) The polar coordinates found in conjunction with the Fourier-slice identity of straight-path tomography, with

$$T(\alpha, \beta) = \begin{bmatrix} \cos \alpha & -\sin \alpha \\ \sin \alpha & \cos \alpha \end{bmatrix} \begin{bmatrix} 0 \\ \beta \end{bmatrix}, \quad \text{for} \quad \begin{cases} 0 \leqslant \alpha < \pi, \\ |\beta| < \infty. \end{cases} \tag{21}$$

(b) The semicircles given in eq. (18) with radius k_0 associated with diffraction tomography. In this case, T is given by

$$T(\alpha, \beta) = \begin{bmatrix} \cos \alpha & -\sin \alpha \\ \sin \alpha & \cos \alpha \end{bmatrix} \begin{bmatrix} -k_0 + \sqrt{k_0^2 - \beta^2} \\ \beta \end{bmatrix} \text{ for } \begin{cases} 0 \leqslant \alpha \leqslant 2\pi, \\ 0 \leqslant |\beta| \leqslant k_0. \end{cases}$$

$$\tag{22}$$

As mentioned previously, $O(\alpha, \beta)$ is usually only available over a finite number of variables (α_n, β_m). The objective of the reconstruction algorithms is to obtain an estimate of $f(r)$ based on the set of available $O(\alpha_n, \beta_m)$. It is at this point that the question of interpolation arises. Those who have favored Fourier domain reconstruction have interpolated $O(\alpha_n, \beta_m)$ onto a uniform grid, using one of the traditional polynomial interpolation functions, to obtain an estimate of $f(\mu_i, \lambda_j)$ [19]. This approach ignores the specific and intimate relation that exists between (μ', λ') and (α, β). In the following, we discuss a technique that is based on the above mentioned relation and operates on the exact integral relations between $f(x, y)$, $O(\alpha, \beta)$ and $\tilde{f}(\mu', \lambda')$. Recall that the Fourier-slice relations are based on Fourier integrals, leading one to continue using the integral relations in formulating the interpolations as well.

Consider the inverse Fourier transform equation for $f(x, y)$:

$$f(x, y) = \frac{1}{4\pi^2} \int_{-\infty}^{\infty} \int_{-\infty}^{\infty} \tilde{f}(\mu', \lambda') \exp[j(\mu'x + \lambda'y)] \, d\mu' \, d\lambda'. \tag{23}$$

With the help of eq. (19), we can make a change of variables in eq. (23) to obtain

$$f(x, y) = \frac{1}{4\pi^2} \int_{-\infty}^{\infty} \int_{-\infty}^{\infty} O(\alpha, \beta) \exp\{j[T_1(\alpha, \beta)x + T_2(\alpha, \beta)y]\}$$

$$\times J(\alpha, \beta) \, d\alpha \, d\beta, \tag{24}$$

where $J(\alpha, \beta)$ is the Jacobian of the transformation, given by

$$J(\alpha, \beta) = \left| \frac{\partial(\mu', \lambda')}{\partial(\alpha, \beta)} \right|. \qquad (25)$$

For data that are obtained at different rotation angles θ, we return to our earlier notation and let $\beta = \nu$ and $\alpha = \theta$. The Jacobian in eq. (25) for the straight-path and diffraction cases are given by

$$J(\theta, \nu) = |\nu|, \qquad \text{straight-path},$$

$$J(\theta, \nu) = \frac{k_0|\nu|}{\sqrt{k_0^2 - \nu^2}}, \qquad |\nu| < k_0, \qquad \text{diffraction}.$$

Equation (24) presents the possibility for two types of reconstruction, each implying interpolation in a different domain. The first one, denoted by backpropagation for diffraction tomography [15], and convolution-backprojection for straight-path tomography [20, 21], performs the effective interpolation in the spatial domain. The second, which has been called unified Fourier reconstruction (UFR) is again based on eq. (25) but does the interpolation in the Fourier space [10].

4.1. Spatial domain reconstruction

We begin by deriving the convolution backprojection algorithm for straight-path reconstruction. Notice that in this case (ν, θ) constitute the polar coordinates in the Fourier space.

Let $[x', y'] \equiv [x, y]\theta$ denote the rotated Cartesian coordinates by $-\theta$. Substituting eq. (21) into eq. (24) results in

$$f(x, y) = \frac{1}{2\pi} \int_0^\pi \left[\frac{1}{2\pi} \int_{-\infty}^\infty |\nu| O(\theta, \nu) \exp(j y' \nu) \right] d\nu \, d\theta. \qquad (26)$$

The term inside [ ] is the convolution of the projection at angle θ and the inverse transform of $|\nu|$, which is backprojected onto the x–y plane. The backprojections are then integrated over all angles to result in $f(\mathbf{r})$. In practice, of course, data are only available over finite number of angles θ_i, $i = 1, \ldots, K$ and in the form of N-point discrete Fourier transforms (DFT). An $N \times N$ reconstruction then becomes

$$\hat{f}(x_k, y_l) = \frac{1}{2\pi} \sum_{i=1}^K \hat{f}'(y_j', \theta_i), \qquad (27)$$

where

$$\hat{f}'(y_j', \theta_i) = \text{IDFT}\big[|\nu_j|O(\theta, \nu_j)\big] \tag{28}$$

and (x_k, y_l), for an angle θ_i and rotated coordinate y_j', are taken as the nearest grid points satisfying

$$x \sin \theta_i + y \cos \theta_i = y_j', \quad j = 1, \dots, N. \tag{29}$$

The computational burden of the above process using the one-dimensional FFT, consists of K N-point FFT's followed by N^2 assignments based on eq. (28). Because of the nature of the assignment of the pixels' values we consider this approach to fall in the category of interpolation in the spatial domain. Most CAT systems use one implementation or other of the convolution-backprojection algorithm. The reconstructions from this algorithm are generally considered to be more accurate than ad hoc polynomial interpolation-based Fourier reconstruction methods. Another advantage of this algorithm is that it can directly be generalized to fan-beam reconstruction [21].

Let us now consider the diffraction counterpart of the above algorithm. This method was originally proposed by Devaney and denoted the backpropagation algorithm. The algorithm can, once again, be developed from the general formulation in eq. (24). Using our earlier notation it follows that

$$f(x, y) = \frac{1}{2\pi} \int_0^{2\pi} \left\{ \frac{1}{2\pi} \int_{-k_0}^{k_0} \frac{k_0|\nu|}{\sqrt{k_0^2 - \nu^2}} O(\theta, \nu) \right.$$

$$\left. \times \exp\big[j(\gamma(\nu)x' + \nu y')\big] \, d\nu \right\} d\theta, \tag{30}$$

where

$$\gamma(\nu) = -k_0 + \sqrt{k_0^2 - \nu^2}.$$

Again, for a discrete set of parameters and sampled image reconstruction, we have

$$\hat{f}(x_k, y_l) = \frac{1}{2\pi} \sum_{i=1}^{K} \hat{f}'(y_j', x_m', \theta_i), \tag{31}$$

where

$$\hat{f}'(y_j', x_m', \theta_i) = \text{IDFT}\left[\frac{k_0|\nu_j|}{\sqrt{k_0^2 - \nu_j^2}} O(\theta_i, \nu_j) \exp\big(\gamma(\nu_j)x_m'\big) \right]. \tag{32}$$

Equation (32) differs from eq. (28) in one major respect. In the present case the partial reconstruction, $f'(y'_j, x'_m, \theta_i)$, is a function of x'_m. Thus for an $N \times N$ reconstruction, it has to be evaluated for N values of x'_m. (x_k, y_l) are then taken as the nearest grid points satisfying

$$[x \quad y]\boldsymbol{\Theta} = [x'_m \quad y'_j], \quad j = 1, \ldots, k; \quad m = 1, \ldots, N. \tag{33}$$

The computational requirements for this algorithm is on the order of (KN) N-point FFT's followed by N^2 assignments in the spatial domain. In summary, the backpropagation reconstruction technique contains the same theoretical elegance for diffraction tomography that backprojection offers straight-path tomography. However, the backprojection method is substantially less computation-intensive than the backpropagation method. We next consider interpolation in the frequency domain.

4.2. Frequency domain reconstruction

As mentioned earlier one may perform the reconstruction efficiently in the Fourier domain by interpolating the values of O from the RCON's onto a uniform two-dimensional grid that is suitable for an FFT operation. Earlier attempts in this regard for both the straight-path and diffraction cases centered around some form of simple polynomial interpolation. The simplest interpolation is nearest-neighbor. An example of a more complex interpolation is the bilinear interpolation method [19] is illustrated in fig. 6. In this case,

$$\tilde{f}(\mu_i, \lambda_k)$$

$$= \frac{\dfrac{O(\alpha_n, \beta_{m-1})}{a} + \dfrac{O(\alpha_n, \beta_m)}{b} + \dfrac{O(\alpha_{n+1}, \beta_m)}{c} + \dfrac{O(\alpha_{n+1}, \beta_{n-1})}{d}}{\dfrac{1}{a} + \dfrac{1}{b} + \dfrac{1}{c} + \dfrac{1}{d}}, \tag{34}$$

where the arguments of $O(,)$ indicate the four nearest neighbors of the grid point at (μ_i, λ_k). We classify this type of interpolation as grid-driven. That is, one assigns a value to a grid point based on some of the data that are closest to it. In the following we discuss a method denoted as a unified frequency domain reconstruction (UFR) method, that derives its interpolation scheme from eq. (24). In order to do so, we temporarily maintain the continuous-space formalism of eq. (24).

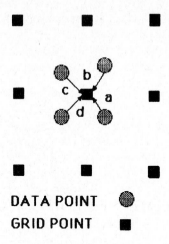

DATA POINT ◉

GRID POINT ■

Fig. 6. Bilinear interpolation assignment.

Let $I(x, y)$ be the indicator function for the support of the region of the image D that we desire to reconstruct. That is, $I(x, y) = 1$, $(x, y) \in D$ and $I(x, y) = 0$, otherwise. D is usually taken as a circle of radius R that is large enough to accommodate the expected support of $f(x, y)$. It can, however, be smaller to focus attention on a certain portion of the image. In any event we define the spatially windowed object function by q. Then

$$q(x, y) = I(x, y)f(x, y), \tag{35}$$

and

$$\tilde{q}(\mu, \lambda) = \tilde{I}(\mu, \lambda) * \tilde{f}(\mu, \lambda) \tag{36}$$

where $*$ indicates the two-dimensional convolution integral. Note that when I includes the support of f, then $q = f$. Then, substituting eq. (24) in eq. (36) we obtain the exact relation at any Cartesian grid point, (μ_i, λ_k), in the frequency domain as

$$\tilde{q}(\mu_i, \lambda_k) = \frac{1}{4\pi^2} \int_{-\infty}^{\infty} \int_{-\infty}^{\infty} \tilde{I}[\mu_i - T_1(\alpha, \beta), \lambda_k - T_2(\alpha, \beta)]$$
$$\times O(\alpha, \beta) J(\alpha, \beta) \, d\alpha \, d\beta. \tag{37}$$

Equation (37) is an exact consequence of the basic relation expressed in eq. (24). At this point we can express the fact that $O(\alpha, \beta)$ is only available over a finite number of points (α_n, β_m), through an ideal sampling representation. That is

$$O(\alpha, \beta) = \sum_{n=1}^{K} \sum_{m=1}^{N} O(\alpha_n, \beta_m)\delta(\alpha - \alpha_n, \beta - \beta_m), \qquad (38)$$

where $\delta(x, y)$ is the two-dimensional Kronecker delta function. Substituting eq. (37) into eq. (38), the exact [*within the sampling errors offered by* (α_m, β_m)] reconstruction formula in the Fourier domain is obtained. We have

$$\tilde{q}(\mu_i, \lambda_k) = \sum_{n=1}^{K} \sum_{m=1}^{N} J(\alpha_n, \beta_m) O(\alpha_n, \beta_m)$$

$$\times \tilde{I}[\mu_i - T_1(\alpha_n, \beta_m), \lambda_k - T_2(\alpha_n, \beta_m)]. \qquad (39)$$

The form of eq. (39) is reminiscent of Nyquists interpolation formula for a sampled signal. The interpolation kernel in the present case has two distinct components. The scale-factor $J(\alpha_n, \beta_m)$ appropriately normalizes the density of the irregularly spaced available data and is crucial to the success of this algorithm. $\tilde{I}(,)$ is the continuous frequency-space interpolation function containing information about the support of the function to be reconstructed. An example of $I(r)$ is a disk of radius R. In this case,

$$\tilde{I}(\mu, \lambda) = \frac{J_1\left(R\sqrt{\mu^2 + \lambda^2}\right)}{R\sqrt{\mu^2 + \lambda^2}},$$

where J_1 is a first-order Bessel function. When R is on the order of half the image dimensions, the main lobe of $\tilde{I}$ has a width on the order of the grid separation in the frequency domain. Thus the main contribution of the kernel $\tilde{I}$ in eq. (39), for a given data point at (α_n, β_m), is to the grid points that are closest to (α_n, β_m). Consequently, for computational efficiency, we approximate $\tilde{q}(\mu_i, \nu_k)$ in terms of a small number of its neighboring data points. This assignment is obviously data-driven. Thus,

$$\tilde{q}(\mu_i, \lambda_k) \simeq \sum_{n} \sum_{m} J(\alpha_n, \beta_m) O(\alpha_n, \beta_m)$$

$$\times I[\mu_i - T_1(\alpha_n, \beta_m), \lambda_k - T_2(\alpha_n, \beta_m)], \qquad (40)$$

(n, m) s.t. (μ_i, ν_k) is one of L nearest neighbors of $T(\alpha_n, \beta_m)$.

The computational requirements for (40) are on the order of LN^2 operations followed by a two-dimensional FFT, where L is typically 4 or

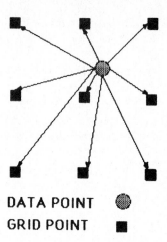

DATA POINT ⬤

GRID POINT ■

Fig. 7. UFR assignment with nine neighbors.

9. The assignment operation indicated in eq. (40) is graphically shown in fig. 7 for nine nearest neighbors. The computation time for the nine neighbor case is slightly higher than the bilinear interpolation algorithm and about one-twentieth of a (128 × 128) backpropagation algorithm. The error of this algorithm is usually substantially less than that for the bilinear method and comparable to that of the convolution-backprojection or backpropagation.

We close this section by giving the UFR formulae explicitly for the straight-path and diffraction case. Substituting the expression for J and T in eq. (21), we obtain

$$\tilde{f}(\mu_i, \lambda_k) \simeq \sum_n \sum_m |\nu| O(\theta_n, \nu_m) \tilde{I}(\mu_i + \nu_m \sin \theta_n, \lambda_k - \nu_m \sin \theta_n),$$

(n, m) s.t. (μ_i, λ_k) is one of L nearest neighbors of $T(\theta_n, \nu_m)$.

Using eq. (22) and the appropriate J for the diffraction case, it follows that

$$\tilde{f}(\mu_i, \lambda_k) \simeq \sum_n \sum_m \frac{2j}{k_0} e^{-j\gamma_m} |\nu_m| O(\theta_n, \nu_m)$$

$$\times \tilde{I}(\mu_i - \cos \theta_n \gamma_m + \sin \theta_n \nu_m, \lambda_k - \sin \theta_n \gamma_m - \cos \theta_n \nu_m),$$

(n, m) s.t. (μ_i, λ_k) is one of L nearest neighbors of $T(\theta_n, \nu_m)$.

Note that in the above two formulations $O(\theta_n, \nu_m)$ is simply the value of the DFT of the projection for an angle θ_n, at a frequency of ν_m.

5. Reconstruction from partial angular views

In the previous section, we presented algorithms to reconstruct the object function when a sufficiently complete set of angular projections were available. There are many circumstances, however, in which physical constraints allow measurements over only limited angular sectors. Clearly, the interpolation methods discussed earlier would fail. This is especially true for Fourier-domain methods. A general solution to this practical problem involves an intermediate projection recovery step that uses certain constraints of the mathematical formulation of the problem. Other approaches to the general limited-data reconstruction problem, that is, cases of either partial angular view and/or partially blocked projections in straight-path tomography, have been the subject of active research [23]–[26]. Many of these methods are variations on the two-dimensional Gerschberg–Papoulis iterative reconstruction method [27, 28] and will not be discussed here. In the following, we present a method first presented in ref. [29] that uses a constrained interpolation step for approximating the missing projections in the Fourier domain followed by the UFR algorithm for reconstruction. The approach is equally applicable to the straight-path and diffraction cases.

Let $P(\theta', \rho')$ be the mapping of $\tilde{f}(\mu', \lambda')$ into the polar spatial frequency coordinator; i.e.,

$$P(\theta', \rho') = \tilde{f}(\mu', \lambda'), \tag{41}$$

where

$$\theta' = \operatorname{arctg}\left(\frac{\lambda'}{\mu'}\right), \qquad \rho' = \sqrt{\mu'^2 + \lambda'^2},$$

and arctg is the four-quadrant inverse tangent operator. Suppose for a constant value of ρ', that $P(\theta', \rho')$ is known over a discrete, finite set of values of θ'. We denote this set by S and its complement by S^c. The constraint used for interpolation is that $P(\theta', \rho')$ is a periodic function of θ' with period 2π. Hence, if $\theta_i' \in S$ then the values of $P(\theta', \rho')$ are also known at $\theta' = \theta_i' + 2\pi k$, with $k = 0, \pm 1, \pm 2, \ldots$.

Without a loss in generality we assume that $\theta' = 0$ is an element of S and fit a periodic interpolating function over $[0, 2\pi]$ that fits $P(\theta_j', \rho')$ for all $\theta_j' \in S$. For any $\theta_i' \in S^c$, the value of $P(\theta_i', \rho')$ can be determined from the periodic interpolating function. In the following we present an interpolation scheme based on trigonometric polynomials.

$P(\theta', \rho')$ can be represented by the following Fourier series expansion:

$$P(\theta', \rho') = \sum_{n=-\infty}^{\infty} p_n(\rho') e^{jn\theta'}. \tag{42}$$

Assume that the object function is contained within a disk of radius R. Thus, the fastest fluctuations of $P(\theta', \rho')$ in the (θ', ρ') domain correspond to the function

$$P_1(\theta', \rho') = e^{j2\pi R\rho'\cos\theta'}. \tag{43}$$

The right-hand side of eq. (43) has the following Fourier series expansion,

$$e^{j2\pi R\rho\cos\theta'} = \sum_{n=0}^{\infty} \varepsilon_n j^n J_n(2\pi R\rho') \cos(n\theta'), \tag{44}$$

where $\varepsilon_0 = 1$, $\varepsilon_n = 2$ for $n \geqslant 1$, and J_n is the Bessel function of the first kind, nth order. $J_n(2\pi R\rho)$ and, as a consequence, the magnitude of the components of the Fourier series in eq. (44) become very small for $n > 2\pi R\rho'$. Hence, the following truncated Fourier series may be used as a good approximation of the original infinite series

$$P_1(\theta', \rho') \approx \sum_{n=0}^{M} \varepsilon_n j^n J_n(2\pi R\rho') \cos(n\theta'), \tag{45}$$

where

$$M \approx 2\pi R\rho'. \tag{46}$$

This implies that the following truncated Fourier series expansion is a good approximation of $P(\theta', \rho')$ with M chosen according to eq. (46):

$$P(\theta', \rho') \approx \sum_{n=-M}^{M} p_n(\rho') e^{jn\theta'}. \tag{47}$$

Let the number of the elements in S be K_1. In practice, $K_1 \geqslant 2M + 1$. Hence, we can write the following set of K_1 linear equations:

$$P(\theta'_i, \rho') \approx \sum_{n=-M}^{M} p_n(\rho') e^{jn\theta'_i}, \quad \theta'_i \in S. \tag{48}$$

Using $\hat{p}_n(\rho')$ as an estimate of $p_n(\rho')$, we can write

$$E\hat{p} = P, \tag{49}$$

where

E is a $K_1 X(2M + 1)$ matrix with elements $\{e^{jn\theta_i'}\}$,

$\hat{p}$ is a column vector of size $(2M + 1)$ of the estimates of the truncated Fourier series coefficients $\{p_n(\rho')\}$, and

P is a column vector of size K_1 of the available data $\{P(\theta_i', \rho')\}$.

The least-squares solution of eq. (49) is given by

$$\hat{p} = (E^T E)^{-1} E^T P. \tag{50}$$

Finally, $P(\theta', \rho')$ can be estimated for any value of θ' with the help of eqs. (47) and (50).

The above results form the basis for restoring the unknown values of the observed function in straight-path and diffraction tomography, as described below.

(a) Straight-path tomography. In this case, we have

$$P(\theta', \rho') = O(\theta, \nu). \tag{51}$$

where

$$\theta' = \theta, \quad \text{and} \quad \rho' = \nu.$$

Thus, the scheme described above is directly applicable to straight-path tomography.

(b) Diffraction tomography. For this problem, we have

$$P(\theta', \rho') = O(\theta, \nu), \tag{52}$$

where

$$\theta' = \theta + \text{arctg}\left(\frac{\nu}{\gamma}\right), \quad \text{and} \quad \rho' = \sqrt{\nu^2 + \gamma^2}.$$

Note that for a constant value of $\pm\nu$, ρ' also assumes a constant value. Thus, for $\rho' = |\nu_m|$, the interpolating points are the $(\theta_i, \pm\gamma_m)$ pairs where the values of $O(\theta_i, \pm\nu_m)$ are known. Therefore, the accuracy of this interpolation method depends on the gap generated in the θ' domain, and not in the θ domain, by the unknown projections. This is an interesting feature in diffraction tomography where it is required that one collects the object profiles over a 360° angular interval.

6. Algebraic reconstruction

Thus far, we have established the reconstruction algorithms within the framework of the projection-slice theorems with resulting interpolations either in the spatial domain or in the Fourier space. In the straight-path case, another possibility is to directly solve the simultaneous collection of integral equations represented by the Radon transform of the object function. Clearly, this has to be done numerically, dictating an immediate discretization of $f(r)$, and the solution of a very large number of simultaneous linear equations. Various approximate solutions to these equations are collectively known as the algebraic reconstruction techniques (ART). In this section we briefly set up the model discussed above and present one possible solution to the problem. Detailed discussion on ART can be found in ref. [28] and references cited therein. We should comment here that algebraic techniques may also, theoretically, be used in conjunction with diffraction tomography. No successful attempt, however, has been made in this regard and this issue will not be addressed.

Let us discretize the object function into $N \times N$ pixels and represent it by an $N^2 \times 1$ array by assuming its value to be a constant over each pixel. Denote the resulting vector by F. Similarly, we sample the projections and denote the one for an angle θ_i by H_{θ_i}. A discrete-space approximation of the line-integral relating F with H_{θ_i} is given by

$$Q_{\theta_i} F = H_{\theta_i}, \quad i = 1, \ldots, K, \tag{51}$$

where $Q_{\theta_i}(N^2 \times N)$ is a sparse matrix (on the order of N non-zero entries in each row. The non-zero elements of Q_{θ_i} may represent the relative lengths of the ray through the pixels that it traverses to generate an element of H_{θ_i}. A simplified model simply uses elemental values of one and zero for Q_{θ_i} depending on whether or not a given ray passes through a corresponding pixel.

The K equations in relation (51) may be appended to each other to give

$$QF = H, \tag{52}$$

where Q is $N^2 \times NK$ and H is $NK \times 1$. The reconstruction problem is now formulated as the solution to eq. (52). Three cases arise depending on the number of projections available. For $K = N$, eq. (52) can be solved exactly. For $K > N$, the system of equations is overdetermined and a least-squares solution can be obtained through the generalized inverse of Q. Finally, if $K < N$ a minimum-norm solution for F may be obtained, again through the use of the appropriate generalized inverse of Q. No matter what the case may be, however, the exact solution of eq.

(52) is often practically impossible (for example, the dimensions of Q for 128×128 reconstruction is on the order of 10^4!). Furthermore, Q is often ill-conditioned resulting in an inversion with noisy data that is highly erroneous.

The methods classified as ART normally use iterative approximate solutions to eq. (52). Within iterations, any prior knowledge such as the support and/or the positivity of $f(r)$ can be used to speed up the convergence and improve the accuracy of reconstruction.

One iterative method for the approximate solution of eq. (52) is the so-called projection method described in ref. [29]. In this technique a "guess" at F is successively projected onto the hyperplanes represented by the equations in relation (52), with an iteration ending with the projection onto the last hyperplane. The process is repeated until some measure of convergence, which is a compromise between the fidelity of reconstruction and possible noise amplification, is obtained. The method is mathematically given by

$$F(i, k) = F(i, k - 1) - \frac{q_k^T F(i, k - 1) - H_k}{q_k q_k^T} q_k \tag{53}$$

where $F(i, k)$ is the approximation for F in the ith iteration following projection onto the kth hyperplane, q_k^T is the kth row of Q, H_k is the kth entry of H, $F(i, 0) = F(i - 1, NK)$ and $F(1, 0)$ is an initial guess based on prior information about F.

In summary, ART is a theoretically viable approach to reconstruction from projections. The algorithms, however, are generally highly computation-intensive and in practice inferior to the methods discussed in the previous sections. A definite advantage of ART is its ability to incorporate prior information about $f(r)$ in its formulation. For example, if refraction is important, iterative corrections to the straight-path model can be made by updating Q according to the approximation obtained for F in a previous iteration. Furthermore, ART can be used in situations where Fourier-slice or backprojection formulations are not possible. An example of this has been given in geophysical imaging [4].

7. Examples

In this section we present a few examples of images reconstructed from actual tomographic measurements or numerical simulations of the projections. Figure 8 shows four backprojection reconstructed slices of the

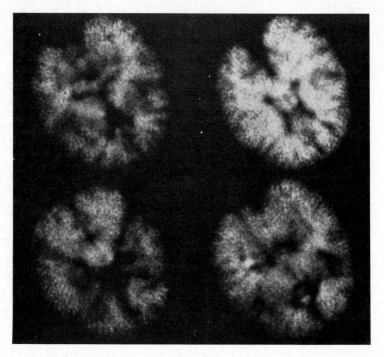

Fig. 8. Straight-path tomograms of excised brain. (Courtesy of J. Greenleaf.)

real index of refraction in an excised human brain. The straight-path model was used with the projection obtained from measurements of time-of-flight of ultrasound pulses through the specimen.

Figure 9 demonstrates the importance of using the correct model or approximation in formulating the reconstruction algorithm. The object is a cylinder with $f(r) = f_r(r) + jf_i(r)$ and $f_i(r) \simeq 0$. The diffraction model was used in conjunction with the Born and Rytov transformations. It is obvious from the example that the Born transformation gives a totally erroneous result even in this relatively simple situation.

Figure 10 shows 64×64 reconstructed images of three cylinders. The data were given as values of $\tilde{f}(\Lambda)$ on 64 diffraction RCON's. Figure 10a uses nearest-neighbor interpolation, fig. 10b uses 4-neighbor UFR and fig. 10c uses backpropagation. The superiority of the latter two methods over the first is obvious. UFR and backpropagation produced nearly identical images, with the latter algorithm requiring about twenty times

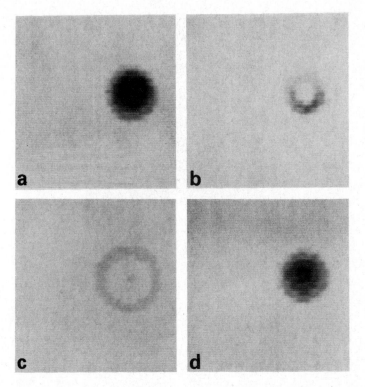

Fig. 9. Diffraction tomograms of 5% saline solution cylinder using ultrasonic measurements. Rytov transformations: (a) f_r, (b) f_i; Born transformations: (c) f_r, (d) f_i.

the computation time of the former. The final example demonstrates the reconstructions based on limited angular data. Sixteen straight-path projections were available within the first quadrant. Figure 11a shows the reconstruction with the method of section 5. Figure 11b shows the four-neighbor UFR reconstruction in which the unavailable projections were set to zero.

8. Summary

The problem of image reconstruction was formulated under both geometrical optics and diffraction models for the response of the object to external excitation. Solutions were developed along three general lines,

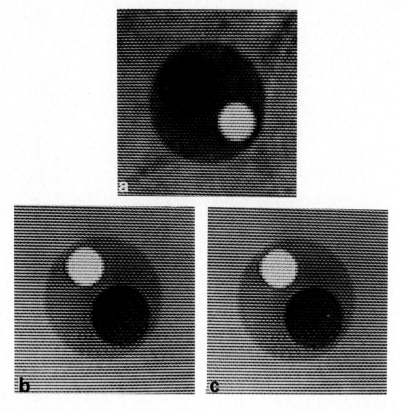

Fig. 10. Diffraction reconstructions of 3 cylinders: (a) nearest-neighbor interpolation, (b) four-neighbor UFR, (c) backpropagation.

not all of which are applicable to all measurement situations. These methods consisted of the spatial-domain interpolation schemes of convolution-backprojection and backpropagation, Fourier-domain interpolation and algebraic reconstruction.

A class of computationally efficient Fourier-domain reconstruction methods, the UFR methods, were discussed in detail and were shown to result in reconstructions of comparable accuracy as the spatial domain methods, but with substantially less computational complexity than the method of backpropagation for diffraction tomography. Algebraic reconstruction was shown to be computationally demanding but well-suited for measurement situations for which other reconstruction formulations

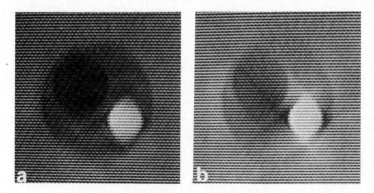

Fig. 11. Limited-angle, straight-path reconstructions of 3 cylinders, with 16 projections in the first quadrant available: (a) interpolated UFR, (b) UFR without interpolation. (Courtesy of M. Soumekh.)

are not possible. Prior information may also easily be included in this class of reconstruction methods.

Future research in the area of image reconstruction is expected to proceed in several distinct but interrelated areas. Of fundamental importance are improved source-medium interaction models, especially for diffraction tomography. Inversion method requiring milder approximations than required by the Born and Rytov transformations are also necessary. Applications and generalizations of the methods of diffraction tomography to cases such as seismic imaging are needed. Much of the current research in straight-path reconstruction is concerned with the practical problem of measurements from limited angular or partially obstructed views. Finally, Fourier-domain interpolation-reconstruction will continue to be of interest, requiring better understanding of the sources of errors in these reconstructions as compared to methods of spatial-domain reconstruction.

References

[1] Special issue on Computerized Tomography, Proc. IEEE 71 (1983) 291–435.
[2] R.N. Bracewell and A.C. Riddle, Inversion of fan-beam scans in radio astronomy, the Astrophys. J. 150 (1967) 427–434.
[3] D.J. DeRosier and A. Klug, Reconstruction of three-dimensional structures from electron micrographs, Nature 217 (1968) 130–134.

[4] K.A. Dines and R.J. Lytle, Computerized geophysical tomography, Proc. IEEE 67 (1979) 1065–1073.

[5] R.K. Mueller, M. Kaveh and G. Wade, Reconstructive tomography and applications to ultrasonics, Proc. IEEE 67 (1979) 567–587.

[6] A.C. Kak, Computerized tomography with X-ray, emission, and ultrasound sources, Proc. IEEE, (1979) 1245–1272.

[7] Z.H. Cho, H.S. Kim, H.B. Song and J. Cumming, Fourier transform nuclear magnetic resonance tomographic imaging, Proc. IEEE 70 (1982) 1152–1173.

[8] M. Soumekh, Image reconstruction techniques in tomographic imaging systems, IEEE Trans. Acoust. Speech & Signal Process. 34 (1986) 952–962.

[9] L. Wang, Cross-section reconstruction with a fan-beam scanning geometry, IEEE Trans. Comput. 26 (1974) 264–270.

[10] M. Kaveh, M. Soumekh and J.G. Greenleaf, Signal processing for diffraction tomography, IEEE Trans. Sonics & Ultrason. 31 (1984) 230–238.

[11] M. Kaveh, R.K. Mueller and R.D. Iverson, Ultrasonic tomography based on a perturbation solution of the wave equation, Comput. Graphics & Image Process. 9 (1979) 105–116.

[12] M. Kaveh, M. Soumekh and R.K. Mueller, A comparison of Born and Rytov approximations in acoustic tomography, in: Acoustical Imaging, Vol. 10, ed. J. Powers (Plenum, New York, 1981) pp. 325–334.

[13] M. Kaveh, M. Soumekh, Z.Q. Lu, R.K. Mueller and J.F. Greenleaf, Further results on diffraction tomography using Rytov's approximation, in: Acoustical Imaging, Vol. 12, eds. E. Ash and C.R. Hill (Plenum, New York, 1982).

[14] M. Kaveh and M. Soumekh, Computer-assisted diffraction tomography, in: Image Recovery: Theory and Applications, ed. H. Stark (Academic Press, New York, 1987) pp. 369–413.

[15] A.J. Devaney, A filtered backpropagation algorithm for diffraction tomography, Ultrason. Imaging 4 (1982) 336–350.

[16] V.I. Tatarski, Wave Propagation in Random Media (McGraw-Hill, New York, 1961).

[17] M. Soumekh and M. Kaveh, A theoretical study of model approximation errors in diffraction tomography, IEEE Trans. Sonics & Ultrason. 33 (1986) 10–20.

[18] J. Radon, On the determination of functions from their integrals along certain manifolds, Ber. Saechs, Akad, Wiss. Leipzig, Math. Phys. K1, 69 (1917) 262–277 (in German).

[19] R.M. Mersereau and A.V. Oppenheim, Digital reconstruction of multidimensional signals from their projections, Proc. IEEE 62 (1974) 1319–1338.

[20] G.N. Ramachandran and A.V. Lakshminarayanan, Three-dimensional reconstruction from radiographs and electron micrographs: application of convolutions instead of Fourier transforms, Proc. Nat. Acad. Sci. USA 68 (1971) 2236–2240.

[21] H.J. Scudder, Introduction to computer-aided tomography, Proc. IEEE 66 (1978) 628–637.

[22] K.C. Tam and V. Peres-Mendez, Tomographical imaging with limited angle input, J. Opt. Soc. Am. 71 (1981) 582–592.

[23] M. Sezan and H. Stark, Image reconstruction by the method of convex projections, IEEE Trans. Med. Imag. 1 (1982) 95–101.

[24] B. Medoff, W. Brody and A. Macovski, The use of a priori information in image reconstruction from incomplete data, in: Proc. IEEE on Acoustics, Speech & Signal Processing ICASSP 83 (Boston, 1983) pp. 131–134.

[25] R.W. Gerchberg, Super-resolution through error energy reduction, Opt. Acta 21 (1974) 709–720.

[26] A. Papoulis, A new algorithm in spectral analysis and band-limited extrapolation, IEEE Trans. Circuits and Systems 22 (1975) 735–742.

[27] M. Soumekh, Image reconstruction from limited projections using the angular periodicity of the radon transform of an object, in: Acoustical Imaging, Vol. 13, eds. M. Kaveh, R. Mueller and J. Greenleaf (Plenum: New York, 1984) pp. 31–42.

[28] G.T. Herman, Image Reconstruction from Projections: The Fundamentals of Computerized Tomography (Academic Press, New York, 1980).

[29] K. Tanabe, Projection method for solving a singular system of equations and its applications, Numer. Math. 17 (1971) 203–214.

PART III

TECHNIQUES IN SIGNAL PROCESSING

COURSE 12

AN INTRODUCTION TO DISPLACEMENT
RANKS AND RELATED FAST ALGORITHMS *

C. GUEGUEN

*Ecole Nationale Supérieure des Télécommunications
46, rue Barrault, 75634 Paris Cédex 13, France*

---

* This work was partially supported by the Office of Naval Research, Mathematical Division, Arlington, VA 22217, under contract 00014-85-K-0256; L. Scharf principal investigator.

J.L. Lacoume, T.S. Durrani and R. Stora, eds.
Les Houches, Session XLV, 1985
Traitement du signal / Signal processing
© *Elsevier Science Publishers B.V., 1987*

Contents

"Toeplitz or not Toeplitz ...
... that is the question"

1. The occurrence of Toeplitz matrices in signal processing

Special kinds of quadratic forms were introduced (around 1905) by the German mathematician Toeplitz [1] to study some properties of spectral representations. Let $f(\varphi)$ be an integrable function over $(-\pi, +\pi)$; its Fourier coefficients are defined as

$$t_k = \frac{1}{2\pi} \int_{-\pi}^{+\pi} f(\varphi) \exp(-jk\varphi) \, d\varphi.$$

Many interesting properties of $f(\varphi)$ may be studied by constructing the Toeplitz matrix T depending on n coefficients t_k:

$$T = \begin{bmatrix} t_0 & t_1 & \cdots & t_n \\ t_1 & t_0 & & \vdots \\ \vdots & & & t_1 \\ t_n & \cdots & t_1 & t_0 \end{bmatrix}. \tag{1.1}$$

This matrix is mainly characterized by the fact that the same element t_k is repeated along the diagonal such that

$$T = \{t_{ij}\}, \qquad t_{ij} = t_{i-j}, \qquad i, j = 0, n.$$

Closely related matrices were introduced, under the name of Hankel matrices, by Frobenius to study the problem of moments and Hurwitz polynomials [2]:

$$H = \begin{bmatrix} h_0 & h_1 & \cdots & h_n \\ h_1 & & & h_{n+1} \\ \vdots & & & \vdots \\ h_n & h_{n+1} & \cdots & h_{2n} \end{bmatrix}. \tag{1.1'}$$

This matrix H is characterized by repeating the same element h_k along the anti-diagonal:

$$H = \{h_{ij}\}, \qquad h_{ij} = h_{i+j}, \qquad i, j = 0, n.$$

707

Toeplitz and Hankel matrices play an outstanding role in the analysis of signals and systems. They provide a useful tool to manipulate the related concepts: stationarity (time-invariance) is described by the Toeplitz character, time-reversibility or causality is translated into symmetry or triangularity, convolution and correlation find their counterparts in matrix formulation.

For motivating a deeper study of Toeplitz matrices and related fast algorithms, we first indicate how the Toeplitz structure shows up in many ways in signal processing and what types of problems involving these matrices have to be solved.

1.1. *Signals and systems in matrix description*

A discrete linear time-invariant dynamical system may be defined by its impulse response h_i, with $h_i = 0$, $i < 0$ for a causal system. The output y_t to a given input u_t is given by the convolution equation

$$y_t = \sum_{i=0}^{t} h_{t-i} u_i, \quad \text{or} \quad y_t = \sum_{i=0}^{t} h_i u_{t-i}. \tag{1.2}$$

These formulas may be set into matrix and vector form according to

$$\begin{bmatrix} y_0 \\ y_1 \\ \vdots \\ y_t \end{bmatrix} = \begin{bmatrix} h_0 & 0 & \cdots & 0 \\ h_1 & h_0 & & \vdots \\ \vdots & & & 0 \\ h_t & \cdots & h_1 & h_0 \end{bmatrix} \begin{bmatrix} u_0 \\ u_1 \\ \vdots \\ u_t \end{bmatrix}, \quad \text{or} \quad y = Hu, \tag{1.3}$$

making a Triangular Toeplitz matrix (TTM), H, to appear. In this form, the causality condition yields a lower triangular matrix, the time-invariance of the impulse response is expressed by the shift of a constant column with parameters h_i. Equation (1.3) gives an alternative description of the first-order dynamical properties of the system.

If the linear system is now time-varying, a two-variables impulse response has to be considered: h_t^i, where i indicates the time of occurrence of the input, and t the time when its contribution is considered. As a consequence, we have the two equivalent formulations

$$y_t = \sum_{i=0}^{t} h_{t-i}^i u_i$$

and

$$
\begin{bmatrix} y_0 \\ y_1 \\ \vdots \\ y_t \end{bmatrix} = \begin{bmatrix} h_0^0 & 0 & \cdots & 0 \\ h_1^0 & h_0^1 & & \vdots \\ \vdots & \ddots & & 0 \\ h_t^0 & \cdots & & h_0^t \end{bmatrix} \begin{bmatrix} u_0 \\ u_1 \\ \vdots \\ u_t \end{bmatrix}. \tag{1.4}
$$

In this new expression (1.4) for a time-varying system the Toeplitz structure of the impulse-response matrix H is lost, but it is still triangular (causality).

Another important appearance of Toeplitz matrices occurs when computing the correlation matrix R of a stationary real stochastic process $\{x_t\}$ with zero mean. Let x_t be a vector containing $(p + 1)$ past samples,

$$
x_t^T = [x_t, x_{t-1}, \ldots, x_{t-p}],
$$

then, the covariance matrix is defined by

$$
R_p \triangleq E[x_t, x_t^T], \quad \text{with} \quad r_{ij} = E[x_{t-i} x_{t-j}].
$$

If the process is stationary, the correlation coefficients r_{ij} are such that

$$
r_{ij} = r_{i-j}, \quad \text{and} \quad r_i = r_{-i},
$$

and R_p is a symmetrical positive definite Toeplitz matrix.

If moreover the process under study is Gaussian, its probability density function $p(x)$ is written as

$$
p(x) = (2\pi)^{-p-1}(\mathrm{Det}\, R_p)^{1/2} \exp\left(-\tfrac{1}{2} x^T R_p^{-1} x\right), \tag{1.5}
$$

where x is any vector of $(p + 1)$ successive samples. This makes the inverse of a Toeplitz correlation matrix to appear as an important ingredient of a Gaussian probability function. The quadratic form in eq. (1.5) has to be computed in many detection and estimation problems (see sections 1.3 and 1.4). Unfortunately, the inverse of a Toeplitz matrix R is not (in general) Toeplitz. In section 2.4 we will see that, even if not Toeplitz, the inverse of R has early been recognized by statisticians as having a nice particular structure.

Along the same lines, this raises the interesting question addressed by Shamsan [3] of associating a spectrum to R^{-1}, or of computing the correlation matrix of the inverse spectrum of the process (is this matrix far from R^{-1}?).

1.2. Second-order description of dynamical systems

In many signal processing applications, a stochastic signal is filtered by passing it through a linear dynamical system. If the input is a (white) Gaussian noise, the output will also have the Gaussian property by the linear transformation (1.3). It is therefore of interest to study how the second-order statistical moments of the input are transferred to the output by a linear dynamical system. This will be realized using the impulse response, also referred to as the first-order description, or using the ARMA description, and will bring new Toeplitz matrices.

Supposing that u_t is a sequence of independent identically distributed variables,

$$u_t(\text{iid}): N(0, \sigma^2), \quad \text{with} \quad \sigma^2 = 1,$$

i.e., stationary white noise with unit power, then the covariance matrix R of the output may be easily computed from the impulse response (1.3):

$$R_t = E(yy^{\mathrm{T}}) = H^{\mathrm{T}}E(uu^{\mathrm{T}})H = H^{\mathrm{T}}H. \tag{1.6}$$

This covariance matrix is a product of TTM, but is not itself Toeplitz. The corresponding process y_t is therefore non-stationary due to the zero initial conditions in the filter. The asymptotic correlation matrix may then be determined by partitioning and rewriting eq. (1.3) as

$$
\begin{bmatrix} y_t \\ \vdots \\ y_{t-p} \end{bmatrix}
=
\begin{bmatrix}
h_0 & \cdots & h_p & \cdots & h_t \\
0 & & & & \vdots \\
\vdots & & \ddots & & \vdots \\
0 & \cdots & 0 & h_0 & \cdots & h_{t-p}
\end{bmatrix}
\begin{bmatrix} u_t \\ u_{t-1} \\ \vdots \\ u_0 \end{bmatrix}.
\tag{1.7}
$$

Multiplying eq. (1.7) by its transpose, taking the expectation, and letting t go to infinity, the output correlation matrix is then

$$
R_{\mathrm{p}} = H_\infty H_\infty^{\mathrm{T}}, \qquad
H_\infty =
\begin{bmatrix}
h_0 & h_1 & \cdots & h_p & \cdots & h_\infty \\
& h_0 & & \ddots & & \vdots \\
& & \ddots & & h_0 & \cdots & h_\infty
\end{bmatrix},
\tag{1.8}
$$

where the (Toeplitz) correlation matrix R_p is computed as a product of infinite-dimensional Toeplitz matrices constructed from the impulse-response coefficients.

In many circumstances, instead of defining the filter by its impulse response, it is more convenient, assuming a rational transfer function, to describe it as an ARMA model of order (p, p):

$$a_0 y_t + a_1 y_{t-1} + \cdots + a_p y_{t-p} = b_0 u_t + \cdots + b_p u_{t-p}. \tag{1.9}$$

The same question of computing the correlation coefficients of the output for a given correlation sequence at the input, will now be addressed in a vector form. For that purpose, the following auto- and intercorrelation are defined:

$$r_i = E(y_t y_{t-i}), \qquad v_i = E(u_t u_{t-i}), \qquad s_i = E(y_t u_{t-i}) \tag{1.10}$$

with

$$r_i = r_{-i}, \qquad v_i = v_{-i}, \qquad s_i = 0, \qquad i < 0.$$

The intercorrelations s_i are zero for negative i due to the causality property. If the input is a δ distribution or an iid sequence, then the s_i coincide with the impulse response h_i of the system. Multiplying the two sides of eq. (1.9) by y_{t+i} and by u_{t+i}, and taking the expectation, we get two sets of matrix equations in Toeplitz form:

(i) Multiplying by y_{t+i}:

$$
\begin{array}{l}
i > 0 \\
i = 0 \\
\\
\\
i = -p \\
i < -p
\end{array}
\begin{bmatrix} r_1 & r_2 & \cdots & r_{p+1} \end{bmatrix}
\begin{bmatrix}
r_0 & r_1 & \cdots & r_p \\
r_1 & r_0 & & \vdots \\
\vdots & & r_0 & r_1 \\
r_p & \cdots & r_1 & r_0
\end{bmatrix}
\begin{bmatrix} a_0 \\ a_1 \\ \vdots \\ a_p \end{bmatrix}
=
\begin{bmatrix}
s_1 & s_2 & \cdots & s_{p+1}
\end{bmatrix}
\begin{bmatrix}
s_0 & s_1 & \cdots & s_p \\
0 & s_0 & & \vdots \\
\vdots & & & s_1 \\
0 & \cdots & 0 & s_0
\end{bmatrix}
\begin{bmatrix} b_0 \\ b_1 \\ \vdots \\ b_p \end{bmatrix}.
$$

$$
\begin{bmatrix} r_{p+1} & \cdots & r_2 & r_1 \end{bmatrix} \qquad \begin{bmatrix} 0 & \cdots & 0 \end{bmatrix}
$$

$$\tag{1.11}$$

The center matrix equation (1.11) is written in closed form by

$$R \qquad a \quad = \qquad S^T \qquad b. \tag{1.12}$$

It is noted that eq. (11) may be extended outside $(0, -p)$, preserving the same Toeplitz structure.

(ii) Multiplying by u_{t+i}:

$$
\begin{matrix} i = 0 \\ \\ \\ i = -p \end{matrix}
\begin{bmatrix} s_0 & & & \\ s_1 & s_0 & & \\ \vdots & & s_0 & \\ s_p & \cdots & s_1 & s_0 \end{bmatrix}
\begin{bmatrix} a_0 \\ a_1 \\ \vdots \\ a_p \end{bmatrix}
=
\begin{bmatrix} v_0 & v_1 & \cdots & v_p \\ v_1 & v_0 & & \vdots \\ \vdots & & & v_1 \\ v_p & \cdots & v_1 & v_0 \end{bmatrix}
\begin{bmatrix} b_0 \\ b_1 \\ \vdots \\ b_p \end{bmatrix},
$$

$$
(1.13)
$$

or

$$
S \qquad a \;=\; V \qquad b \qquad (1.14)
$$

Finally, equations (1.12) and (1.14) can be gathered in a same block-Toeplitz matrix equation:

$$
\begin{bmatrix} R & S \\ S & V \end{bmatrix}\begin{bmatrix} a \\ b \end{bmatrix} = 0. \tag{1.15}
$$

The AR and MA parts may then be decoupled, assuming the correlation matrices R and V to be positive definite (invertible):

$$
b = V^{-1}Sa, \quad \text{and} \quad [R - S^T V^{-1} S]a = 0, \tag{1.16}
$$
$$
a = R^{-1}S^T b, \quad \text{and} \quad [SR^{-1}S^T - V]b = 0. \tag{1.17}
$$

An interesting special case of eq. (1.16) is the white-noise input where

$$
V = I, \quad S = H, \quad \text{and} \quad [R - HH^T]a = 0 \tag{1.18}
$$

This shows that, given first- and second-order information H and R, the AR parameters are a solution of a Toeplitz system R, perturbed by a product of two TTM matrices H. This situation will appear as a generic case in the discussion of close-to-Toeplitz matrices (see section 3.2).

Another approach to eqs. (1.11) and (1.12) and eqs. (1.13) and (1.14) is to try to eliminate the intercorrelation matrix S and get a direct relationship between R and V for some a and b parameters. This will be made by expanding eqs. (1.11) and (1.13) beyond p, making a full use of the shift property of Toeplitz matrices. For instance, eq. (1.11) may be repeated several times and gives rise to the more general system:

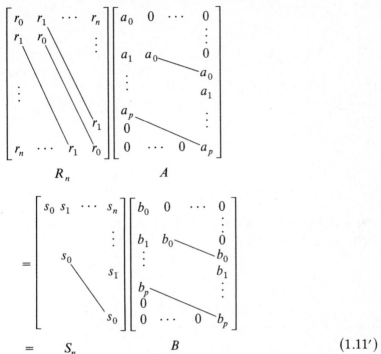

$$= \quad S_n \qquad\qquad\qquad B \qquad\qquad\qquad\qquad (1.11')$$

The matrices A and B are Toeplitz operators of a special parallelogram structure built on the coefficients a_i and b_i, with convenient dimensions. Eliminating S_n from eq. (1.11′) and the corresponding eq. (13′) leads to an important matrix relation

$$A^T R_n A = B^T V_n B, \qquad\qquad\qquad\qquad (1.19)$$

which expresses the input-output relationship for an ARMA model in terms of correlation matrices. Two cases are of interest.

(i) Strictly MA case. The output correlation matrix R_n is explicitly computed as

$$R_n = B^T V_n B. \qquad\qquad\qquad\qquad (1.20)$$

It is easily shown that, if the system is of finite order p, the resulting matrix R_n is a Toeplitz band matrix with width $(2p + 1)$.

(ii) Strictly AR case. The output correlation matrix R_n is now implicitly defined by

$$A^T R_n A = V_n. \tag{1.21}$$

Knowing the correlation coefficients of the input v_i, and the system parameters a_i, the computation of the corresponding output correlation sequence is no longer a trivial problem. We will later see that this can be efficiently achieved using linear prediction devices (see sections 5.3 and 6.5).

1.3. Empirical correlation matrix estimates

Given a relatively-short snapshot of a time series $\{x_t\}$, a safe way to preserve positive definiteness (and symmetry) of R is to build it as a product of a data matrix and its transpose. Various solutions may be chosen depending on the data length and the impact of initial and final conditions. Defining the following data matrix X, several useful partitions may be defined, depending on the window W of interest.

$$X^T =
\begin{bmatrix}
x_0 & x_1 & \cdots & x_p & \cdots & x_t & \cdots & x_N & 0 & \cdots & 0 \\
0 & x_0 & & \vdots & & \vdots & & \vdots & & & \vdots \\
\vdots & & & & & & & & & & 0 \\
0 & \cdots & 0 & x_0 & \cdots & x_{t-p} & \cdots & & x_{N-p} & \cdots & x_N
\end{bmatrix}$$

$$\tag{1.22}$$

The corresponding estimate of R will be defined by the empirical covariance of the columns of X for each window W:

$$\hat{R}_W = X_W^T X_W. \tag{1.23}$$

(i) $W = 0$, $N + p$ (Correlation case of linear prediction). The corresponding R_{COR} matrix is Toeplitz by construction. If the number of samples N is relatively large ($\sim 10\,p$), then the initial ($t \sim 0$) and final conditions ($t \sim N$) can be neglected and R_{COR} is a reasonable estimate of the correlation matrix. If the number of samples is small, then R_{COR} is still Toeplitz but may be a poor estimate.

(ii) $W = p, N$ (Covariance case of linear prediction). The corresponding R_{COV} is not Toeplitz (but is the product of two Toeplitz matrices). The initial and final conditions (unmeasured data replaced by zeros in eq. 24) are now excluded from the estimate. As a consequence R_{COV} is, in general, more reliable for small numbers of samples ($\sim 3p$).

(iii) $W = 0, t$ (Prewindow case of linear prediction). An interesting intermediate includes initial conditions but no final condition triangle. The corresponding R_{PRW} is not Toeplitz but has a simple structure (it is a Toeplitz matrix corrected by a product of TTM from final conditions). It is the natural choice for time-recursive implementations and converges to the Toeplitz correlation matrix when $t \rightarrow \infty$.

A related approach can be found in the book by T.W. Anderson [5] in his "method of quadratic forms". Many detection or estimation problems are linear in terms of correlation coefficients r_i. It is therefore interesting to estimate r_i as a scalar by a quadratic form in terms of the given samples $\{x_i\}$, $i = 0, N$:

$$r_i = x^T Q_i x, \tag{1.24}$$

where the matrix Q_i may be taken as

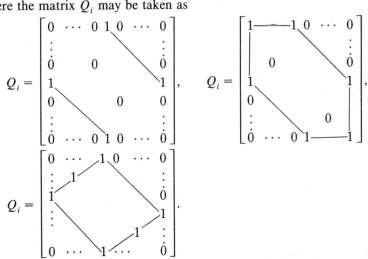

In all these Q_i matrices the two diagonals of ones are at a distance i from each other. An interesting feature about the Q_i is that they are all described as polynomials of the corresponding matrix Q_1 only, and are therefore diagonal in the same basis. This concept may be extended to matrices: the different choices of Q_i amount to variations on the initial and final conditions. The corresponding matrix R is given by

$$R = X^T X, \quad \text{with} \quad X^T = \begin{bmatrix} x_0 & \cdots & x_N \\ & x_0 & \cdots & x_N \end{bmatrix},$$

where the two triangular matrices are repeated values of $(x_0, \ldots, x_p)$ and $(x_N, \ldots, x_{N-p})$, corresponding to periodic extensions with, or without, time reversal.

Estimation of structured correlation matrices.
An alternative way of deriving a correlation matrix for spectral estimation or other purposes, is to deduce the structure and value of the best estimate of this matrix from a given error criterion. For this purpose, a distance from a measured R (empirical correlation matrix for instance), and a model M selected from a given class, has to be defined: Mean Squares (MS), Maximum likelihood (ML), Divergence (JD), ... In many cases, and, at least for large samples, the structure of the estimate can be chosen to be Toeplitz. This gives rise to the problem of estimating a patterned correlation matrix, addressed by Anderson [6], and more recently by Burg et al. [7].

The MS criterion for matrices is the weak (Hilbert–Schmidt) norm of the difference between the measured matrix R and its Toeplitz model M,

$$d_{\mathrm{MS}}(R, M) = \mathrm{Tr}\left[(R - M)(R - M)^T\right]. \tag{1.25}$$

It amounts to the sum of squares of the error elements in $(R - M)$. It is easily shown that the best Toeplitz estimate of R is obtained by averaging the elements of R along the various diagonals:

$$m_k = \frac{\left(\sum_{i=0}^{p-k} r_{i, i+k}\right)}{(p + 1 - k)}.$$

Assuming the underlying process $\{x_i\}$ to be Gaussian, with covariance matrix M, and, moreover, assuming the n observed $(p + 1)$-dimensional vectors to be independent (rather rough assumption for a time series), the ML distance is expressed by

$$d_{\mathrm{ML}}(R, M) = \log(\mathrm{Det}\, M) + \mathrm{Tr}[M^{-1}R], \tag{1.26}$$

where R is the empirical covariance matrix and M is constrained to be Toeplitz:

$$R = \sum_0^N x_i x_i^T, \quad \text{and} \quad M = \sum_{i=0}^P m_i Q_i.$$

The above minimization problem has no explicit solution. It has been studied in some depth by Lecadre et al. [8] who took a steepest-descent approach. Weak and strong convergence conditions have been exhibited, including the positive definiteness constraint for the solution.

Note. As for the optimal structure of the model matrix M, there is no general evidence of a Toeplitz constraint to be optimal in finite dimension. This is due to the fact that matrix norms (distances) can be translated into eigenvalue properties, but in turn, as will be stressed below in section 2.5, the Toeplitz structure does not imply special features for the eigenvalues.

1.4. Spectral estimation in matrix form

Spectral estimation is one of the most common and powerful methods in signal processing. Assume an underlying stationary stochastic process $\{x_t\}$ with zero mean, (true) correlation coefficients r_i $i = 0, \infty$ and a corresponding spectrum $S(\varphi)$. For a given finite snapshot $[x_0, x_1, \ldots, x_N]$ of this process, the problem is to build a reasonable estimate $\hat{S}(\varphi)$ of the spectrum. The problem can be solved by estimating some convenient correlation coefficients $\hat{r}_i$, or directly on a first rough estimate of the spectrum $P_n(\varphi)$ (periodogram). In both cases, we will show that Toeplitz forms come into play.

Spectral estimation by quadratic forms I.
A convenient estimator of the spectrum for a truncated data sequence is the average periodogram.

The periodogram is defined in vector form by (square modulus of the sample Fourier transform)

$$P_n(\varphi) = e^*(xx^T)e, \tag{1.27}$$

with

$$x^T = [x_0, x_1 \ldots, x_n], \quad \text{and} \quad e^T = [1, e^{j\varphi}, \ldots, e^{jn\varphi}].$$

The ensemble average of $P_n(\varphi)$ is therefore

$$S_{AP}(\varphi) = E[P_n(\varphi)] = e^*R_n e, \tag{1.28}$$

and $S_{AP}(\varphi)$ is therefore a quadratic form in the Toeplitz matrix R_n. Many other well-known estimators (maximum likelihood, maximum entropy, etc.) may also be expressed as ratios of quadratic forms comprising Toeplitz matrices (or their inverses):

$$S_{\mathrm{ML}} = \frac{1}{eR^{-1}e}, \qquad S_{\mathrm{ME}} = \frac{a^{\mathrm{T}}Ra}{e*aa^{\mathrm{T}}e} \cdots S_{\mathrm{LA}} = \frac{e*R^{q}e}{e*R^{q-1}e}.$$

As a consequence, an attractive idea is to replace R in one of the previous definitions by some estimate $\hat{R}$. Such an estimate is also of outstanding importance in other problem including linear prediction (see section 1.4).

A naive solution for estimating $\hat{R}$ would be to estimate the correlation coefficients r_i individually (and for that, we have many biased, unbiased solutions) and plug them into a Toeplitz structure. Unfortunately in this case, the positive-definite property will not be ensured in general for $\hat{R}$ and this may result in a non-positive spectrum estimate $\hat{S}$. Consequently, the R matrix should be estimated as a whole, according to one of the previous schemes of section 1.3.

Spectral estimation by quadratic forms II
In the definition of the periodogram (1.27), the role of the sample vector x and of the frequency vector e (wave number) may be interchanged:

$$P_n(\varphi) = e*(xx^{\mathrm{T}})e = x^{\mathrm{T}}(ee*)x = x^{\mathrm{T}}E(\varphi)x. \tag{1.29}$$

The idea of estimating $S(\varphi)$ as a quadratic form in frequency, follows and leads to define

$$\hat{S}(\varphi) = x^{\mathrm{T}}Q(\varphi)x, \tag{1.30}$$

where $Q(\varphi)$ is an Hermitian matrix to be chosen.

A good way to select $Q(\varphi)$ is to evaluate and minimize the bias and variance of such an estimator, as advocated by Clergeot [9].

Assuming a Gaussian process and recalling that

$$E[x^{\mathrm{T}}Qx] = E[\mathrm{Tr}(Qxx^{\mathrm{T}})] = \mathrm{Tr}(QR),$$

the following values are easily obtained:

$$E[\hat{S}] = \mathrm{Tr}(QR),$$

$$\mathrm{Var}[\hat{S}] = \mathrm{Tr}[(QR)^{2}].$$

Using the orthonormal matrix U:

$$\Lambda = \mathrm{Diag}\,\lambda_i = URU^{\mathrm{T}}, \qquad \tilde{Q} = UQU^{\mathrm{T}},$$

the bias and variance are expressed as

$$E[\hat{S}] = \text{Tr}(QR) = \sum_i \tilde{q}_{ii}\lambda_i,$$

$$\text{Var}[\hat{S}] = \text{Tr}(QR)^2 = \sum_i \left(\tilde{q}_{ii}^2\lambda_i^2 + \sum \mu_i\mu_j\tilde{q}_{ij}^2 \right).$$

It then follows that the variance is minimized by setting $q_{ij}^2 = 0$ and the orthonormal matrix U diagonalizes R and Q simultaneously for all values of φ.

Moreover, seeking for an unbiased (or fixed bias) estimate of $S(\varphi)$, leads to set

$$E[S(\varphi)] = S(\varphi) = \sum_{i=0}^{n-1} \tilde{q}_{ii}\lambda_i,$$

where n is the rank of Q (N is the sample size).

Minimizing the variance under this constraint, provides the Lagrangian

$$\mathscr{L} = \sum_{i=0}^{n-1} \tilde{q}_{ii}^2\lambda_i^2 - \mu\left[S(\varphi) - \sum_{i=0}^{n-1} \tilde{q}_{ii}\lambda_i \right],$$

the optimality conditions are

$$\mu = \frac{2S(\varphi)}{n}, \quad \text{and} \quad \tilde{q}_{ii} = \frac{S(\varphi)}{n\lambda_i},$$

and, therefore, the optimal matrix Q^* has to be proportional to R^{-1}.

For a stationary time-series with R being Toeplitz, its inverse is generally non-Toeplitz. However, for a large enough time-window (compared to the correlation time), R^{-1} is asymptotically Toeplitz (see section 2.3) and a Toeplitz quadratic form should be selected for $Q(\varphi)$. Coming back to the variance the optimized value is

$$\text{Var}[\hat{S}(\varphi)] = \text{Tr}[(Q^*R)^2] = \sum_{i=0}^{n-1} \frac{S^2(\varphi)}{n^2} = \frac{S^2(\varphi)}{n},$$

and it is then advisable to let the rank n of $Q(\varphi)$ increase with N. (This is not the case for the rough periodogram where $n = 1$.)

1.5. Linear prediction and other modelling problems

Instead of trying to estimate the spectrum directly from the data, another approach consists in inferring from the measures a time domain model of the generating process of the signal. This takes back the problem in the domain of system identification, but with some particular features (e.g. no access to the input excitation signal).

The most celebrated technique along these lines is the linear prediction procedure [10]. The model is a simple order p autoregressive (AR) system on the observed output

$$a_0 y_t + a_1 y_{t-1} + \cdots + a_p y_{t-p} = u_t, \qquad (1.31)$$

where the unknown input u_t is supposed to be a zero-mean sequence of iid variables. Choosing $a_0 = 1$, the model may be turned into a one-step-ahead linear predictor, based on the last p past samples,

$$\hat{y}_t = - \sum_i a_i y_{t-i}. \qquad (1.32)$$

A convenient error value ε_t is then defined as

$$\varepsilon_t = y_t - \hat{y}_t, \qquad (1.33)$$

and ε_t is nothing but an estimate of the noisy input in (1.31). This error is evaluated on a given time window W by a mean-squares criterion,

$$MS = \sum_{t \in W} \varepsilon_t^2, \qquad (1.34)$$

which is then minimized with respect to the unknown a_i.

The problem is cast in vector form using the above Toeplitz matrix Y.

$$
\begin{bmatrix}
y_0 & 0 & \cdots & 0 \\
 & & & \vdots \\
\vdots & & & 0 \\
y_p & \cdots & & y_0 \\
\vdots & & & \vdots \\
y_t & \cdots & & y_{t-p} \\
\vdots & & & \vdots \\
y_N & \cdots & & y_{N-p} \\
0 & & & \vdots \\
\vdots & & & \\
0 & \cdots & 0 & y_N
\end{bmatrix}
\begin{bmatrix}
a_0 \\
a_1 \\
\vdots \\
a_p
\end{bmatrix}
=
\begin{bmatrix}
\varepsilon_0 \\
\vdots \\
\varepsilon_p \\
\vdots \\
\varepsilon_t \\
\vdots \\
\varepsilon_N \\
\vdots \\
\varepsilon_{N+p}
\end{bmatrix}, \qquad (1.35)
$$

$$Y \qquad\qquad a \;=\; \varepsilon.$$

This leads to a classical quadratic minimization problem with constraint,

$$\underset{a}{\mathrm{Min}}\,(a^{\mathrm{T}} R a), \quad \text{with} \quad a_0 = 1, \quad R = Y^{\mathrm{T}} Y. \qquad (1.36)$$

Under this general framework, many variation of the problem may be discussed:

(i) Time-window. The MS criterion (1.34) may be defined for various time-windows W, depending on the relative sizes of p and N. They will give rise for matrix R to the different correlation matrix estimates, already discussed in section 1.3. Depending on the initial and final conditions, it will be close to a Toeplitz matrix or not, and will be a good or bad estimate of the true correlation matrix of $\{ y_t \}$.

(ii) Normalization. In the classical definition of a linear predictor, the normalization imposed on vector a is $a_0 = 1$. This constraint is, by no means, the only reasonable normalization, and many others, either linear or quadratic, may be considered:

Linear constraint.

$$\text{Min} (a^\mathrm{T} R a), \quad \text{with} \quad u^\mathrm{T} a = 1, \tag{1.37}$$
$$\phantom{\text{Min}}_{a}$$

where u is a given vector. The following classical problems fall under this case.

Forward one-step-ahead
predictor: $u^\mathrm{T} = [1, 0, \ldots, 0]$, $u^\mathrm{T} a = 1$,
Backward one-step: $u^\mathrm{T} = [0, \ldots, 0, 1]$, $u^\mathrm{T} a = 1$,
ith position smoother: $u^\mathrm{T} = [0, \ldots, 0, 1, 0, \ldots 0]$, $u^\mathrm{T} a = 0$,
q-steps-ahead predictor: $u^\mathrm{T} = [0, 1, \ldots, 1, 0, \ldots, 0]$, $u^\mathrm{T} a = 0$.

This type of constraint will lead to a general type of normal equations, associated to the Choleski factorization of R.

Quadratic constraint.

$$\text{Min} (a^\mathrm{T} R a) \quad \text{with} \quad a^\mathrm{T} W a = 1, \tag{1.38}$$
$$\phantom{\text{Min}}_{a}$$

where W is a given weight matrix.

Such a quadratic constraint has an energetic meaning and appears in two basic situations:

–Noise cancellation: $\text{Min}_a [a^\mathrm{T} (R - \sigma^2 W) a]$.

W is the correlation matrix of the noise, with unknown SNR. This leads to the Prony–Pisarenko type of techniques.

–Matched filter: $\text{Min}_a [a^T R a / a^T W a]$.

W is the correlation of the signal to be rejected. A discriminant analysis (SNR, rayleigh quotient, likelihood, ...) criterion is extremized. This type of constraint will lead to generalized eigenvalue problems, associated to the Karhunen–Loeve expansion of R.

Choleski factorization of R.
If we first consider the linear constraint (1.37) on the parameters, the solution is obtained through the Lagrangian

$$\mathscr{L} = a^{\mathrm{T}} R a + \mu(1 - a^{\mathrm{T}} u).$$

Setting the derivatives to zero:

$$2 R a + \mu u, \quad \text{and} \quad \mu = \tfrac{1}{2}(a^{\mathrm{T}} R a),$$

two cases are of particular interest, for special values of u. The forward linear predictor is obtained by imposing $a_0 = 1$, and leads to the well-known normal or Yule–Walker equations

$$
\begin{bmatrix}
r_0 & r_1 & \cdots & r_p \\
r_1 & r_0 & & \vdots \\
\vdots & & \ddots & r_1 \\
r_p & \cdots & r_1 & r_0
\end{bmatrix}
\begin{bmatrix}
a_0 \\ a_1 \\ \vdots \\ a_p
\end{bmatrix}
=
\begin{bmatrix}
\alpha_p \\ \vdots \\ 0 \\ \vdots \\ 0
\end{bmatrix}. \tag{1.39}
$$

The successive order increasing forward predictors produce the following Choleski factorization:

$$
\begin{bmatrix}
r_0 & r_1 & \cdots & r_p \\
r_1 & r_0 & & \vdots \\
& & & r_1 \\
r_p & \cdots & r_1 & r_0
\end{bmatrix}
\begin{bmatrix}
a_0^p & & & \\
\vdots & & A & \ddots \\
a_p^p & & \cdots & a_0^0
\end{bmatrix}
$$

$$
=
\begin{bmatrix}
s_0^p & \cdots & s_p^0 \\
 & H^{\mathrm{T}} & \vdots \\
 & \ddots & \\
 & & s_0^0
\end{bmatrix}, \tag{1.40}
$$

where $a_0^i = 1$ and $s_0^i = \alpha_i$.

Taking into account the symmetry and positive-definiteness of R,

$$RA = A^{-T}D, \quad \text{where} \quad D = \text{Diag}[\alpha_i]. \tag{1.41}$$

This qualifies $(AD^{-1/2})$ as the lower Cholesky factor of R^{-1}, while $(A^{-T}D^{1/2})$ is the upper Cholesky factor of R.

The backward linear predictor, associated to the constraint $a_p = 1$, will produce the lower-upper (LU) factorization of R, instead of the above UL factorization,

$$RB = B^{-T}D,$$

where B is an upper triangular matrix of successive order backward predictors.

Moreover, if R is a symmetric real Toeplitz matrix, it is also centro-symmetric:

$$R = JRJ, \quad \text{where} \quad J = \begin{bmatrix} 0 & & \cdots & 0 & 1 \\ \vdots & & & & 0 \\ 0 & 1 & & & \vdots \\ 1 & 0 & & \cdots & 0 \end{bmatrix}. \tag{1.42}$$

In this case there is a simple relationship between the forward and backward predictors:

$$B = JAJ, \quad \text{or} \quad b = Ja.$$

Karhunen–Loeve factorization of R.
If we now consider the quadratic constraint (1.38), the extremization of the Lagrangian

$$\mathcal{L} = a^T Ra + \mu(1 - a^T Wa)$$

leads to the generalized eigenvalue problem

$$(R - \mu W)a = 0,$$

and the optima a are the eigenvectors of the pencil of matrices (R, W).

A special case of interest is the case where $W = I$, and a is the eigenvector associated to the minimum eigenvalue of R (Pisarenko,

Prony). This extremization problem is connected to the Karhunen–Loeve expansion of R. Let U be the orthonormal matrix of the eigen vectors u_i, we then have the factorization

$$R = U\Lambda U^T, \quad \text{where} \quad U\Lambda U^T = \text{Diag}[\lambda_i]. \tag{1.43}$$

If, moreover, R is a Toeplitz matrix with distinct eigenvalues, then the centro-symmetry property (1.42) implies that

$$u_i = \pm J u_i,$$

and the eigenvectors of R are either symmetrical or anti-symmetrical.

Note. The two above factorization problems are very fundamental in signal processing. They are two ways of obtaining uncorrelated data from a given correlated process by means of linear transformations. In other words, they define linear, but perhaps time-varying filters, which whiten the corresponding given process. In the case of Cholesky factorization, a causality constraint is imposed and the resulting transformed sequence is the innovation sequence. In the case of Karhunen–Loeve factorization, an orthonormal constraint is imposed (which results into a linear-phase filter in the stationary Toeplitz case).

The two factorization problems can be related taking into account the fact that in case of a deterministic signal, the prediction error energy α is zero and the linear predictor is also the eigenvector of R associated to the zero eigenvalue. This provides some ways to approach the computation of eigenvectors of Toeplitz matrices through linear prediction fast algorithms.

References

[1] O. Toeplitz, Zur Theorie der quadratischen Formen und bilinearen Formen von unendlichvielen Veränderlichen. I. theorie der L-formen, Mathematiche Annalen 70, (1911) 351–376.

[2] G. Frobenius, Uber das Trägheitsgesetz der quadratischen Formen, S.B. Deutsch Akad, Wiss. Berlin Math. Nat. Kl. (1896) 7–16.

[3] P. Shaman, An approximate inverse for the covariance matrix of moving average and autoregressive processes, The Annals of Statistics 3-2 (March 1975) 532–538.

[4] P. Shaman, Approximations for stationary covariance matrices and their inverses with application to ARMA models, The Annals of Statistics 4-2 (March 1976) 292–301.

[5] T.W. Anderson, Serial correlations, in: The Statistical Analysis of Time-Series, Series in Probability and Mathematical Statistics (Wiley, New York, 1958) ch. 6.

[6] T.W. Anderson, Asymptotically efficient estimation of covariance matrices with linear structure, The Annals of Statistics 1-1 (January 1973) 135–141.

T.N. Anderson, Statistical inference for covariance matrices with linear structure, in: Multivariate Analysis II, ed. P.R. Krishnaiah WSV, Dayton (Academic Press, New York, 1969).

M.E. Shaikin, Optimal estimates for covariance matrices, Auto. & Remote Control 34-1/1 (January 1973) 53–60.

[7] J.P. Burg, D. Luenberger and D. Wenger, Estimation of structured covariance matrices, Proc. IEEE 70-9 (September 1982) 863–974.

R. Nitzberg, An alternative derivation of the maximum likelihood of a covariance matrix, Proc. IEEE, 63-8 (November 1975) 16123–1624.

[8] J.P. Lecadre, P. Lopez, Estimation d'une matrice interspectrale de structure imposée. Applications, Traitement du Signal 1-1 (October 1984) 3–17.

[9] H. Clergeot, Choix entre différentes méthodes quadratiques d'estimation spectrale, Ann. Télécommun. 39-3/4 (1984) 113–128.

[10] J. Makhoul, Linear prediction; a tutorial review, Proc. IEEE 63-4 (April 1975) 561–580.

A. Papoulis, Maximum entropy and spectral estimation; a review, IEEE Trans. Acoust. Speech & Signal Process. 29-6 (December 1981) 1176–1186.

J.A. Cadzow, Spectral estimation: an overdetermined rational model equation approach, Proc. IEEE 70-9 (September 1982) 907–939.

2. Some properties of Toeplitz matrices

Toeplitz matrices arise in many signal and system related problems. They provide alternative descriptions of properties that can be approached by other formulations (polynomials, geometric projections). However, at the same time, they have specific properties as matrices [1, 2].

2.1. Polynomials and Toeplitz matrices

The usual analysis of linear systems make a wide use of frequency domain transforms, defining rational transfer functions for describing their dynamical behavior. There are strong connections between the theories of matrices and of rings of polynomials (minimum polynomials, canonical forms, etc.). The connection is even stronger with Toeplitz matrices, and there is an isomorphism between usual algebra on polynomials and Toeplitz matrices.

Let $A(z)$ and $B(z)$ be two polynomials (in z^{-1}) with degrees p and q, then the resulting product $C(z)$ of degree $(p + q)$ can be expressed by arranging the coefficients in columns, using the Toeplitz matrix A,

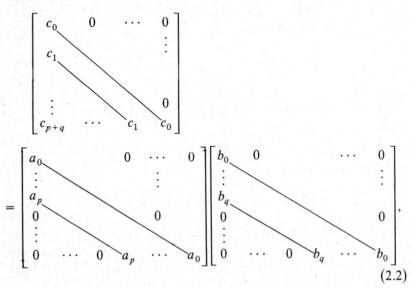

$$
\begin{bmatrix} c_0 \\ c_1 \\ \vdots \\ c_q \\ \vdots \\ c_p \\ \vdots \\ c_{p+q} \end{bmatrix}
=
\begin{bmatrix} a_0 & & & \\ a_1 & a_0 & & \\ \vdots & & a_0 & \\ a_p & & a_1 & \\ & & & \vdots \\ & & & a_p \end{bmatrix}
\begin{bmatrix} b_0 \\ b_1 \\ \vdots \\ b_q \end{bmatrix},
\tag{2.1}
$$

or

$$c = Ab.$$

In eq. (2.1), there is no point in giving a particular role to a column description and the equation can be restated in rows. Mixing the two descriptions, make triangular Toeplitz matrices (TTM) to appear as a useful device:

$$
\begin{bmatrix}
c_0 & 0 & \cdots & 0 \\
c_1 & & & \vdots \\
& & & 0 \\
\vdots & & & \\
c_{p+q} & \cdots & c_1 & c_0
\end{bmatrix}
$$

$$
=
\begin{bmatrix}
a_0 & & 0 & \cdots & 0 \\
\vdots & & & & \vdots \\
a_p & & & & 0 \\
0 & & & & \\
\vdots & & & & 0 \\
0 & \cdots & 0 & a_p & \cdots & a_0
\end{bmatrix}
\begin{bmatrix}
b_0 & 0 & & \cdots & 0 \\
\vdots & & & & \vdots \\
b_q & & & & 0 \\
0 & & & & \\
\vdots & & & & 0 \\
0 & \cdots & 0 & b_q & \cdots & b_0
\end{bmatrix},
\tag{2.2}
$$

or

$$\mathscr{C}(c) = \mathscr{C}(A) \cdot \mathscr{C}(B),$$

where the matrices have convenient sizes (extended by zeros). It is noted that commutativity of polynomials thus imply commutativity of TTMs: $\mathscr{C}(\cdot)$. More precisely, the inverse of a (lower) TTM being also TTM, the matrices $\mathscr{C}(\cdot)$ form a field. This is associated to the modulo polynomial equation

$$[A(z) \cdot B(z)]\operatorname{Mod} z^{-n} = 1,$$

where $A(z)$ is a polynomial and $B(z)$ is the series expansion $A^{-1}(z)$ truncated at degree n

Examples.

$$\mathscr{C}(A) = \begin{bmatrix} 1 & & 0 & & \cdots & & 0 \\ -\alpha & & & & & & \vdots \\ 0 & & & 1 & & & \\ \vdots & & & -\alpha & & & 0 \\ 0 & & \cdots & & 0 & -\alpha & 1 \end{bmatrix},$$

$$\mathscr{C}^{-1}(A) = \begin{bmatrix} 1 & 0 & & \cdots & & 0 \\ \alpha & 1 & & & & \vdots \\ \alpha^2 & \alpha & & & & \\ \vdots & & & & & 0 \\ \alpha^n & \cdots & & \alpha^2 & \alpha & 1 \end{bmatrix}. \tag{2.3}$$

Many other polynomial properties can be transposed into vector and matrix form. An important instance is the inner product of $A(z)$ and $B(z)$, expressed in Toeplitz form

$$\langle A(z), B(z)\rangle = \frac{1}{2\pi j}\int A(a)R(z)B(z^{-1})\frac{\mathrm{d}z}{z} = a^{\mathrm{T}}Rb,$$

which makes the connection between the orthogonal (Szegö) polynomial formulation of linear prediction and Toeplitz matrices [3].

Example. Another example of interest is the Bezout identity for relatively prime polynomials $A(z)$ and $B(z)$ [4, 5].

$$A(z)U(z) + B(z)V(z) = 1.$$

This identity, translated in matrix form, gives

$$\mathscr{C}(A) \cdot \mathscr{C}(U) + \mathscr{C}(B)\mathscr{C}(V) = \mathscr{C}(1),$$

which is known as the resolvant, or Sylvester identity.

2.2. Circulant and k-circulant matrices

A special, and important, case of a Toeplitz matrix is the circulant matrix defined by [6]

$$
c = \begin{bmatrix} c_0 & c_n & & \cdots & c_1 \\ c_1 & c_0 & c_n & & \vdots \\ c_2 & c_1 & & c_0 & \\ \vdots & & & & c_n \\ c_n & & \cdots & c_1 & c_0 \end{bmatrix}, \quad
Q_1 = \begin{bmatrix} 0 & & \cdots & 0 & 1 \\ 1 & & & & \\ 0 & & & & \vdots \\ \vdots & & & & \\ 0 & \cdots & & 0 & 1 & 0 \end{bmatrix}. \quad (2.4)
$$

It is easily checked that the eigenelements of C are given by

$$x_m^{\mathrm{T}} = [1, e^{j\varphi}, \ldots, e^{jn\varphi}],$$

$$\mu_m = \sum_{i=0}^{n} c_i e^{ji\varphi}, \qquad \text{where} \quad \varphi = -\frac{2m\pi}{(n+1)}.$$

In particular, the eigenvalue μ_m is recognized as the DFT of the sequence c_i. A straightforward generalization of eq. (2.4) is the k-circulant matrix, where the upper triangular corner in C is weighted by a complex coefficient k. Correspondingly, the isolated 1 in Q is replaced by k. These matrices are characterized by

$$C_k Q_k = Q_k C d k.$$

The eigenvectors of C_k are still the same x_m, associated with

$$\mu_m = \sum_i c_i (k)^{i/n+1} e^{ji\varphi},$$

which corresponds to the DFT of the exponentially weighted sequence.

It is possible to show that k-circulant matrices form a field (even the pseudo-inverse is k-circulant if $|k| = 1$). An even more specialized case is the TTM which is obtained for $k = 0$.

The most important result about circulant matrices in our context, is an asymptotic equivalence property between Toeplitz and circulant matrices.

2.3. Asymptotic properties of Toeplitz matrices

Given a Toeplitz matrix with elements $t_0, \ldots, t_n$, it may be extented by zeros for $n < i < N$ to provide a band Toeplitz matrix T_N.

$$
T_N = \begin{bmatrix}
t_0 & t_1 & \cdots & t_n & 0 & \cdots & 0 \\
t_1 & & & & & & \vdots \\
\vdots & & & & & & 0 \\
t_n & & & & & & t_n \\
0 & & & & & & \\
\vdots & & & & & & \vdots \\
& & & & & & t_1 \\
0 & \cdots & 0 & t_n & \cdots & t_1 & 0
\end{bmatrix}.
\tag{2.5}
$$

Correspondingly, a circulant matrix C_N may be constructed:

$$
C_N = \begin{bmatrix}
t_0 & \cdots & t_n & 0 & t_n & \cdots & t_1 \\
\vdots & & & & & & t_n \\
t_n & & & & & & 0 \\
0 & & & & & & t_n \\
t_n & & & & & & \vdots \\
\vdots & & & & & & \\
t_1 & \cdots & t_n & 0 & t_n & \cdots & t_0
\end{bmatrix}.
\tag{2.5'}
$$

Defining the weak norm as a distance between sequences of (supposed uniformly bounded) matrices, it is shown that [7]

$$
\lim_{N \to \infty} |T_N - C_N| \to 0.
$$

As a consequence, all continuous functions of the eigenvalues λ_i^N of T_N will converge to the same continuous function of the eigenvalues μ_i^N of C_N. The latter eigenvalues and eigenvectors being known, many interesting asymptotic results relating the spectrum $S(\varphi)$ and the corresponding infinite Toeplitz correlation matrix were deduced by Szegö [3]:

$$
S(\varphi) = \sum_{-N}^{+N} r_i e^{ji\varphi}, \qquad r_i = \frac{1}{2\pi} \int_{-\pi}^{+\pi} S(\varphi) e^{-ji} d\varphi.
\tag{2.6}
$$

Let R_N be the correlation matrix with eigenvalues λ_i^N, and

$$
\operatorname{Tr} R_N = \sum \lambda_i^N, \qquad \operatorname{Det} R_N = D_N = \pi \lambda_i^N,
$$

then the following properties hold [8, 9]:

$$\lim_{N \to \infty} \frac{1}{N} \sum (\lambda_i^N)^k = \frac{1}{2\pi} \int_{-\pi}^{+\pi} S^k(\varphi) \, d\varphi, \tag{2.7}$$

$$\lim_{N \to \infty} \frac{D_N}{D_{N-1}} = \exp\left[\frac{1}{2} \int_{-\pi}^{+\pi} S(\varphi) \, d\varphi\right] \lim_{N \to \infty} \frac{D_N}{N} = \frac{1}{2} \int_{-\pi}^{+\pi} \log S(\varphi) \, d\varphi. \tag{2.8}$$

Moreover, arranging the eigenvalues in increasing value order, we have

$$m < \lambda_0^N < \lambda_1^N \cdots < \lambda_N^N < M,$$

where $m = \text{Min}_\varphi \, S(\varphi)$, and $M = \text{Max}_\varphi \, S(\varphi)$,

$$\lambda_0^N \to m, \quad \text{and} \quad \lambda_N^N \to M, \quad \text{as } N \to \infty. \tag{2.9}$$

Roughly speaking, asymptotically, the eigenvalues of R_∞ behave like the spectrum, while the eigenvalues tend to DFT vectors with an infinite number of roots equally spaced on the unit circle.

Unfortunately, the above results are not useful for the finite-dimensional cases that appear in most of the problems.

2.4. Toeplitz matrix inversion formulas

From the above results, it appears that the inverse of a Toeplitz matrix is asymptotically Toeplitz. Even if this is no longer the case for finite-dimensional matrices, it has been recognized that the inverse matrix had interesting properties. This has been proven by Siddiqui [10] and Durbin [11] using a stochastic argument, but the result is algebraically true.

Consider the stochastic time series $\{y_t\}$ generated by an AR model of order p with coefficients a_i, when excited by an iid sequence u_t. A given sample y_N may be segmented into

$$y_N^T = [y_0, y_1, \ldots, y_p | y_{p+1}, \ldots, y_N] = [y_0^T, y_1^T].$$

For a Gaussian input $u_t \sim N(0, \sigma^2)$, the output density is

$$p(y_N) = k(\text{Det } R_N)^{-1} \exp - y_N^T R_N^{-1} y_N,$$

using the conditional probabilities by Bayes rule,

$$p(y_N) = p(y_0/y_1)p(y_1) = p(y_1/y_0)p(y_0). \tag{2.10}$$

In eq. (2.10) the partial densities may be expressed as

$$p(y_0) \sim y_0^T R_p^{-1} y_0,$$

$$p(y_1/y_0) \sim \frac{1}{\sigma^2} u^T I u,$$

where

$$\boldsymbol{u} = \begin{bmatrix} 0 & & a_p & \cdots & a_0 & 0 & 0 \\ 0 & 0 & & a_p & \cdots & a_0 & 0 \\ 0 & & & 0 & a_p & \cdots & a_0 \end{bmatrix} y_N,$$

and, as a consequence,

$$R_N^{-1} = \begin{bmatrix} 0 & \cdots & & 0 \\ a_p & & & 0 \\ \vdots & & & \\ a_0 & & & a_p \\ 0 & & & \\ \vdots & & & \vdots \\ 0 & \cdots & 0 & a_0 \end{bmatrix}^2 - \begin{bmatrix} 0 & & \cdots & & 0 \\ & & & & \vdots \\ \vdots & & & & \\ & & 0 & \cdots & 0 \\ & & \vdots & R_p^{-1} & \\ 0 & \cdots & 0 & & \end{bmatrix},$$

$$\tag{2.11}$$

where $[\cdot]^2$ indicates the product of the matrix $[\cdot]$ by its transposed.

The same procedure applies for the left-hand side of (2.10) and equating the two corresponding expressions (2.11) for R_N^{-1}, the following (Gohberg–Semencul)

$$R_p^{-1} = \begin{bmatrix} a_0 & & & \\ a_1 & a_0 & & \\ \vdots & & & \\ a_p & \cdots & & a_0 \end{bmatrix}^2 - \begin{bmatrix} 0 & & \cdots & 0 \\ a_p & & & \vdots \\ \vdots & & & \\ a_1 & \cdots & a_p & 0 \end{bmatrix}^2, \tag{2.12}$$

where the predictor is normalized, $a_0 = 1/\sigma$.

Any inverse of a Toeplitz matrix may then be expressed using TTMs built on the coefficients of the corresponding (normalized) linear predictor. This result will be generalized in section 3.4 for close-to-Toeplitz matrices.

2.5. Eigenvectors of Toeplitz matrices

Given a symmetric real Toeplitz matrix R, the centro-symmetry property has already been noted [16, 17]:

$$R = JRJ,$$

where J is the order reversal operator ($J^2 = I$). This implies some particular properties for the Karhunen–Loève expansion of R, defined in terms of eigenvectors u_i (assuming simple eigenvalues),

$$Ru_i = \lambda_i u_i, \qquad R = U\Lambda U^T, \qquad (2.13)$$

where U is the orthonormal matrix of the eigenvectors.

Structural properties [18–24].
Supposing the dimension of R to be even ($n = 2p + 1$) to simplify the notation, the matrix may be partitioned as

$$R_n = \begin{bmatrix} T & JH \\ HJ & T \end{bmatrix}, \quad \text{where} \quad \begin{array}{l} T \text{ is Toeplitz on } \quad r_0, \ldots, r_p, \\ H \text{ is Hankel on } \quad r_1, \ldots, r_{p+1}. \end{array}$$

We have the following decomposition:

$$\begin{bmatrix} I & J \\ J & -I \end{bmatrix}\begin{bmatrix} T & JH \\ HJ & T \end{bmatrix}\begin{bmatrix} I & J \\ J & -I \end{bmatrix} = \begin{bmatrix} J(T + H)J & 0 \\ 0 & (T - H) \end{bmatrix}, \qquad (2.14)$$

and the eigenvectors of R may be deduced by the inverse transformation from those of $(T + H)$ or of $(T - H)$. As previously announced, these eigenvectors have the property

$$Rx = \lambda x, \quad \text{with} \quad x = \pm Jx. \qquad (2.15)$$

The $(p + 1)$ symmetric eigenvectors are associated to $(T + H)$, the $(p + 1)$ anti-symmetric eigenvectors are associated to $(T - H)$.

Note 1. For Hermitian (complex) matrices, we have

$$JRJ = \overline{R} \neq R, \quad \text{and} \quad Jx = e^{j\varphi}\overline{x},$$

and symmetrical elements in x have the same modulus.

Note 2. For multiple eigenvalues, for instance $\lambda = 0$, the corresponding null subspace is spanned by columns of the X matrix,

$$RX = 0, \quad \text{where} \quad X^T = \begin{bmatrix} x_0 & x_1 & \cdots & x_p \\ & x_0 & & \\ & & x_0 & x_1 & \cdots & x_p \end{bmatrix}.$$

Any vector in this null-space, is then expressed as: $x = Xc$, where c is any vector of convenient dimension. This explains the spurious frequencies often notices when applying Pisarenko's method at an over-determined order.

Stability properties [25–29].

An interesting connection between linear prediction vectors and eigenvectors (Choleski and Karhunen–Loeve factorizations) is made through the singular case of linear prediction. A perturbed matrix $\tilde{R}_p$ is defined by

$$\tilde{R}_n(\lambda) = R_n - \lambda I,$$

with the following linear prediction results

$$R_n a = \alpha_n u, \quad \text{with} \quad \alpha_n = \frac{\text{Det } R_n}{\text{Det } R_{n-1}}.$$

The location of the zeros of $A(z)$ as a function of λ is then analyzed. The matrix $\tilde{R}_n$ being no longer positive definite, the roots of $A(z)$ may lie outside the unit circle ("unstable" roots), or on the unit circle (but by reciprocal pairs).

The signature of $\tilde{R}_n$ (number of positive squares π, and negative squares ν) may be computed using Choleski or Karhunen–Loeve decompositions:

$$(n + 1) = \pi + \nu, \quad \sigma = \pi - \nu.$$

(i) If $\lambda \neq \lambda_i$ (λ_i eigenvalue of R), then the number q of unstable roots of $A(z)$ is given by a Schur–Cohn criterion:

$$q = \#(\alpha_i < 0, i = 0, n - 1), \quad \text{if } \alpha_n > 0,$$
$$q = \#(\alpha_i > 0, i = 0, n - 1), \quad \text{if } \alpha_n < 0,$$

or equivalently,

$$q = \nu, \quad \text{if } \alpha_n > 0, \quad \text{and} \quad q = \pi, \quad \text{if } \alpha_n < 0.$$

(ii) If $\lambda = \lambda_i$ (singular case of linear prediction), then the following holds

$$a = u_i, \quad |k_n| = 1, \quad \alpha_n = 0, \quad Ja = \pm a,$$

and the corresponding "predictor" (eigenmodel) is a linear phase MA (FIR) filter. It has been shown by Gueguen [2], that

$U_i(z)$ has at least $(n - 2i)$ roots on the unit circle

This also proves, as particular cases ($i = 0, i = n$), that all the roots of the eigen models associated with the extreme eigenvalues lie on the unit circle. This result has been extended to multiple eigenvalues by Delsarte et al. [28].

Computation of eigenvalues

The above singular case of linear prediction (ii) can be used as an efficient procedure for computing the eigenvectors of a correlation matrix (or, more generally, of an Hermitian interspectral matrix). In this computation scheme, a fast linear prediction algorithm is imbedded in a recursive search loop for the eigenvalues. The error criterion is given by one or several conditions (ii), location of a particular estimate λ among the eigenvalues is directed by (i). In usual cases (not too close eigenvalues) this type of algorithm converges to one eigenvector within 10 iterations with a satisfactory precision (10^{-3}). The procedure may be extended to close-to-Toeplitz matrices (covariance case).

References

[1] P. Roebuck and S. Barnett, A survey of Toeplitz and related matrices, Int. J. Syst. Sci. 9-8 (August 1978) 921–934.

[2] H. Widom, Toeplitz matrices, in: Studies in Real and Complex Analysis, ed. I.I. Hirschman (Prentice Hall, London, 1965).

[3] U. Grenander and G. Szego, Toeplitz Forms and their Applications (University of California Press, Berkeley, 1958).

[4] E.I. Jury, The theory and application of Inners, Proc. IEEE 63 (July 1975) 1044–1068.

[5] B.D.O. Anderson and E.I. Jury, Generalized Bezoutian and Sylvester matrices IEEE Trans. Autom. Control 21-8 (August 1976) 551–556.

[6] R. Cline, R. Plemmons and G. Worms, Generalized inverses of certain Toeplitz matrices, Linear Algebra & Appl. 8 (1974) 25–33.

[7] J. Pearl, Asymptotic equivalence of spectral representations, IEEE Trans. Acoust. Speech & Signal Proc. 23-6 (December 1975) 547–551.

[8] R.M. Gray, On the asymptotic eigenvalue distribution of Toeplitz matrices, IEEE Trans. Inf. Theory 18-6 (November 1972) 725–730.

[9] M. Kac, W. Murdock and G. Szego, On the eigenvalues of certain Hermitian forms, J. Rat. Mech. Anal 2 (1953) 767–800.

[10] M.M. Siddiqui, On the inversion of the sample covariance matrix in a stationary auto-regressive process, Annals of Math. Stat. 29 (1958) 585–588.

[11] J. Durbin, Efficient estimation of parameters in MA models, Biometrika 46 (1959) 306–316.

[12] R.F. Galbraith and J.I. Galbraith, On the inverses of some patterned matrices arising in the theory of stationary time series, J. Appl. Probab. 11 (1974) 63–71.

[13] I. Gohberg and G. Hajnig, On the inverses of Toeplitz matrices (in Russian) Rev. Roum. Math Fures Appl. 29-5 (1974) 629–663.

[14] L.M. Kitikov, The structure of matrices which are the inverse of the correlation matrix of a random vector process, USSR Comput. Math. & Math. Phys. 7-4 (1967) 58–71.

[15] A. Viera and T. Kailath, An another approach to the Schur–Cohn criterion, IEEE Trans. Circuits & Syst. 24-4 (April 1977) 218–220.

[16] I.J. Good, The inverse of a centro-symmetric matrix, Technometrics 12-4 (November 1970) 925–928.

[17] W. Pye, T. Bouillon and T. Atchison, The pseudo-inverse of a centro-symmetric matrix, Linear Algebra & Appl. 6 (1973) 201–204.

[18] W.D. Ray and R. Driver, Further decomposition of the Karhunen–Loeve series representation of a stationary random process, IEEE Trans. Inf. Theory 16-6 (November 1970) 663–668.

[19] C. Gueguen, The modified linear prediction, a factor analysis approach to speech analysis, Report UCLA-ENG 7540 (University of California, System Science Dept, Los Angeles, 1975).

[20] A. Cantoni and P. Butler, Properties of the eigenvectors of persymmetric matrices with applications to communication theory, IEEE Trans. Commun. 24-8 (August 1976) 804–809.

[21] J. Makhoul, On the eigenvectors of symmetric Toeplitz matrices, IEEE Trans. Acoust. Speech & Signal Process. 29-4, (August 1981) 868–872.

[22] L. Datta and S. Morega, Comments and corrections to "On the eigenvectors of symmetric Toeplitz matrices", IEEE Trans. Acoust. Speech & Signal Process. 32-2 (April 1984) 440, 441.

[23] L. Andrews, Further comments on "On the eigenvectors of symmetric Toeplitz matrices, IEEE Trans. Acoust. Speech & Signal Process. 33-4 (August 1985) 1013.

[24] E. Robinson, Statistical Communication and Detection, ed. Griffin (London, 1967) pp. 269–272.
S. Treitel and T. Ulrych, A new proof of the minimum phase property of the unit prediction error operator, IEEE Trans. Acoust. Speech & Signal Process. 27-1 (February 1979) 99, 100.

[25] G. Carayannis and C. Gueguen, The factorial linear modelling; a Karhunen–Loeve approach to speech analysis, Proc. IEEE Int. Conf. on Acoustics, Speech and Signal Processing, Philadelphia, PA (April 1976) pp. 489–492.

[26] S.S. Reddi, Eigenvector properties of Toeplitz matrices and their application to spectral analysis of time series, Signal Process. 7 (1984) 45–56.

[27] C. Gueguen, Linear prediction in the singular case, Proc. IEEE Int. Conf. on Acoustics, Speech and Signal Processing, Vol. 2, Atlanta, GA (April 1981) pp. 881–884.

[28] P. Delsarte, Y. Genin and Y. Kamp, Parametric Toeplitz Systems Circuits, Signal Process. 3-2 (1984) 208–223.

[29] F. Giannella and C. Gueguen, Extraction des vecteurs propres des matrices de Toeplitz, 8ième Colloque GRETSI, Nice, France (Juin 1981) pp. 215–219.

3. Displacement ranks and the distance to Toeplitz

Toeplitz matrices constitute a class of highly-structured matrices with important properties which are, in general, associated to linearity and stationarity. Unfortunately, in most of the cases, due to some intrinsic non-stationarity, or due to the observation process (short snapshot from a stationary time series), the matrices of interest are no longer Toeplitz, but related to Toeplitz by some properties (e.g. R_{COV} is a product of two Toeplitz matrices). We will show in this section how the concept of displacement rank measures the complexity of a given matrix in terms of its "distance" to Toeplitz, and how it provides a useful representation of this matrix in terms of TTMs.

3.1. Definitions of displacement ranks

Let R be any given (symmetrical) matrix, for which two partitions are constructed:

$$R = \begin{bmatrix} r_0 & r^T \\ r & R_1 \end{bmatrix} = \begin{bmatrix} R_0 & s \\ s^T & s_0 \end{bmatrix}, \qquad Z = \begin{bmatrix} 0 & & \cdots & & 0 \\ 1 & & & & \vdots \\ 0 & 1 & & & \\ \vdots & & \ddots & & \\ 0 & \cdots & 0 & 1 & 0 \end{bmatrix}.$$

$$(3.1)$$

The displacement rank δ is defined by computing the difference of R and its shifted version along this diagonal [1]. Using the shift operator Z, the rank of the following residual matrices M and N are computed:

$$\nabla R: \quad R - ZRZ^T = M \qquad \delta_- = \text{Rank}\,[M],$$
$$\Delta R: \quad R - Z^TRZ = N, \qquad \delta_+ = \text{Rank}\,[N],$$

where δ_- is the lower displacement rank, and δ_+ the upper displacement rank of R. Using the above partitions, for instance in the lower displacement, gives

$$\begin{bmatrix} r_0 & r^T \\ r & R_1 \end{bmatrix} - Z \begin{bmatrix} R_0 & s \\ s^T & s_0 \end{bmatrix} Z^T = \begin{bmatrix} r_0 & r^T \\ r & P \end{bmatrix},$$

where $P = R_1 - R_0$. For a Toeplitz matrix, the diagonal shift gives $P = 0$, the upper corner residual generally has rank 2, and $\delta_- = 2$. For a non-Toeplitz matrix in general $P \neq 0$, and $(\delta - 2)$ may be regarded as a "distance" to the Toeplitz structure.

In order to explicitate the rank of the residual matrix, the δ generators of R are defined by the diadic decomposition,

$$M = \sum_{i=1}^{\delta} \pm g_i g_i^T, \quad \text{or} \quad M = G\Sigma G^T,$$

where the matrix G has columns g_i and Σ is a signature matrix. It is noted that the g_i are not uniquely defined. Moreover, due to the signature matrix Σ, M is not a grammian matrix of linear dependence. As a matter of fact, the question of the minimal index δ will be ignored in this introduction.

By simple inspection, it is easily shown that many estimates of the correlation matrix R, as given in section 1.3, have a low displacement rank (no matter their dimension):

R_{COR} is Toeplitz, and $\delta_+ = \delta_- = 2$,

R_{PRW} has displacement rank $\delta_- = 3$,

R_{COV} has displacement rank $\delta_+ = \delta_- = 4$.

The definition of displacement ranks are easily generalized to non-symmetric matrices.

3.2. Representation theorem

The most important consequence of the definition of displacement ranks is a representation theorem of any matrix R in terms of TTMs [1].

Suppose, for instance, that the lower displacement rank of R is one and is defined by a single generator h,

$$R - ZRZ^T = hh^T, \qquad h^T = [h_0, h_1, \ldots, h_n]. \tag{3.2}$$

A system of equations is deduced from (3.2) by multiplying both sides by successive powers of Z. Summing up these equations and noting that Z is a nihilpotent matrix ($Z^{n+1} = 0$), we have

$$R - ZRZ^T = hh^T$$

$$ZRZ^T - Z^2RZ^{2T} = Zhh^TZ^T$$

$$\cdots \qquad \cdots \qquad \cdots$$

$$R = \sum_{i=0}^{n} Z^ihh^TZ^{iT}$$

As a consequence, the matrix R is recognized as a product of two lower triangular Toeplitz matrices,

$$R = HH^T = \begin{bmatrix} h_0 & & & \\ h_1 & h_0 & & \\ \vdots & & \ddots & \\ h_n & \cdots & h_1 & h_0 \end{bmatrix}^2, \qquad H = \mathscr{C}(h). \tag{3.3}$$

In the general case, it is easily shown that the matrix R may be represented as the sum of δ terms as previously defined,

$$R = \sum_{i=1}^{\delta} \pm G_i G_i^{\mathrm{T}}, \quad \text{where} \quad G_i = \mathscr{C}(g_i). \tag{3.4}$$

An analogous development is realized using the upper displacement rank and leads to a representation by a sum of δ_+ products of upper TTMs. In some cases, a mixed representation (lower and upper) is also of interest [2].

As a matter of fact, the displacement rank one matrix (3.3) appears as the generic case. It is easily interpreted as the covariance matrix of a time-invariant system starting at time $t = 0$ from zero initial conditions, as in section 1.1. Although the system is time-invariant, the resulting covariance is non-stationary due to initial conditions.

In the general case (3.4), the representation is more difficult to interpret physically. Some elements in the sum may have a negative sign and cannot represent real covariances. The interpretation should appeal to complex signals or to reactances. An amazing situation is the case of a truly stationary signal with a Toeplitz correlation matrix, which may be interpreted as combining two ($\delta = 2$) non-stationarities in such a way that the overall signal appears to be stationary.

3.3. Related displacement ranks

A question of interest for practical use of displacement ranks, is the invariance of the lower or upper displacement ranks after some transformation of the matrix.

Upper-lower displacement ranks.
By definition of the lower displacement rank, and multiplying both sides of (3.2) by Z^{T} and Z:

$$\begin{aligned} R - ZRZ^{\mathrm{T}} &= M, \\ Z^{\mathrm{T}}RZ - Z^{\mathrm{T}}ZRZ^{\mathrm{T}}Z &= Z^{\mathrm{T}}MZ, \end{aligned} \qquad \text{Rank}[M] = \delta_-,$$

the matrix $(Z^{\mathrm{T}}Z)$ is easily recognized to be almost the identity matrix

$$Z^{\mathrm{T}}Z = \begin{bmatrix} 1 & 0 & \cdots & 0 \\ 0 & 1 & & \vdots \\ \vdots & & 1 & 0 \\ 0 & \cdots & 0 & 0 \end{bmatrix}, \qquad Z^{\mathrm{T}}ZRZ^{\mathrm{T}}Z = R - S,$$

$$S = \begin{bmatrix} 0 & s \\ s^{\mathrm{T}} & s_0 \end{bmatrix}.$$

As a consequence, we have

$$N = R - Z^T R Z = -Z^T M Z - S, \qquad \text{Rank}[N] = \delta_+,$$

and

$$|\delta_-(R) - \delta_+(R)| \leqslant 2. \tag{3.5}$$

A useful example of this property is the prewindow case of linear prediction. Given a lower rank one representation, the problem consists in computing an upper rank three $(2 + 1)$ representation, thus providing the Choleski factors.

Displacement rank of the inverse.
Considering the lower displacement of R, and the upper displacement of R^{-1},

$$R - Z R Z^T = M, \qquad R^{-1} - Z^T R^{-1} Z = N,$$

these two equations are put into a canonical form by multiplying both sides by the convenient square-root (Choleski factor) of R:

$$I - (R^{-1/2} Z R^{1/2})(R^{T/2} Z^T R^{-T/2}) = R^{-1/2} M r^{-T/2} = I - W W^T,$$

$$I - (R^{T/2} Z^T R^{-T/2})(R^{-1/2} Z R^{1/2}) = R^{T/2} N R^{1/2} = I - W^T W.$$

Using a singular value decomposition for W, it is easily shown that the ranks of M and N are the same and

$$\delta_-(R) = \delta_+(R^{-1}). \tag{3.6}$$

This important property will allow us to establish the representation of the inverse in terms of TTMs in section 3.4.

Displacement rank of the Schur complement.
The (lower) Schur complement of a matrix R is defined by discarding the first row and column from R using a diadic product. When applied to R and R^{-1} for instance, we have

$$R = \frac{1}{r_0} \begin{bmatrix} r_0 \\ r \end{bmatrix} \begin{bmatrix} r_0 & r^T \end{bmatrix} + \begin{bmatrix} 0 & \cdots & 0 \\ \vdots & S(R) & \\ 0 & & \end{bmatrix}, \tag{3.7}$$

where $(r_0 r^T)$ is the just row of R,

$$R^{-1} = \frac{1}{\alpha} \begin{bmatrix} a_0 \\ a \end{bmatrix} \begin{bmatrix} a_0 & a^T \end{bmatrix} + \begin{bmatrix} 0 & \cdots & 0 \\ \vdots & S(R^{-1}) & \\ 0 & & \end{bmatrix}, \tag{3.8}$$

where $(a_0 a^T)$ is the prediction vector with $a_0 = 1$. A remarkable property is that the displacement rank of a given matrix R is invariant under Schur complement.

The first column of R may always be selected as one possible generator:

$$R - ZRZ^T = \frac{1}{r_0}\begin{bmatrix} r_0 \\ r \end{bmatrix}[r_0 \quad r^T] + \sum_{i=2}^{\delta} g_i g_i^T. \tag{3.9}$$

Doing so, the subsequent generators g_i have a zero first coordinate by construction. Considering the Schur complement

$$\tilde{R} = R - \frac{1}{r_0}\begin{bmatrix} r_0 \\ r \end{bmatrix}[r_0 \quad r] = \begin{bmatrix} 0 & & 0 \\ 0 & S & 0 \end{bmatrix}.$$

Equation (3.9) is then rewritten to express the displacement rank of R:

$$\tilde{R} - Z\tilde{R}Z^T = -\frac{1}{r_0}Z\begin{bmatrix} r_0 \\ r \end{bmatrix}[r_0 \quad r^T]Z^T + \sum_{i=2}^{\delta} g_i g_i, \tag{3.10}$$

which has consequently the same displacement rank δ as R (note that in eq. 3.10 all the matrices have a zero first column and first row).

3.4. Inverses of close-to-Toeplitz matrices

For a Toeplitz matrix R with displacement rank $\delta_- = \delta_+ = 2$, the inverse matrix R^{-1} has the same property from (3.6). Applying the representation theorem (3.4) explains the particular TTM structure of the earlier derived inversion formula (2.12). The problem addressed in this section is to generalize this formula for close-to-Toeplitz matrices [3].

For any inverse matrix R^{-1} the lower and upper Schur complements are

$$R^{-1} = \frac{1}{\alpha}\begin{bmatrix} a_0 \\ a_1 \\ \vdots \\ a_p \end{bmatrix}[a_0, a_1, \ldots, a_p] + \begin{bmatrix} 0 & \cdots & 0 \\ \vdots & R_1^{-1} & \\ 0 & & \end{bmatrix}, \tag{3.11}$$

where a_i is the forward predictor with $a_0 = 1$, and α is the energy of predictor error.

$$R^{-1} = \frac{1}{\beta}\begin{bmatrix} b_p \\ \vdots \\ b_1 \\ b_0 \end{bmatrix}[b_p, \ldots, b_1, b_0] + \begin{bmatrix} R_0^{-1} & & 0 \\ & & \vdots \\ 0 & \cdots & 0 \end{bmatrix}, \tag{3.12}$$

where b_i is the backward predictor with $b_0 = 1$, and β is the energy of predictor error. Using these two expressions in the lower displacement of R^{-1} leads to

$$R^{-1} - ZR^{-1}Z^T = \frac{1}{\alpha}aa^T - \frac{1}{\beta}Z(bb^T)Z^T + \begin{bmatrix} 0 & 0 \\ 0 & R_1^{-1} - R_0^{-1} \end{bmatrix}.$$

(3.13)

Introducing the previous matrix P,

$$P = R_1 - R_0, \qquad R_1^{-1}PR_0^{-1} = R_1^{-1}(R_1 - R_0)R_0^{-1} = R_0^{-1} - R_1^{-1},$$

we can expand P in an orthogonal sum of diadics, and define

$$P = \sum_{i=1}^{\delta-2} v_i v_i^T, \qquad R_0^{-1}v_i = c_i, \qquad R_1^{-1}v_i = \frac{1}{\gamma_i}c_i.$$

Now we have

$$R_1 - R_0 = \sum_i R_1^{-1}v_i v_i R_0^{-1} = \sum_i \frac{1}{\gamma_i}c_i c_i^T,$$

and

$$R^{-1} - ZR^{-1}Z^T = \frac{1}{\alpha}aa^T - \frac{1}{\beta}Zbb^TZ^T + \sum_i^{\delta-2} \frac{1}{\gamma_i}c_i c_i^T.$$

(3.14)

From eq. (3.14), the representation theorem gives an explicit formula in terms of TTMs for a displacement rank δ matrix. In the Toeplitz case, it coincides with the Gohberg–Semencul formula (2.12). In the non-Toeplitz case, the role of the forward and backward predictors is preserved but with additional vectors c_i. These latter vectors may be interpreted as "Kalman gains" and show up in time recursive fast algorithms.

Note 1. Displacement ranks are dedicated to exhibit a particular structure of the data matrix relating to shifts. As already mentioned, this shift property appears in a very natural way in many problems dealing with discrete time-series. But this structure is mainly an a priori property of the usual correlation estimates $R_{COR}, R_{COV}, R_{PRW}, \ldots$, which turn out to have low displacement ranks. There is no evidence that a naive estimate will exhibit interesting properties from this point of view.

Note 2. Without going further in the story of fast algorithms, the representation theorem (3.4) in itself is of practical interest. Many problems require an efficient way of computing Toeplitz or close-to-

Toeplitz quadratic forms. This computation can be realized by passing the signal through a bank of linear causal filters, squaring (quadration) and algebraically summing the outputs [4].

Note 3. A useful alternative representative of the inverse of a Toeplitz matrix R is the following:

$$R^{-1} = HVH^{\mathrm{T}},$$

where V is Toeplitz and H is an impulse-response matrix (TTM).

Note 4. The concept of displacement rank is very general in nature. The Z operator in (3.1) associated to a shift may be replaced by any well-chosen operator in order to fit the special structure of a given matrix. Various authors have shown that fast algorithms (for inversion or solution of linear sets of equations) may be found for other types of matrices as well [5-7]. This question will be developed in the concluding section.

References

[1] T. Kailath, S.Y. Kung and M. More, Displacement ranks of a matrix, Bull. Am. Math. Soc. 5 (September 1979).
[2] J.M. Delosme and M. Morf, Mixed and minimal representations for Toeplitz and related systems, in: Proc. 14th Asilomar Conf. Control Systems and Communication, Monterey, CA (November 1980) pp. 19–24.
[3] B. Friedlander, M. Morf, T. Kailath and L. Ljung, New inversion formulas for matrices classified in terms of their distance from Toeplitz matrices. Linear Algebra & Appl. 27 (1979) 31–60.
[4] T. Kailath, B. Levy, L. Ljung and M. Morf, Fast time-invariant implementation of Gaussian signal detectors, IEEE Trans. Inf. Theory 24-4 (July 1978) 469–477.
[5] G. Carayannis, N. Kalouptsidis and D. Manolakis, Fast recursive algorithms for a class of linear equations, IEEE Trans. Acoust. Speech & Signal Process. 30-2 (April 1982) 227–239.
[6] B. Picinbono, Properties and applications of stochastic processes with stationary increments, J. Appl. Probab. 6 (1974) 512–523.
[7] C. Gueguen, Sur une mesure du caractère non-stationnaire, 9ième Colloque GRETSI, Vol. 1, Nice, France (Mai 1983) pp. 27–33.

4. Slow and fast factorization algorithms

4.1. Introduction

A straightforward solution of optimal filtering problems is often obtained by decomposing a given spectrum into sum or product components. For instance, the realizable Wiener filter solution for predicting an output signal y_t from a given input u_t, in the mean-square sense, is given by

$$\xrightarrow{u_t} \boxed{H(z)} \xrightarrow[y_t]{\hat{y}_t} \underset{\varepsilon_t}{\otimes} \xrightarrow{} \quad \text{where} \quad H(z) = \frac{1}{R^+} \frac{S}{[R^-]_+} \qquad (4.1)$$

where $R(z)$ is the (auto-) spectrum of u_t, $S(z)$ the inter-spectrum (u_t, y_t), $[\cdot]^+$ is a product decomposition into causal, stable and minimum phase components, and $[\cdot]_+$ is a sum decomposition into causal and stable components. In this decomposition, $1/R^+$ is understood as a prewhitening filter on u_t, which alleviates the main difficulties of the optimal filter design. This spectral factorization can be performed in the frequency domain for stationary signals by factoring $R(z)$ or its inverse $R^{-1}(z)$ as rational transfer functions, i.e. ratios of symmetrical polynomials.

For a given real symmetrical polynomial $P(z)$ with simple roots, the following decomposition holds:

$$P(z) = M(z)M^*(z)Q(z), \qquad Q(z) = N(z) + N^*(z),$$

where $M(z)$ and $N(z)$ have roots strictly inside the unit circle, and where $M^*(z)$ and $N^*(z)$ are the corresponding reciprocal polynomials. The three terms in $P(z)$ have their roots inside, outside and on the unit circle. $P(z)$ is said to be factorable iff $Q(z)$ has degree zero. This factorization may evidently be conducted by discriminating among the roots of $P(z)$ but the computational cost is very high.

The approach advocated in the following will rely on the corresponding matrix decomposition making use of (i) sum versus product decomposition, (ii) AR (or IIR) versus MA (or FIR) models and (iii) forward versus backward prediction.

(i) *Sum versus product decomposition.* Let R be an admissible correlation matrix ($R > 0$), we then have

(a) *Sum decomposition.*

$$R = H_+ + H_-,$$

where $H_- = H_+^{\mathrm{T}}$, and H_+ is triangular (see eq. 4.2).

In this decomposition the correlation sequence has been split into causal and anti-causal components associated to the triangular character of H_+ and H_-. This decomposition is related to the Szegö polynomials of the second kind, but it will not be emphasized here. It is worth noting, however, that if R is definite positive then the corresponding polynomial $H_+(z)$ has poles inside the unit circle.

(b) *Product decomposition.*

$$R = H^+ H^-,$$

where $H^- = (H^+)^T$, and H^+ is triangular (see eq. 4.2).

In this decomposition, causality is, once more, associated to triangularity. The factor H^+ is identified to a special square root of R, namely the lower Choleski factor. The R matrix is always factorable if definite positive (excluding zero determinant for any principal minor). This decomposition is related to linear prediction (see section 1.5) and to the usual Szegö orthogonal polynomials of the first kind:

$$H_+ = \begin{bmatrix} r_{00/2} & 0 & \cdots & 0 \\ r_{10} & r_{11/2} & & \vdots \\ \vdots & & \ddots & 0 \\ r_{p0} & \cdots & r_{pp-1} & r_{pp/2} \end{bmatrix},$$

$$H^+ = \begin{bmatrix} h_0^0 & & & \\ h_1^0 & h_0^1 & & \\ \vdots & & \ddots & \\ h_p^0 & \cdots & h_1^{p-1} & h_0^p \end{bmatrix}. \tag{4.2}$$

(ii) *AR versus MA models.* The above decomposition may be applied to R or on its inverse R^{-1}, and we have the following FIR synthetizers and whiteners.

(a) *FIR Synthetizers.* Factoring R gives

$$R = \begin{bmatrix} h_0^0 & & & \\ h_1^0 & h_0^1 & & \\ \vdots & & \ddots & \\ h_p^0 & \cdots & h_1^{p-1} & h_0^p \end{bmatrix}^2, \quad \text{and}$$

$$\begin{bmatrix} y_0 \\ y_1 \\ \vdots \\ y_p \end{bmatrix} = \begin{bmatrix} h_0^0 & & & \\ h_1^0 & h_0^1 & & \\ \vdots & & \ddots & \\ h_p^0 & \cdots & h_1^{p-1} & h_0^p \end{bmatrix} \begin{bmatrix} \varepsilon_0 \\ \varepsilon_1 \\ \vdots \\ \varepsilon_p \end{bmatrix}, \tag{4.3}$$

where $[\cdot]^2$ denotes the product of a matrix by its transposed.

This defines a sequence of time-varying order-increasing synthezisers as finite impulse responses h^i_t(FIR) with increasing lengths, for generating the output data y_t with covariance R from some white input noise sequence ε_t. In this expression (4.3), each FIR sequence is normalized by a time varying gain h^t_0.

This triangular factorization solves the inverse problem of section (1.2).

(b) *FIR whiteners*. Factoring R^{-1} gives

$$R^{-1} = \begin{bmatrix} a^p_0 & a^1_1 & \cdots & a^p_p \\ 0 & a_0 & & \vdots \\ \vdots & & \ddots & a^{p-1}_1 \\ 0 & \cdots & 0 & a^p_0 \end{bmatrix}, \quad \text{and}$$

$$\begin{bmatrix} \varepsilon_0 \\ \varepsilon_1 \\ \vdots \\ \vdots \\ \varepsilon_p \end{bmatrix} = \begin{bmatrix} a^0_0 & 0 & \cdots & 0 \\ a^1_1 & a^1_0 & & \vdots \\ \vdots & & \ddots & 0 \\ a^p_p & \cdots & a^{p-1}_1 & a^p_0 \end{bmatrix} \begin{bmatrix} y_0 \\ y_1 \\ \vdots \\ y_p \end{bmatrix}. \tag{4.4}$$

This defines a sequence of time-varying order-increasing analyzers, as MA models with coefficients a^p_i and increasing orders for whitening the input data sequence y_t with covariance R and producing a white output residual ε_t. In this expression (4.4), each MA model has been normalized by a time varying gain a^i_0 to generate a normalized constant energy innovation sequence. From comparing eqs. (4.3) and (4.4), it is now clear that

–The two lower triangular factors with coefficients a^j_i and h^j_i are of each other's inverse.

–The MA analyser with coefficients a^j_i may be inverted to produce an AR synthetizer.

–If R has a lower-upper factorization, then R^{-1} is upper-lower.

Note that in the above, we have preserved some distinction between a time-varying FIR filter (with column-wise indexes) and a time-varying MA filter (with row-wise indexes).

(iii) *Forward versus backward models*. A given definite positive matrix R has two Choleski decompositions according to the prescribed order of the factors: upper-lower versus lower-upper. They are associated to causal (lower) and anti-causal (upper) impulse responses. Connecting the two descriptions of the same matrix R is a meaningful problem (see below). A first attempt is to use a time-reversal argument, implemented through pre- and post-multiplication by the order-reversal operator J:

$$\check{R} = JRJ, \quad \text{with} \quad R = HH^T, \quad H \text{ lower triangular.}$$

This results into

$$\check{R} = (JHJ)(JH^TJ), \quad \text{where } (JHJ) \text{ is upper triangular.}$$

A forward causal model for R is thus connected to a backward anti-causal model for $\check{R}$. If, moreover, the process is stationary, then R is Toeplitz and $\check{R} = R$. In the latter case, the relation between the forward and backward models is very simple.

These factorization concepts may be extended to non-symmetrical matrices, with Toeplitz structure or not, i.e. intercorrelation matrices. In the Toeplitz case, given an intercorrelation sequence, the problem of selecting the convenient index on the diagonal of R is of special importance. This non-symmetric factorization procedures will, however, not be investigated here.

In the following sections, the problem of factoring a symmetrical covariance matrix will be examined in some details. The relevance of factoring R instead of R^{-1} will be advocated. The procedures recalled in section 4.2 are referred to as "slow algorithms", because they do not take into account the special (close-to-Toeplitz) structure of the covariance estimate. They consequently reach a computation cost of $\mathcal{O}(n^3)$ multiplications, where n is the dimension of R. In section 4.3 we introduce the Toeplitz constraint and we will, consequently, reduce the computations to $\mathcal{O}(n^2)$, emphasizing an important algorithm due to Bareiss [6], which was, later, independently developed by several authors [7–11].

4.2. Slow factorization procedures

Triangular matrix factorization is a standard procedure in numerical analysis for solving a linear set of equations (so-called LU factorization),

$$Mx = y, \quad \text{with} \quad M, y \text{ given, } x \text{ unknown.}$$

If R is factored into

$$M = LU, \quad \text{with } L \text{ and } U \text{ lower and upper triangular,}$$

then x is computed using an efficient and numerically stable backsubstitution method,

$$Lz = y, \quad \text{and} \quad Ux = z.$$

There are two basic approaches to extract triangular factors.

4.2.1. Direct triangularization

The Gauss elimination procedure constructs a sequence of left elementary operations on the rows of M. At a given stage i, R_i is supposed to be triangular up to column i. A lower triangular matrix P_i with ones on the diagonal and v as ith column, where

$$v^T = \begin{bmatrix} \cdots & -m_{ji}/m_{ii} & \cdots 1\, 0 & \cdots & 0 \end{bmatrix}$$

is then computed in order to bring a new lower column of zeros in ith position, thus giving M_{i+1}. The divisor m_{ii} is the ith pivotal element. There is no breakdown in the procedure if all the pivots are non-zero. This is the case when M is positive definite (all principal minors positive). Moreover in this case, the Gaussian elimination is conducted without requiring any row interchange.

If n is the order of M (dimension $n + 1$, including index 0), then the overall computation load is of the order of

$$n^3/3 \text{ multiplies} \quad \text{and} \quad \text{no square root.}$$

Assuming the R matrix to be symmetric and definite positive, another well-known triangularization procedure is the classical Choleski decomposition. This is a dimension-increasing technique where the principal minor M_{j-1} of M is supposed to be already factored:

$$M_{j-1} = L_{j-1}L_{j-1}^T, \quad \text{where } L_{j-1} \text{ is lower triangular.}$$

The matrix M_j is therefore obtained by bordering M_{j-1} by m_j and m_{jj},

$$M_j = \begin{bmatrix} M_{j-1} & m_j \\ m_j^T & m_{jj} \end{bmatrix}, \quad \text{where} \quad \begin{array}{l} m_j \quad \text{is a vector,} \\ m_{jj} \quad \text{is a scalar.} \end{array}$$

The corresponding bordering elements in L_j, say l_j and l_{jj}, are computed by straightforward identification:

$$L_{j-1}l_j = m_j, \qquad l_j^T l_j + l_{jj}^2 = m_{jj}.$$

If the matrix is symmetric, but not positive definite, some row interchange may be needed as above, but the stability of the numerical procedure is no longer insured. The computational cost is

$$n^3/6 \text{ multiplies,} \quad \text{and} \quad (n-1) \text{ square roots.}$$

The Choleski procedure is considered as robust in the sense that the errors on the triangular factors L may be less than the rounding error on the M matrix itself.

Among the interesting features of the Choleski factorization, is the fact that the decomposition of a perturbed M is well understood [2]. For instance, if M, with Choleski factors L, is replaced by $\tilde{M}$:

$$\tilde{M} = M + vv^T, \qquad M = LL^T,$$

then, the Choleski factors $\tilde{L}$ of the perturbed matrix $\tilde{M}$ are

$$\tilde{M} = \tilde{L}\tilde{L}^T = L(I + L^{-1}vv^T L^{-T})L^T,$$

and $\tilde{L}$ is obtained from L by factoring the center matrix only. This may be achieved by a $\mathcal{O}(4n^2)$ procedure.

4.2.2. Orthogonalization
Another approach to computing triangular factors is provided through sequential orthogonalization of the columns of M (so-called QR factorization).

The Gram–Schmidt procedure determines an orthonormal matrix Q with columns q_i, and an upper triangular matrix U with elements u_{ij}, to triangularize the given matrix M with columns m_i:

$$QM = U, \quad \text{where} \quad \begin{matrix} Q = [q_0, \ldots, q_n], \qquad M = [m_0, m_1, \ldots, m_n], \\ U \text{ is upper triangular.} \end{matrix}$$

The decomposition is essentially unique if M is non-singular. By direct identification, we get the following set of equations,

$$m_1 = u_{11}q_1, \qquad m_2 = u_{12}q_1 + u_{22}q_2, \qquad m_3 = \ldots, \text{etc.},$$

and the q_i and u_{ij} may be computed sequentially, using the order of

n^3 multiplies, and n square roots.

However, this sequential orthogonalization procedure is known to give poor numerical results. In particular, due to cumulative rounding errors, the successive columns q_i do not remain orthogonal in practice [3].

To overcome this difficulty, the matrix Q may be computed as a product of elementary orthonormal matrices. This is achieved by the Householder transformation. Let M_i be the partially triangularized matrix deduced from M; an elementary orthogonal matrix Q_i is defined by

$$Q_i = \begin{bmatrix} I & 0 \\ 0 & I - 2vv^T \end{bmatrix},$$

where the first identity matrix has dimension i, and

$$v^\mathrm{T} = \frac{1}{2\beta}\left[\; \cdots \; m_{ji} \; \cdots \; \alpha \; 0 \; \cdots \; 0\right].$$

Selecting convenient values for α and β by

$$\gamma^2 = \sum_{j=i}^{n} m_{ji}^2, \qquad 2\beta^2 = \gamma(\gamma \pm m_{ii}), \qquad \alpha = m_{ii} \pm \gamma,$$

where the $\pm$ sign is adjusted to insure numerical stability, we then have

$$Q^\mathrm{T} = Q, \qquad v^\mathrm{T}v = 1, \qquad Q^2 = QQ^\mathrm{T} = \mathrm{I}.$$

This Householder procedure gives good numerical results for a computational cost of

$2n^3/3$ multiplies, and n square roots.

An alternate way of implementing the above orthonormal transformations is to consider a product of plane rotations, giving the so-called Givens method. In that case, the non-diagonal elements in the ith column of M_i are set to zero, element by element, instead of as a partial column. The associated Q_{ij} matrix is

$$Q_{ij} = \begin{bmatrix} 1 & & & & & & \\ & \ddots & & & & & \\ & & \cos\varphi & \cdots & -\sin\varphi & & \\ & & \vdots & & \vdots & & \\ & & \sin\varphi & \cdots & \cos\varphi & & \\ & & & & & \ddots & \\ & & & & & & 1 \end{bmatrix},$$

where φ is the angle of rotation in the i, j plane, and the corresponding cost of computation reaches:

$4n^3/3$ multiplies, and $n^2/2$ square roots.

It is generally admitted that Givens' method is slightly more sensitive than Householder's. Surprisingly enough, it can be shown that the Householder matrix used to produce a partial column of zeros, is not the product of individual Givens matrices designed to set the corresponding elements to zero.

4.3. Slow versus fast factorization algorithms

The main hypothesis underlying the above procedures is that the R matrix to be factored is (symmetrical) positive definite. They make no use of the possible internal structure, e.g. Toeplitz structure, to improve the efficiency of the computations. This section will introduce this extra hypothesis, to elaborate the so-called "fast" algorithms.

4.3.1. The Bauer algorithm
The factorization algorithm due to Bauer [4] makes use of the Toeplitz structure, but still remains in the slow class. It computes an MA model coefficients from a given output Toeplitz correlation matrix. Let b_i be the coefficients of an MA model of order q:

$$y_t = \sum_{i=0}^{q} b_i u_{t-i}, \quad \text{with} \quad u_t(\text{iid}): N(0, \sigma^2 = 1).$$

Then, we have the following output correlation matrix (see section 1.2):

$$R = B^T B, \quad \text{with} \quad B^T = \begin{bmatrix} b_0 & b_1 & \cdots & b_p & 0 & \cdots & 0 \\ & & & & & & \vdots \\ 0 & b_0 & & & & & 0 \\ \vdots & & & & & & \\ 0 & \cdots & 0 & b_0 & b_1 & \cdots & b_q \end{bmatrix}.$$

$$(4.5)$$

Due to the Toeplitz structure of R, this non-linear set of equations is indeed redundant and may be reduced to the first column of R, i.e.

$$\begin{bmatrix} r_0 \\ r_1 \\ \vdots \\ \vdots \\ r_q \end{bmatrix} = \begin{bmatrix} b_0 & b_1 & \cdots & b_q \\ 0 & b_0 & & \vdots \\ \vdots & & & b_1 \\ 0 & \cdots & 0 & b_0 \end{bmatrix} \begin{bmatrix} b_0 \\ b_1 \\ \vdots \\ b_q \end{bmatrix}.$$

$$(4.6)$$

The problem of computing the b_i from the r_i $(i = 0, q)$ may be approached through a fixed point algorithm. For that purpose, the diagonal term b_0 in eq. (4.6) may be discarted to generate

$$
\begin{bmatrix} r_0 \\ r_1 \\ \vdots \\ r_q \end{bmatrix} = \begin{bmatrix} 0 & b_1 & \cdots & & b_q \\ 0 & 0 & & & \vdots \\ \vdots & & & & b_1 \\ 0 & \cdots & & 0 & 0 \end{bmatrix} \begin{bmatrix} 0 \\ b_1 \\ \vdots \\ b_q \end{bmatrix} + b_0 \begin{bmatrix} b_0 \\ b_1 \\ \vdots \\ b_q \end{bmatrix}
\tag{4.7}
$$

Equation (4.7) can be used to produce an iterative scheme. The estimate b_i at stage n will be estimated by s_i^n:

$$
\begin{bmatrix} r_0 \\ r_1 \\ \vdots \\ r_q \end{bmatrix} - \begin{bmatrix} 0 & s_1^{n-1} & \cdots & & s_1^{n-q} \\ 0 & 0 & & & \vdots \\ \vdots & & & & s_1^{n-1} \\ 0 & \cdots & & & 0 \end{bmatrix} \begin{bmatrix} 0 \\ s^{n-1} \\ \vdots \\ s_q^n \end{bmatrix} = s_0^n \begin{bmatrix} s_0^n \\ s_1^n \\ \vdots \\ s_q^n \end{bmatrix}
\tag{4.8}
$$

Applying this scheme recursively is equivalent to factoring an infinite band Toeplitz matrix (as in note 1). Convergence of this scheme has been proven in the literature. Moreover, it was shown that at each intermediate step n, the corresponding MA model is minimum phase. See ref. [5] for an extensive survey.

4.3.2. The Bareiss algorithm

In a widely overlooked paper which appeared in the numerical analysis literature, E.H. Bareiss introduced a clever algorithm for solving the following general non-symmetric Toeplitz set of equations [6]:

$$Tx = y, \quad \text{where } y \text{ is any known vector.}$$

The solution is obtained by constructing a sequence of matrices T_i and T_{-i}, where T_i (resp. T_{-i}) have zeros on the i first upper (lower) diagonals:

$$
T_0 = T = \begin{bmatrix} t_0 & t_{-1} & \cdots & t_{-n} \\ t_1 & t_0 & & \\ \vdots & & t_0 & t_{-1} \\ t_n & \cdots & t_1 & t_0 \end{bmatrix}, \quad T_i = \begin{bmatrix} 0 & & U_i \\ & \ddots & \\ L_i & & 0 \end{bmatrix},
$$

$$
T_{-i} = \begin{bmatrix} 0 & & U_{-i} \\ & \ddots & \\ L_{-i} & & 0 \end{bmatrix}.
$$

The recurrence relationships are linear combinations:

$$T_{i+1} = T_i + k_{i+1}Z^{i\mathrm{T}}T_{-i},$$

$$T_{-i-1} = T_{-i} + k_{-i-1}Z^iT_i,$$

where Z and Z^T are the usual lower and upper shift operators.

In these equations, the scalar weight factors k_{i+1} (resp. k_{-i-1}) are adjusted so as to generate a new diagonal of zeros in T_{i+1} (resp. T_{-i-1}). This means that the lower corner of U_{-i} in T_{-i} is used, after convenient shift and weight adjustment, to cancel the main diagonal in U_i. In this process, the first i rows of T_i remain Toeplitz, while the last rows (non-Toeplitz) in L_i will remain unchanged later on.

At the end-stage, T_n and T_{-n} are clearly, respectively lower and upper triangular (the transformation matrix applied on the left of $T_0 = T$ may be computed back from the k_i, $i = 1, n$).

The total number of operations to achieve both triangularizations is

$2n^2 - n$ multiplications, n divisions, no square roots,

which compares to the $n^3/3$ multiplications of the Gauss elimination procedure.

4.3.3. The Cybenko algorithm

As previously mentioned, an alternative approach to Toeplitz matrix triangularization is to perform orthogonalization procedure. D.R. Sweet proposed to use the rank one perturbation of Gill et al. [2] to compute the Q and U matrices in $25n^2$ operations (and 4.5 n^2 for U alone) [12]. More recently, G. Cybenko has attacked this problem using an inner product formulation [13], thus generalizing previous results of De Meersman [14].

We first consider the following rectangular Toeplitz matrix T with bordering zeros:

$$T = \begin{bmatrix} t_{-n} & \cdots & t_{-1} & t_0 & t_1 & \cdots & t_n & 0 & \cdots & 0 \\ 0 & t_{-n} & & t_{-1} & t_0 & & t_1 & t_n & & 0 \\ \vdots & & & & & & & & & \\ 0 & \cdots & 0 & t_n & \cdots & t_{-1} & t_0 & t_1 & \cdots & t_n \end{bmatrix}$$

$$= \begin{bmatrix} t_0^\mathrm{T} \\ t_1^\mathrm{T} \\ \vdots \\ t_n^\mathrm{T} \end{bmatrix}$$

The columns t_i of T are then shifted versions of the first t_0 column,

$t_i = Z^i t_0$,   where Z is the down-shift operator.

Notice that the center matrix in T is the usual non-symmetric Toeplitz matrix. The successive columns may be easily orthogonalized with respect to the usual Euclidean inner product (i.e. (x, y) is $x^T y$). Two sequences of vectors q_i and p_i are computed such that

$$\begin{aligned} q_{i+1} &= Zq_i + k_{i+1}p_i, \\ p_{i+1} &= p_i + k_{i+1}Zp_i, \end{aligned} \quad \text{where} \quad k_{i+1} = -\frac{(Zq_i, p_i)}{(p_i, p_i)},$$

with initialization $p_0 = q_0 = t_0$, and for $i = 0, n$.
These vectors have the following properties:

$$q_i, p_i \in E_i, \qquad (q_i, q_j) = 0, \quad \text{for } i \neq j,$$

and

$$E_i = \text{Span } t_0, \ldots, t_i, \qquad (p_i, x) = 0, \quad \text{for } x \in \{ E\theta t_0 \},$$

and, consequently, T is triangularized:

$$Q^T T = U, \quad \text{where } U \text{ is upper triangular}$$

This procedure will later be related to the standard signal lattice algorithm.

At this stage, the procedure is not complete because it solves the Toeplitz orthogonalization problem only with special zero-boundary conditions (the T matrix is thus cyclic). In a recent paper [15], Cybenko extends its previous work to general non-symmetric Toeplitz matrices, i.e. the center part of T. This is done by orthogonalizing the vectors in T, with respect to a non-Euclidean inner product, defined by a window W, such that

$$(x, y) = x^T W y, \quad \text{with} \quad W = \text{Diag}[0, \ldots, 0, 1, 1, \ldots 1, 0, \ldots, 0],$$

where W has $(n + 1)$ ones on the diagonal.

Applying this inner product of shifted vectors will make some boundary conditions to appear. The final algorithm uses a nested vector relationship as above, but will encompass two extra vectors to take the boundary conditions into account (this has to relate to the displacement rank two of a Toeplitz matrix).

This method uses $\mathcal{O}(47/2 \, n^2)$ multiplications. It turns back to a sequential Gram–Schmidt procedure, but nevertheless remains efficient. This is due to some fundamental properties of the lattice structure which still have to be more deeply understood [16].

References

[1] J.H. Wilkinson, The Algebraic Eigenvalue Problem, Monographs in Numerical Analysis (Clarendon Press, Oxford, 1965) ch. 4.

[2] P. Gill, G. Golub, W. Murray and M. Saunders, Methods for modifying factorizations, Math. Comput. 28-216 (April 1974) 505–535.

[3] A. Bjorck, Solving linear least-squares problems by Gram–Schmidt orthogonalization, BIT 7 (1967) 1–21.

[4] F.L. Bauer, A direct iterative process for a Hurwitz decomposition of a polynomial. Arch. Electron. & Uebertragung stech. 9 (June 1955) 285–290.

F.L. Bauer. Beitrage zur Entwicklung numerischer Verfahren für programmgesteuerte Rechenanlagen: Direkte faktorisierung eines Polynoms, Sitzung. Ber. Bayer. Akad. Wiss (1956) 163–203.

[5] J. Leroux, Sur les algorithmes de factorisation spectrale dans le cas des signaux stationnaires échantillonnés, Thèse de doctorat d'Etat (Université de Nice, Octobre 1985).

[6] E.H. Bareiss, Numerical solution of linear equations with Toeplitz and vector Toeplitz matrices. Numer. Math. 13 (1969) 404–424.

[7] J. Rissanen, Algorithms for triangular decomposition of block Hankel and Toeplitz matrices with application to factoring positive matrix polynomials, Math. Comput. 27-121 (January 1973) 147–154.

[8] J. Leroux and C. Gueguen, A fixed-point computation of parcor coefficients, IEEE Trans. Acoust. Speech & Signal Process. 25-3 (June 1977) 257–259.

[9] R. Brent, F. Gustavson and D. Yun, Fast solution of Toeplitz systems of equations and computation of Pade approximants, J. Algorithms 1 (1980) 259–295.

[10] R. Brent and F. Luk, A systolic array for the linear-time solution of Toëplitz systems of equations, J. of VLSI and Computer Systems 1-1 (1983) 1–21.

[11] S.Y. Kung and Y.H. Hu, A highly-concurrent algorithm and pipelined architecture for solving Toeplitz systems, IEEE Trans. Acoust. Speech & Signal Process. 31-1 (February 1983) 66–76.

[12] D.R. Sweet, Fast Toeplitz orthogonalization, Numer. Math. 43 (1984) 1–24.

[13] G. Cybenko, A general orthogonalization technique with applications to time-series analysis and signal processing, Math. Comput. 40-161 (January 1983) 323–336.

[14] R. de Meersman, A method for least-squares solution of systems with a cyclic rectangular coefficient matrix, J. Comp. Appl. Math. 1-1, (1975) 51–54.

[15] G. Cybenko, Fast Toeplitz orthogonalization using inner products, submitted for publication.

[16] G. Cybenko, The numerical stability of the Levinson–Durbin algorithm for Toeplitz systems of equations, SIAM J. Sci. Stat. Comp. 1-3 (September 1980) 303–319.

5. Fast Choleski factorization of R^{-1}

Solving the linear prediction problem in the mean-square sense (with normalization $a_0 = 1$) is nothing but computing the first column of the inverse covariance matrix R^{-1}. Moreover, as noted in section 1, if the solution is seeked for in increasing order of the predictors, this amounts to constructing a Choleski factor of R^{-1}.

If R is Toeplitz, a celebrated algorithm for solving the corresponding normal correlation equations was introduced by Levinson [1], then reformulated by Durbin [2]. This Levinson–Durbin (LD) algorithm, not only has a lower computational cost of $\mathcal{O}(n^2)$ [instead of $\mathcal{O}(n^3)$ as for slow algorithms, see section 4], but exhibits interesting features in relation to fundamental properties of least-square problems.

When short-time windows are considered, the covariance matrix estimate is no longer Toeplitz. In early work, this covariance case of linear prediction was solved, ignoring the special structure of the matrix [13–15]. More recently, fast algorithms were discovered for solving the prewindow and covariance case [11, 12]. The notions of displacement rank, introduced in section 3, now give a unified framework to imbed this type of fast algorithms in the same class of "generalized Levinson" procedures.

5.1. The Levinson–Durbin (LD) algorithm

The LD algorithm is an efficient order recursive way to solve the normal equations in the symmetrical Toeplitz (correlation) case [1, 2]. The principle relies on the shift property and centro-symmetry of R. Assuming the solution to be known at order n, the predictor vector a^n satisfies

$$R_n a^n = \alpha_n u, \quad \text{with} \quad u^{\mathrm{T}} = [1 \ 0 \ \cdots \ 0]. \tag{5.1}$$

The order $(n + 1)$ solution is then constructed by combining a^n and its reverted and shifted version (backward predictor) with a linear weight factor k_{n+1}, applied to R_{n+1}. Under vector form:

$$
\left(
\begin{bmatrix}
r_0 & r_1 & \cdots & r_n & r_{n+1} \\
r_1 & r_0 & & & r_n \\
r_n & & r_0 & & \vdots \\
r_{n+1} & r_n & \cdots & r_1 & r_0
\end{bmatrix}
\right)
\left(
\begin{bmatrix}
a_0 \\
a_1 \\
\vdots \\
\vdots \\
a_n \\
0
\end{bmatrix}
+ k_{n+1}
\begin{bmatrix}
0 \\
a_n \\
\vdots \\
a_1 \\
a_0
\end{bmatrix}
\right)
$$

$$
=
\left(
\begin{bmatrix}
\alpha_n \\
0 \\
\vdots \\
0 \\
\omega_n
\end{bmatrix}
+ k_{n+1}
\begin{bmatrix}
\omega_n \\
0 \\
\vdots \\
0 \\
\alpha_n
\end{bmatrix}
\right) \tag{5.2}
$$

In eq. (5.2), ω_n is the (non-zero) residual obtained when applying the order n predictor on the $(n + 1)$ order matrix. The reflection coefficient k_{n+1} is then selected as $k_{n+1} = -\omega_n/\alpha_n$, in order to annihilate the last non-zero coordinate in the right-hand side of (5.2), giving rise to the order $(n + 1)$ normal equations,

$$R_{n+1}a^{n+1} = \alpha_{n+1}u, \quad \text{with} \quad u^T = \begin{bmatrix} 1 & 0 & \cdots & 0 \end{bmatrix}.$$

The recurrence relationships on the scalar coordinates a_i^n of the predictors are then summarized by the well-known LD formulas

$$\begin{cases} k_{n+1} = -\omega_n/\alpha_n \\ \alpha_{n+1} = \left(1 - k_{n+1}^2\right)\alpha_n & \text{for} \quad 0 < i \leqslant n. \\ a_i^{n+1} = a_i^n + k_{n+1}a_{n-i+1}^n \end{cases} \tag{5.3}$$

Note that in (5.3) the range of index i can be extended to any value by

$$a_0^n = 1, \qquad a_n^n = k_n,$$
$$a_i^n = 0, \quad \text{for} \quad i < 0 \quad \text{and} \quad i > n.$$

However, in any case, the information used in (5.3) at stage n, is restricted to $r_0, \ldots, r_n$ even if the next correlation coefficients of higher order are already available.

The reflection coefficients k_i have a very useful lattice filter interpretation and are computed sequentially. Along with r_0, they retain for $i = 1, n$, all the information contained in $r_0, \ldots, r_n$. Moreover, we have the useful determinantial equations

$$\text{Det } R_n = \alpha_0, \alpha_1 \ldots, \alpha_n, \qquad \alpha_n = \frac{\text{Det } R_n}{\text{Det } R_{n-1}}, \tag{5.4}$$

with $\alpha_{i+1} = (1 - k_{i+1}^2)\alpha_i$ and $\alpha_0 = r_0$.
A direct consequence is that if R_n is positive definite, all the R_i, for $i \leqslant n$ have this same property, and

$$\alpha_i > 0, \qquad |k_i| < 1, \qquad \text{for} \quad \forall i. \tag{5.5}$$

Many other important properties and interpretations in relation with the LD algorithm have been introduced in early literature [3–6].

5.2. Levinson–Durbin algorithm in vector form

The principle of the LD algorithm has been introduced as an order recursive procedure. As a consequence, the prediction vectors under consideration, increase in size as the algorithm is operated. For implementation on vectorized computers it is more suitable to deal with a maximum length (or order), say p. The previous equations can then be written in vector form and arranged in two triangular matrices containing the successive forward and backward predictors:

$$A = \begin{bmatrix} a_0^0 & a_0^1 & \cdots & a_0^p \\ 0 & a_1^1 & & a_1^p \\ & & \ddots & \vdots \\ 0 & & 0 & a_p^p \end{bmatrix}, \quad B = \begin{bmatrix} a_0^0 & a_1^1 & \cdots & a_p^p \\ 0 & a_0^1 & & a_{p-1}^p \\ & & \ddots & \vdots \\ 0 & & 0 & a_0^p \end{bmatrix}.$$

$$\text{with columns } a_n \qquad\qquad \text{with columns } b_n$$

$$(5.6)$$

Note that the forward matrix A has ones in the first row and k_i on the diagonal. The backward matrix B has k_i on the first row and ones on the diagonal.

As a matter of fact, multiplying R_p by B is nothing but constructing the (lower–upper) factorization of R_p:

$$\begin{bmatrix} r_0 & r_1 & \cdots & r_p \\ r_1 & r_0 & & \vdots \\ \vdots & & \ddots & \\ r_p & \cdots & r_1 & r_p \end{bmatrix} \begin{bmatrix} a_0^0 & \cdots & a_p^p \\ 0 & \ddots & \vdots \\ \vdots & & \vdots \\ 0 & \cdots & 0 & a_0^p \end{bmatrix} = \begin{bmatrix} \alpha_0 & 0 & \cdots & 0 \\ * & \alpha_1 & & \vdots \\ \vdots & & \ddots & 0 \\ * & & * & \alpha_p \end{bmatrix} \triangleq H$$

$$R_p \qquad\qquad B \qquad\qquad\qquad H$$

$$(5.7)$$

The zeros in the triangular form of H explicitate the orthogonality conditions in the classical normal equations for successive orders (the product of R_p by A will be used below). The last ingredient to be used is the Z matrix to operate the desired shift. As a sequence, the LD algorithm is equivalent to the vector form nested recursions

$$\begin{cases} a_{n+1} = a_n + k_{n+1} Z b_n \\ b_{n+1} = Z b_n + k_{n+1} a_n \end{cases} \quad \text{with} \quad Z = \begin{bmatrix} 0 & \cdots & & 0 \\ 1 & \ddots & & \vdots \\ \vdots & & \ddots & \\ 0 & \cdots & 1 & 0 \end{bmatrix}, \quad (5.8)$$

where

$$a_n^T = [a_0^n, a_1^n, \ldots, a_n^n, 0, \ldots, 0],$$
$$b_n^T = [a_n^n, a_{n-1}^n, \ldots, a_0^n, 0, \ldots, 0]$$

and

$$a_0^T = [1, 0, \ldots, 0], \qquad b_0^T = [0, \ldots, 0, 1].$$

5.3. Inverse Levinson–Durbin algorithm

As already mentioned, there is a one-to-one correspondence between a set of correlation coefficients r_i $(i = 0, n)$, a set (r_0, k_i) $(i = 1, n)$ of reflection coefficients, a set $(\alpha_n, a_i^n)(i = 1, n)$ of predictor coefficients and the associated residual energy (this is even clearer if the energy of the signal is normalized to $r_0 = 1$). This will be proven by inverting the LD algorithm from end-informations at order n. Two main connections will be considered: a_i to k_i (AITOKI), and a_i to r_i (AITORI).

AITOKI

The above vector recursions (5.3) may be easily reverted from $n + 1$ to n. This is achieved by combining the two equations in (5.8) with weights $(1, -k_{n+1})$ and $(-k_{n+1}, 1)$, respectively. The coefficient k_n is adjusted so as to annihilate the last coordinate in the resulting combination, thus retrieving an order n predictor (recall that the last coordinate of a_{n+1} is precisely k_{n+1}). The inverse LD algorithm is therefore written under vector form as

$$\begin{cases} \left(1 - k_{n+1}^2\right)a_n = a_{n+1} - k_{n+1}b_{n+1}, \\ \left(1 - k_{n+1}^2\right)2b_n = b_{n+1} - k_{n+1}a_{n+1}. \end{cases} \tag{5.9}$$

From these equations, it is clear that Zb_n may be shifted back with no loss of information, before going to the next recursion step. This inverse algorithm reconstructs the matrices A and B entirely from their last columns, and compute the set of reflection coefficients k_i. It can be shown that this algorithm coincides with the well-known Schur–Cohn criterion for testing the stability of a discrete linear system. The k_i coincide with the Schur–Cohn parameters, and if

$$|k_i| < 1, \quad \text{for} \quad \forall i = n, r,$$

the polynomial $A_n(z)$ with coefficients a_i^n has all its zeros inside the unit circle (the transfer function $1/A_n(z)$ is stable).

Note. The case where $|k_i| = 1$ has been excluded here. It corresponds to the singular case of linear prediction, where $\alpha_i = 0$, Det $R_i = 0$ (deterministic signals).

AITORI

The second problem consists in computing the correlation coefficients r_i from the higher-order predictor coefficient a_i. Equivalently, after applying the previous step AITOKI, the problem is to compute the correlation matrix from its Choleski factor, according to eq. (5.7). In this equation the prediction error energies α_i are known from r_0 and k_i by

$$\alpha_{i+1} = \left(1 - k_{i+1}^2\right)\alpha_i, \qquad \alpha_0 = r_0.$$

As a consequence, a variable length recurrence relationship with coefficients a_i^n may be used to compute the r_i, with initial condition r_0,

$$\begin{cases} a_0^0 = \alpha_0 = r_0 & r_0, \\ a_1^1 + r_1 a_0^1 = 0 & r_0, \\ \cdots \qquad \cdots \\ a_n^n + r_1 a_{n-1}^n + \cdots + r_n a_0^n = 0 & r_0. \end{cases} \qquad (5.10)$$

In this triangular system, the r_i are easily computed taking $a_0^n = 1$ into account.

The problem of an efficient computation of the correlation coefficients generated by a given AR system has been attacked by various authors [8, 9, 10]. The above algorithm gives a fast (but not the only) way to compute these coefficients using linear prediction concepts.

5.4. Generalized Levinson algorithms

The fast algorithms introduced so far, are restricted to the (symmetrical) Toeplitz case. As discussed earlier, in many cases of interest (short-time windows for instance), the R matrix no longer exhibits the shift and centro-symmetry properties on which the LD recursions are based. The question is then how the fast algorithms can be extended to a broader class of matrices, including close-to-Toeplitz matrices, using the concept of displacement rank. More precisely, we will instead consider the distance of the given matrix R to Toeplitz by assuming that R is a Toeplitz matrix perturbed by a (small) number of triangular Toeplitz factors. This assumption is not necessary, but will create a clearer connection to the LD algorithm and its forward and backward predictors.

The first authors who understood the significance of perturbating a Toeplitz matrix by triangular factors, although without displacement rank interpretation, were Mullis and Roberts [11]. They dealt with first-(impulse response) and second-(correlation) order information in designing ARMA models. They solved the linear prediction problem for (displacement rank 3)

$$R = T - HH^{\mathsf{T}},$$

where T is symmetrical (correlation) and H is triangular (impulse response) Toeplitz. An important contribution was made by Morf et al.

[12] by introducing a fast algorithm for solving the covariance case (displacement rank 4).

The generalized Levinson algorithm will be developed here following the lines of the classical correlation case, but updating a number of vectors (forward, backward predictors, auxiliary vectors) equal to the displacement rank of R. The description will be restricted to displacement rank 3 (prewindow case) for instance, but may be easily extended.

Let R be a covariance matrix of the form

$$R = T + SS^T = \begin{bmatrix} t_0 & t_1 & \cdots & t_n \\ t_1 & t_0 & & \vdots \\ \vdots & & \ddots & t_1 \\ t_n & \cdots & t_1 & t_0 \end{bmatrix}$$

$$+ \begin{bmatrix} s_0 & 0 & \cdots & 0 \\ s_1 & s_0 & & \vdots \\ \vdots & & \ddots & 0 \\ s_n & \cdots & s_1 & s_0 \end{bmatrix} \begin{bmatrix} s_0 & s_1 & \cdots & s_n \\ 0 & s_0 & & \vdots \\ & & \ddots & s_1 \\ 0 & 0 & & s_0 \end{bmatrix}.$$

$$(5.11)$$

It is easily shown that the upper and lower corner matrices R_0 and R_1 in R, have the following property,

$$R_1 = R_0 + ss^T, \quad \text{where} \quad s^T = [s_1, s_2, \ldots, s_n], \tag{5.12}$$

which is a direct consequence of the displacement rank 3 (i.e. $P = ss^T$). As an initial stage, we will assume that the linear equations in R_0 have been solved:

$$R_0 a_0 = \alpha_0 [1, 0, \ldots, 0]^T, \qquad R_0 b_0 = \beta_0 [0, \ldots, 0, 1]^T,$$
$$R_0 c_0 = s, \tag{5.13}$$

where the vectors a_0 and b_0 are recognized as the order $(n-1)$ forward and backward predictors (b_0 is no longer the reversed vector a_0) and c_0 is an auxiliary vector relating to the perturbation. The computation of the same quantities a, b and c for the higher-order matrix R will be achieved in 3 main steps.

Step 1: Solve the linear prediction problem for R_1.
R_1 is obtained from R_0 by a diadic perturbation. The perturbed solution is easy to compute, and give $a_1 b_1$ such that

$$R_1 a_1 = \alpha_1 [1, 0, \ldots, 0]^T, \qquad R_1 b_1 = \beta_1 [0, \ldots, 0\ 1]^T.$$

This computation is achieved using the auxiliary vector c_0.

Step 2: Solve the linear prediction problem for R.

This step is almost identical to the usual Levinson procedure, but uses the convenient $a_0b_0 a_1b_1$ to build the needed reflection coefficients. The forward predictor a of order n is, for instance, given by

$$\left[\begin{array}{c} R_0 \\ R_1 \end{array}\right] \left(\left[\begin{array}{c} 1 \\ a_0 \\ 0 \end{array}\right] + k \left[\begin{array}{c} 0 \\ b_1 \end{array}\right]\right)$$

$$= \left(\left[\begin{array}{c} \alpha_0 \\ 0 \\ \vdots \\ 0 \\ \gamma \end{array}\right] + k \left[\begin{array}{c} \gamma \\ 0 \\ \vdots \\ 0 \\ \beta_0 \end{array}\right]\right) = \left[\begin{array}{c} \alpha \\ 0 \\ \vdots \\ 0 \\ 0 \end{array}\right]. \tag{5.14}$$

A similar equation in $b_0 a_1$ gives the backward predictor b.

Step 3: Update the auxiliary vector c.

This auxiliary vector c is constructed using a reflection coefficient between the former c_0 and the backward predictor b.

$$\left[\begin{array}{c} R_0 \\ R_1 \end{array}\right] \left(\left[\begin{array}{c} c_0 \\ 0 \end{array}\right] + 1\left[\begin{array}{c} b \end{array}\right]\right)$$

$$= \left(\left[\begin{array}{c} s_1 \\ \vdots \\ s_{n-1} \\ \mu \end{array}\right] + 1\left[\begin{array}{c} 0 \\ \vdots \\ 0 \\ \beta \end{array}\right]\right) = \left[\begin{array}{c} s_1 \\ \vdots \\ s_{n-1} \\ s_n \end{array}\right], \tag{5.15}$$

where 1 in the above equation has been computed in order to adjust the last coordinate to s_n.

Step 1 is often referred to as the time-update step due to the particular initial (final) condition meaning of the perturbation vector in the pre-window (post window) case. Step 2 is the order update.

It must be noted that the algorithm uses δ (displacement rank) vectors and δ reflection coefficients at each recursion. As a consequence, the computational cost of this generalized algorithm, evaluated in terms of multiplies is approximately

$\mathcal{O}(\delta p^2)$ multiplies.

Increasing the displacement rank results in a linear increase of the complexity and the generalized LD algorithm remains advantageous for small values of δ. If the matrix under study has no particular shift structure, then $\delta = p$ and the cost of the Choleski factorization reaches the usual $\mathcal{O}(p^3)$ figure.

References

[1] N. Levinson, The Wiener RMS error criterion in filter design and prediction, J. Math and Phys. 25-1 (February 1946) 261–278.

[2] J. Durbin, The fitting of time-series models, Rev. Int. Statist. Inst. 28 (1960) 233–244.

[3] F. Itakura and S. Saito, Analysis synthesis telephony based on the maximum likelihood principle, Rep. 6th Int. Congr. on Acoustics, Budapest, ed. Y. Kohasi (August 1968) C17–20.

[4] M. Pagano, An algorithm for fitting auto-regressive schemes, J. of Roy. Stat. Soc., series C Applied Stat. 21-3 (1972) 274–281.

[5] J. Markel and A. Gray, On autocorrelation equations as applied to speech analysis, IEEE Trans. on Audio and Electroacoustics, 21-2 (April 1973) 69–79.

[6] J. Makhoul, Linear prediction, a tutorial review, Proc. IEEE 63-4 (April 1975) 561–579.

[7] M. Marden, The geometry of the zeros of a polynomial in a complex variable, Math. Surveys III (American Mathematical Society, New York, 1949) ch. 10.

[8] J.P. Dugre, A. Beex and L. Scharf, Generating covariance sequences and the calculation of quantization and rounding error variances in digital fitlers, IEEE Trans. Acoust. Speech & Signal Proc. 28-1 (February 1980) 102–104.

[9] D. Dzung, Generation of cross covariance sequences, IEEE Trans. Acoust. Speech & Signal Proc. 29-4 (August 1981) 922–923.

[10] B. Friedlander, Efficient computation of the covariance sequence of an autogressive process, IEEE Trans. Autom. Control 28-1 (January 1983) 97–99.

[11] C. Mullis and R. Roberts, The use of second-order information in the approximation of discrete-time linear systems, IEEE Trans. Acoust. Speech & Signal Proc. 24-3 (June 1976) 226–238.

[12] M. Morf, B. Dickinson, T. Kailath and A. Vieira: Efficient solution of covariance equations for linear prediction, IEEE Trans. Acoust. Speech & Signal Proc. 25-5 (October 1977) 429–433.

[13] J. Markel and A. Gray, Linear Prediction of Speech (Springer-Verlag, New York, 1976), ch. 3.

[14] B. Dickinson and J. Turner, Relfexion coefficient estimation using Choleski decomposition, IEEE Trans. Acoust. Speech & Signal Proc. 27-2 (April 1979) 146–149.

[15] G. Carayannis, An alternative formulation of the recursive solution of the covariance and auto-correlation equations, IEEE Trans. Acoust. Speech & Signal Proc. 25-6 (December 1977) 574–577.

[16] J.M. Delosme, Algorithms for Finite Shift Rank Processes, Ph.D. dissertation, (EE Dpt., Stanford University, CA, April 1982).

6. Fast Choleski factorization of *R*

The classical LD algorithm manipulates the linear prediction AR coefficients a_i to construct a Choleski factor of R^{-1}. In the mean time, according to eq. (5.7), a Choleski factor of R is constructed in the right-hand side, but it is only considered as a result. In the following, the emphasis will be set on constructing the Choleski factors of R itself (instead of R^{-1}), and the variable involved in the computations will therefore have the meaning of estimated impulse responses (MA or FIR filters).

When R is Toeplitz, this computation, theoretically equivalent to Bauer's method [1, 2], is performed efficiently (in terms of operations) by the Bareiss algorithm introduced in section 4 [3]. The same type of algorithm, generalized to block Toeplitz matrices, was independently developed by Rissanen [4]. More recently, using a different approach and emphasizing fixed point computation, Leroux and Gueguen proposed an impulse response algorithm (LG) quite similar in its formulation to the LD algorithm [5]. Finally, this class of algorithms was traced back to Schur [6] and extended to close-to-Toeplitz matrices, under the name of "generalized Schur" algorithms.

The formulation developed here will follow the lines of the above LG algorithm to stress the impulse-response meaning of the variables.

6.1. The impulse-response algorithm (LG)

Applying an order n AR model on the given time series y_t, leads to prediction residuals u_t^n, estimating the true input u_t,

$$a_0^n y_t + a_1^n y_{t-1} + \cdots + a_n^n y_{t-n} = u_t^n. \tag{6.1}$$

Multiplying both sides of eq. (6.1) by y_{t+i}, and taking the expectation, leads to

$$a_0^n r_i + a_1^n r_{i-1} + \cdots + a_n^n r_{i-n} = s_i^n. \tag{6.2}$$

If the order was correct, the input-output correlation s_i would coincide with the impulse response of the AR model h_i, with a causality condition

$$h_i = 0, \quad \text{for} \quad i < 0.$$

If the correct order is not reached, then the s_i^n estimate the impulse response. Causality is just partially satisfied by the orthogonality conditions. More precisely, the range of index i may be divided in 3 segments:

$$0 \leqslant i < N, \qquad s_i^n \qquad \text{estimate the impulse response } h_i,$$
$$-n \leqslant i < 0, \qquad s_i^n = 0 \quad \text{orthogonality conditions,}$$
$$-N < i < -n, \qquad s_i^n \qquad \text{auxiliary parameters.}$$

$$(6.3)$$

In vector form, the equations are identical to eq. (5.2) in the LD algorithm, but are extended beyond n, making use of $r_{n+1}, \ldots, r_N$, even when dealing with the order n predictor. As in the LD algorithm, the orthogonality condition may be expanded to the higher order by shifting, reverting and combining with a reflection coefficient, the estimated impulse-response vector with coordinates s_i^n. Thus the zero gap in (6.3) is gradually extended. A nice interpretation in the case of layered media modeling for seismic data is given by E. Robinson and S. Treitel in ref. [7].

The corresponding LG algorithm can then be written under scalar recurrence equations form

$$\begin{cases} k_{n+1} = -s_{n-1}^n/s_0^n, \\ s_0^{n+1} = \left(1 - k_{n+1}^2\right)s_0^n, \\ s_i^{n+1} = s_i^n + k_{n+1}s_{-n-i-1}^n, \quad \forall i. \end{cases}$$

$$(6.4)$$

These formulas are very close to eq. (5.3) with which they share common variables:

$$s_0^n = \alpha_n \qquad \text{energy of the prediction error,}$$
$$s_{-n-1}^n = \omega_n \quad \text{and } k_{n+1} \text{ reflection coefficient,}$$

but the main difference is the range of index i and the use of MA variables s_i^n (intercorrelations) instead of AR variables a_i^n. A direct, and rather beneficial, consequence is that in this algorithm all the variables are bounded:

$$|k_i| \leqslant 1, \qquad s_i^n \leqslant s_0^n r_0,$$

$$(6.5)$$

and this may be exploited for fixed-point implementation [5]. This is not the case for the LD algorithm which works out the Choleski factor of R^{-1} (instead of R) and has difficult scaling problems relating to the size of the determinants.

The following full-size matrix equations (6.6) and (6.7) give an illustration in terms of structure and notations for expanding the scalar equation (6.4) in the two indexes n and i. The backward description (6.7) is particularly useful and exhibits the Choleski factors of R (and R^{-1}).

Forward model description

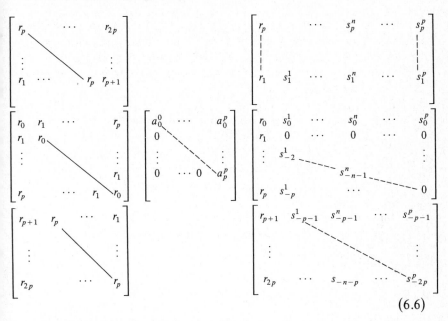

$$(6.6)$$

Backward model description

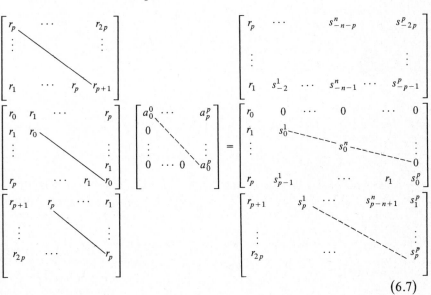

$$(6.7)$$

6.2. First vector form by columns

As previously done for the LD algorithm, the s_i^n in eq. (6.4) may be rearranged into convenient vector forms. A first vector form, obtained by sliding a constant length impulse response along the diagonal in eq. (6.7), will amount to a particular formulation of the Chandrasekar equations for computing Kalman gains [8].

The interpretation of Choleski factors as a set of time-varying synthetizers, generating the given covariance R, has been emphasized in the introduction of this section. The usual device for computing the innovation sequence from the observation y_t is a Kalman filter under state-space representation. We will show, using this first vector form, how the Kalman gains can be computed and "read out" from the Choleski factors [9, 10].

The LG algorithm can be set in vector form by columns by stacking the s_i^n up to a maximum allowed size p. Equations (6.4) are shared into two sets on index i:

Impulse response, $i \geq 0$

$$\begin{bmatrix} s_0^{n+1} \\ s_1^{n+1} \\ \vdots \\ s_p^{n+1} \end{bmatrix} = \begin{bmatrix} s_0^n \\ s_1^n \\ \vdots \\ s_p^n \end{bmatrix} + k_{n+1} \begin{bmatrix} s_{-n-1}^n \\ s_{-n-2}^n \\ \vdots \\ s_{-n-p-1}^n \end{bmatrix} \rightarrow s_0^{n+1} = s_0^n \left(1 - k_{n+1}^2 \right)$$

$$(6.8)$$

Auxiliary vectors, $i < -n$

$$\begin{bmatrix} s_{-n-1}^{n+1} \\ s_{-n-2}^{n+1} \\ \vdots \\ s_{-n-p-1}^{n+1} \end{bmatrix} = \begin{bmatrix} s_{-n-1}^n \\ s_{-n-2}^n \\ \vdots \\ s_{-n-p-1}^n \end{bmatrix} + k_{n+1} \begin{bmatrix} s_0^n \\ s_1^n \\ \vdots \\ s_p^n \end{bmatrix} \rightarrow s_{-n-1}^{n+1} = 0 \qquad (6.9)$$

Discarting the first rows, eqs. (6.8) and (6.9) may be written in condensed form by introducing two vectors v_n and w_n, and a shift operator $\mathscr{Z}$,

$$v_n = \left[s_{-n-1}^n, s_{-n-2}^n, \ldots, s_{n-p}^n \right]^{\mathrm{T}}, \qquad w_n = \left[s_1^n, s_2^n, \ldots, s_p^n \right],$$

$$\mathscr{Z} v_n = \left[s_{-n-2}^n, \ldots, s_{-n-p}^n, s_{-n-p-1}^n \right]^{\mathrm{T}}. \qquad (6.10)$$

The resulting vector form is

$$\begin{cases} w_{n+1} = w_n + k_n \mathcal{L} v_n, \\ v_{n+1} = \mathcal{L} v_n + k_n w_n \end{cases} \tag{6.11}$$

The only problem in eq. (6.10) is implementing the shift operator $\mathcal{L}$ and, more precisely, including the missing information s^n_{-n-p-1}. Two procedures are available:

(i) *Kalman filter.* When computing the innovation sequence from y_t by a Kalman filter, the generating system for y_t is supposed to be known, and consequently the true AR parameters are given. The shift operator $\mathcal{L}$ can then be implemented by the classical companion form matrix $\mathcal{A}$, which is used to extend the impulse-response sequence according to

$$s^n_{-n-p-1} = - \sum_{i=1}^{p} a_i s^n_{-n-p-1+i}$$

$$\begin{bmatrix} 0 & 1 & 0 & \cdots & 0 \\ 0 & 0 & 1 & & \vdots \\ \vdots & & & & \vdots \\ 0 & & & & 1 \\ -a_p & \cdots & & & -a_1 \end{bmatrix} \begin{bmatrix} s^n_{-n-1} \\ s^n_{-n-2} \\ \vdots \\ s^n_{-n-p} \end{bmatrix} = \begin{bmatrix} s^n_{-n-2} \\ \vdots \\ s^n_{-n-p-1} \end{bmatrix},$$

$$\text{or} \quad \mathcal{L} v_n = \mathcal{A} v_n. \tag{6.12}$$

Therefore, eqs. (6.10) may be rewritten as

$$\begin{cases} w_{n+1} = w_n + k_n \mathcal{A} v_n, \\ v_{n+1} = \mathcal{A} v_n + k_n w_n, \end{cases} \tag{6.13}$$

and can be recognized as the Chandrasekar equations [8]. The vector w_n is the Kalman gain vector (except for a multiplicative constant α_n).

(ii) *Extended equations.* Another way to implement the shift operator $\mathcal{L}$ is to extend the computations to a larger initial vector v_0 of dimension $2p$. This will need an extra p correlation coefficient to compute a p-dimensional Kalman gain. Instead of relying on the AR parameters (often unknown in many problems), the impulse-response sequence will be extended using the extra-correlation measures and the (implicit) estimated AR parameters at stage n. Some spurious terms will be included in the bottom of the $2p$-dimensional v_n, but the useful part will not be contaminated.

6.3. Second vector form by columns

If the only available information is r_i ($i = 0, p$), it is interesting to define the recurrence vectors so as to include no spurious term. This will be accomplished by moving the impulse-response vector row-wise, while sliding the auxiliary vector along the diagonal in eq. (6.7). This second vector form has recently been developed by Scharf and Demeure [11].

The following order p (dimension $p + 1$) vectors will now be defined:

$$\boldsymbol{h}_n^{\mathrm{T}} = \left[0, \ldots, 0, s_0^n, s_1^n, \ldots, s_{p-n}^0 \right],$$

$$\boldsymbol{g}_n = \left[0, \ldots, 0, 0, s_{-n-1}^n, \ldots, s_{-n-p}^n \right]. \tag{6.14}$$

These vectors appear in a natural way, when applying the forward and backward matrices A and B to the given correlation matrix. They are columns of the corresponding resulting G and H matrices:

$$
\begin{bmatrix}
r_0 & r_1 & \cdots & & r_p \\
r_1 & r_0 & & & \\
\vdots & & \ddots & r_1 & \\
r_p & & \cdots & r_1 & r_0
\end{bmatrix}
\begin{bmatrix}
a_0^0 & \cdots & a_p^p \\
\vdots & \ddots & \vdots \\
0 & \cdots & a_0^p
\end{bmatrix}
=
\begin{bmatrix}
h_0^0 & 0 & \cdots & & 0 \\
h_1^0 & h_0^1 & & & \\
\vdots & & \ddots & & 0 \\
h_p^0 & & \cdots & h_1^{p-1} & h_0^p
\end{bmatrix},
$$

$$\tag{6.15}$$

$$R_p \qquad\qquad B \quad \triangleq \qquad H \ \uparrow\!\!\!\!\_\text{with columns } \boldsymbol{h}_n$$

and

$$
\begin{bmatrix}
r_0 & r_1 & \cdots & & r_p \\
r_1 & r_0 & & & \\
\vdots & & \ddots & r_1 & \\
r_p & & \cdots & r_1 & r_0
\end{bmatrix}
\begin{bmatrix}
a_0^0 & \cdots & a_0^p \\
\vdots & \ddots & \vdots \\
0 & \cdots & a_p^p
\end{bmatrix}
=
\begin{bmatrix}
g_0^0 & g_0^1 & \cdots & g_0^p \\
g^0 & 0 & \cdots & 0 \\
\vdots & g^1 & \ddots & \vdots \\
g_{-p}^0 & \cdots & g_{-1}^{p-1} & 0
\end{bmatrix}
$$

$$\tag{6.16}$$

$$R_p \qquad\qquad A \quad \triangleq \qquad G \ \uparrow\!\!\!\!\_\text{with columns } \boldsymbol{g}_n$$

It is noted that eq. (6.13) is a more explicit formula for the Choleski factor H in eq. (6.7). The matrix G in eq. (6.14) is triangular except for the first row. This row may be cancelled by discarting the first row in R_p.

Taking the recurrence relationship into account, the following vector form is deduced:

$$
\begin{cases}
\boldsymbol{h}_{n+1} = Z\boldsymbol{h}_n + k_{n+1}\boldsymbol{g}_n, \\
\boldsymbol{g}_{n+1} = \boldsymbol{g}_n + k_{n+1}Z\boldsymbol{h}_n,
\end{cases}
\tag{6.17}
$$

with initial conditions,

$$\mathbf{h}_0^{\mathrm{T}} = [r_0, r_1, \ldots, r_p], \quad \text{and} \quad \mathbf{g}_0^{\mathrm{T}} = [0, r_1, \ldots, r_p].$$

In the vector forms (5.2) and (6.7), the LD and LG algorithms have a striking similarity. The role played by the backward predictor $\mathbf{b}_n$ is now played by its right-hand counterpart $\mathbf{h}_n$, and respectively for $\mathbf{a}_n$ and $\mathbf{g}_n$.

However, it must be noted that the LD proceeds from an empty vector $\mathbf{a}_0$ to a full vector $\mathbf{a}_n$, while the converse is done when using the LG algorithm.

6.4. Vector form by rows

Another useful way of defining vector recursions for the LG algorithm is to consider the rows of H and the diagonals of G. This algorithm by rows has been introduced from a different (lattice filter) point of view by Friedlander [12].

Let us consider the following ($p + 1$)-dimensional vectors:

$$\mathbf{e}_n^{\mathrm{T}} = [0, \ldots, 0, s_0^n, s_1^{n-1}, \ldots, s_n^0]$$

$$\mathbf{f}_n^{\mathrm{T}} = [0, \ldots, 0, s_{-n-1}^n, s_{-n-1}^{n-1}, \ldots, s_{-n-1}^0]. \tag{6.18}$$

The usual recursion formulas (6.4) may be arranged as

$$
\begin{bmatrix} s_{-n-1}^{n+1} \\ s_{-n-1}^n \\ \vdots \\ s_{-n-1}^1 \end{bmatrix}
=
\begin{bmatrix} s_{-n-1}^n \\ s_{-n-1}^{n-1} \\ \vdots \\ s_{-n-1}^0 \end{bmatrix}
+
\begin{bmatrix} k_{n+1} & 0 & \cdots & 0 \\ 0 & k_n & & \vdots \\ \vdots & & \ddots & 0 \\ 0 & \cdots & 0 & k_1 \end{bmatrix}
\begin{bmatrix} s_0^n \\ s_1^{n-1} \\ \vdots \\ s_n^0 \end{bmatrix}
$$

$$\rightarrow s_{-n-1}^{n+1} = 0 \tag{6.19}$$

and also as

$$
\begin{bmatrix} s_0^{n+1} \\ s_1^n \\ \vdots \\ s_{n+1}^1 \end{bmatrix}
=
\begin{bmatrix} s_0^n \\ s_1^{n-1} \\ \vdots \\ s_n^0 \end{bmatrix}
+
\begin{bmatrix} k_{n+1} & 0 & \cdots & 0 \\ 0 & k_n & & \vdots \\ \vdots & & \ddots & 0 \\ 0 & \cdots & 0 & k_1 \end{bmatrix}
\begin{bmatrix} s_{-n-1}^n \\ s_{-n-1}^{n-1} \\ \vdots \\ s_{-n-1}^0 \end{bmatrix}
$$

$$\rightarrow s_0^{n+1} = (1 - k_{n+1}^2) s_0^n \tag{6.20}$$

(the zero terms have been excluded from eqs. (6.19) and (6.20) for convenience of notation). The definition (6.18), along with the usual shift matrix Z, give the following vector form:

$$\begin{cases} Zf_n = f_n + Ke_n, \\ Ze_{n+1} = e_n + Kf_n, \end{cases} \tag{6.21}$$

where K is a diagonal matrix comprising the reflection coefficients k_{p+1} to k_1. It is noted that in fact each reflection coefficient k_n may be computed step by step by annihilating s_{-n-1}^{n+1} in eq. (6.19). Moreover, the way the new information merges into the equation, is by setting the last coordinate s_{n+1}^0 of e_{n+1} (which is not computed in eq. 6.20) to r_{n+1} (which in turn is read from the data).

Equations (6.21) have a quite convenient form since no recursion is operated on f_n (it is replaced by a simple shift). As a consequence, we can get a rather explicit formula for computing the e_n:

$$Ze_{n+1} = \left[I - K(I - Z)^{-1}K \right] e_n = \mathcal{H}e_n. \tag{6.22}$$

In a more detailed form, eq. (6.35) is written as

$$\begin{bmatrix} s_0^{n+1} \\ s_1^n \\ \vdots \\ s_n^1 \end{bmatrix} = \begin{bmatrix} \left(1 - k_{n+1}^2\right) & 0 & \cdots & 0 \\ -k_{n+1}k_n & \left(1 - k_n^2\right) & & \vdots \\ \vdots & & \ddots & 0 \\ -k_{n+1}k_1 & \cdots & -k_2k_1 & \left(1 - k_1^2\right) \end{bmatrix} \begin{bmatrix} s_0^n \\ s_1^{n-1} \\ \vdots \\ s_n^0 \end{bmatrix}.$$

$$\mathcal{H} \tag{6.23}$$

Equation (6.22) is self-contained as soon as the required reflection coefficients are computed (note that $\mathcal{H}$ is triangular), and it is applied from the initial conditions

$$e_0^{\mathrm{T}} = [0, \ldots, 0, r_0].$$

Moreover, informations r_i are fed in at each step.

Note. Equation (6.23) may be extended by one last row, assuming for convenience of notation that $k_i = 1$, for $i = 0$. This leads to eq. (6.24) in a straightforward manner.

6.5. Inverse impulse-response algorithm

It has been noted that the LD algorithm may be inverted in the sense that starting from end-informations a_i^n ($i = 1, n$) (last column of B) or from the k_i ($i = 1, n$) (first row of B), the whole Choleski factor B of R^{-1} may be reconstructed. As matrix H (Choleski factor of R) is the inverse transpose of B, except for a diagonal weight matrix, the same question may be raised for H. Computing H from its first column has already been realized through the direct LG algorithm. The question is therefore to compute back H from its last row or from the k_i. If available, this algorithm would also be a way to compute the correlation coefficient r_i from an impulse response h_t.

Supposing that the reflection coefficients and the last row of H are known, there is no difficulty in inverting eq. (6.21) in order to reconstruct H row by row. An important formula is deduced from the first equation in (6.21), by taking the last row

$$(I - Z) f_n = K e_n, \qquad f_n = (I - Z)^{-1} K e_n,$$

noting that the last element in f_n is nothing but r_{n+1}:

$$r_{n+1} = -e_n^T k, \quad \text{where} \quad k^T = [k_{n+1}, k_n, \ldots, k_1]. \tag{6.24}$$

As a consequence, for different values of n, we have

$$\begin{bmatrix} r_1 \\ r_2 \\ \vdots \\ r_{n+1} \end{bmatrix} = - \begin{bmatrix} h_0^0 & 0 & \cdots & 0 \\ & & & \vdots \\ h_1^0 & h_0^1 & & 0 \\ \vdots & & \ddots & \\ h_n^0 & \cdots & h_1^{n-1} & h_0^n \end{bmatrix} \begin{bmatrix} k_1 \\ k_2 \\ \vdots \\ k_{n+1} \end{bmatrix} = -Hk, \tag{6.25}$$

and the reflection coefficients k_i appear to be the input of the time-varying AR synthetizer (Choleski factor) to generate the correlation sequence as an output.

Using the usual normal equations, another relationship is established:

$$\begin{bmatrix} k_1 \\ k_2 \\ \vdots \\ k_{n+1} \end{bmatrix} = D^{-1} \begin{bmatrix} h_1^0 & & \cdots & h_n^0 \\ 0 & h_0^1 & & \vdots \\ \vdots & & \ddots & \\ 0 & \cdots & 0 & h_0^n \end{bmatrix} \begin{bmatrix} a_1 \\ a_2 \\ \vdots \\ a_{n+1} \end{bmatrix},$$

$$D = \text{Diag}[\alpha_0, \ldots, \alpha_n] \tag{6.26}$$

Note. An alternate expression for eqs. (6.25) and (6.26) includes $k_0 = 1$ in the k vector, and $a_0 = 1$ in the a vector.

When the k_i are not known, but the h^i_{n-i} (last row e_n of H) are given, no tractable inverse algorithm has been established at the best of our knowledge.

6.6. The Schur algorithm

As the displacement rank of R^{-1} was directly related to the complexity of finding the Choleski factors of R^{-1} (successive AR models) by generalized LD algorithms, the same is true, and even simpler, for the displacement rank of R. We will show that a clever algorithm that can be attributed to Schur provides a generalization of the LG algorithm for close-to-close Toeplitz matrices [13–21].

The principle behind the algorithm is to make a full use of the representation theorem in terms of the product of TTM's. This representation can indeed be understood as a first decisive step towards Choleski factorization, where the TTMs will be aggregated in one single non-Toeplitz triangular factor.

Suppose for instance that R has the $\delta = 2$ representation,

$$R = AB \pm CD.$$

The idea is to drive CD to zero by manipulation on $ABCD$ without changing R. For this purpose, some matrix identities can be used:

$$(1 - k^2)(AB - CD) = (A + kC)(B + kD)$$
$$- (kA + C)(kB + D),$$
$$(1 - k^2)(AB + CD) = (A - kC)(B - kD)$$
$$+ (kA + C)(kB + D). \tag{6.27}$$

The scalar k (reflection coefficient) will be chosen to introduce zeros in the new C matrix, however, the TTM components of R are redundant in terms of parameters. A simpler implementation will be provided using the corresponding displacement rank generators.

The Schur algorithm can then be described as a two-step procedure. In order not to obscure the essential matter, we start with the $\delta = 2$ case,

$$\nabla R = R - ZRZ^T = hh^T - gg^T. \tag{6.28}$$

Step 1. Modify the generators in eq. (6.28) by the matrix identities (6.27) in order to bring a zero on the first place in g:

$$R = (h + kg)(h + kg)^T - (kh + g)(kh + g)^T, \tag{6.29}$$

where

$$h^{\mathrm{T}} = (h + kg)^{\mathrm{T}} = [h_0 h'], \quad \text{and} \quad g^{\mathrm{T}} = (kh + g)^{\mathrm{T}} = [0 \ g'],$$

and k is a reflection coefficient.

Step 2. Reduce R by Schur's complement. Define

$$\tilde{R} = R - hh^{\mathrm{T}}, \quad \text{and} \quad \nabla\tilde{R} = \tilde{R} - Z\tilde{R}Z^{\mathrm{T}},$$

then

$$\nabla\tilde{R} = -Zhh^{\mathrm{T}}Z - gg^{\mathrm{T}}, \tag{6.30}$$

and in R the first row and column are zero so that step 1 can be applied to a lower-dimensional matrix.

After applying this procedure p times, one gets a sequence of columns h_n and g_n, with n zeros as first elements. As can be seen from eq. (6.30) and step 1, they follow the vector recursion

$$\begin{cases} h_{n+1} = Zh_n + k_n g_n, \\ g_{n+1} = g_n + k_n Zh_n, \end{cases} \tag{6.31}$$

with initial conditions defined, for instance, by selecting the following generators:

$$h_0^{\mathrm{T}} = [r_0, r_1, \ldots, r_p], \qquad g_0^{\mathrm{T}} = [0, r_1, r_2, \ldots, r_p].$$

This coincides with the vector form (6.15) of section 6.3.

However, the above procedure can be carried out for higher displacement rank close-to-Toeplitz matrices. Step 1 can be applied to each generators in sequence, annihilating the first coordinate of these vectors before applying step 2. But this can also be done by a single matrix step 1. Define Q as a Σ orthonormal matrix such that

$$Q\Sigma Q^{\mathrm{T}} = \Sigma,$$

where Σ is a signature matrix. Step 1 may then be replaced by selecting Q on G such that [16]

$$GQ = [hg_1 \cdots g_{\delta-1}]Q = \begin{bmatrix} h_0 & 0 & \cdots & 0 \\ h' & g_1' & \cdots & g_{\delta-1}' \end{bmatrix}.$$

This matrix step 1 is nothing but applying an Householder transformation on the columns of G so as to bring a first row of zeros, except for the first element h_0. However, this Householder transformation has to be generalized so as to encompass the signature matrix Σ. As a matter of fact, the generators with positive and negative signs may be dealt with separately, and eventually combined.

In this global Householder approach, the Q matrix comprises a set of $(\delta - 1)$ reflection coefficients. The previous generator-by-generator approach is associated with a sequence of Given's plane rotations, each providing a single reflection coefficient (the selected order matters). If the composed generators have the same sign in the signature matrix Σ, then it amounts to an ordinary $(\sin \varphi, \cos \varphi)$ rotation; if they have opposite signs, then it is an hyperbolic $(\operatorname{sh} \varphi, \operatorname{ch} \varphi)$ rotation. The latter case is well-known to be ill-conditioned from a numerical point of view, and for this reason, dealing first with the same sign generators is often preferred.

The over-all computational cost of the Schur algorithm is consequently of the order of

$$(\delta - 1)n \text{ multiplies}, \quad \text{and} \quad \delta \text{ divisions}$$

at each n; $n = p, 0$.

References

[1] F.L. Bauer, A direct iterative process for a Hurwitz decomposition of a polynomial, Arch. Electron. & Uebertragungstech. 9 (June 1955) 285–290.

[2] F.L. Bauer, Beitrage zur Entwicklung numerischer Verfahren für programmgesteuerte Rechenanlagen: Direkte faktorisierung eines Polynoms, Sitzungs Ber. Bayer. Akad. Wiss. xx (1956) 163–203.

[3] E.H. Bareiss, Numerical solution of linear equations with Toeplitz and vector Toeplitz matrices, Numer. Math. 13 (1969) 404–424.

[4] J. Rissanen, Algorithms for triangular decomposition of block Hankel and Toeplitz matrices with application to factoring positive matrix polynomials, Math. Comput. 27-121 (January 1973) 147–154.

[5] J. Leroux and C. Gueguen, A fixed-point computation of parcor coefficients, IEEE Trans. Acoust. Speech & Signal Proc. 25-3 (June 1977) 257–259.

[6] J. Schur, Über Potenzreichen, die im Innern des Einheitskreise beschränkt sind. (On series which are bounded in the unit circle). Z. für die heine und Angewandte Math 147-4 (1917) 205–232.

[7] E. Robinson and S. Treitel, Maximum entropy and the relation of the partial autocorrelation, IEEE Trans. Acoust. Speech & Signal Proc. 28-2 (April 1980) 224–228.

[8] M. Morf, G. Sidhu and T. Kailath, Some new algoithms for recursive estimation on constant linear discrete systems, IEEE Trans. Autom. Control 19-8 (August 1974) 315–323.

[9] C. Gueguen and L. Scharf, Exact maximum likelihood identification of ARMA models; a signal processing perspective, in: Signal Processing: Theories and Applications (Proc. 1st Eur. Conf. EUSIPCO-80, Lausanne, Switzerland, 16–18 September 1980) eds. M. Kunt and F. De Coulon (North-Holland, Amsterdam, 1981) pp. 759–769.

[10] J.P. Dugre, Parametric Spectrum Analysis of Stationary Random Sequences, Ph.D. dissertation (EE Dpt., Colorado State University, Fort Collins, CO, June 1981).

[11] L. Scharf and C. Demeure, Internal variables in the lattice for Choleski factorization, Kalman filtering, and model identification, Techn. Rep. (Office of Naval Research, December 1984).

[12] B. Friedlander, Lattice filters for adaptive processing. Proc. IEEE 70-8 (August 1982) 829–867.

[13] J.M. Delosme, Y. Genin, M. Morf and P. Van Dooren, Σ Contractive embeddings and interpretation of some algorithms for recursive estimation, Proc. 14th Asilomar Conf. on Control Systems and Communications (November 1980) 25–28.

[14] J.M. Delosme and M. Morf, Mixed and minimal representations for Toeplitz and related systems, Proc. 14th Asilomar Conf. on CS and C 19–24, Monterey, Nov. 1980.

[15] M. Morf, Fast Algorithms for Multivariable Systems, Ph.D. dissertation (EE Dpt, Stanford University, CA, August 1974).

[16] M. Morf and J.M. Delosme, Matrix decompositions and inversions via elementary signature-orthogonal transformations, Proc. Int. Symp. on Mathematical Models, San Francisco, May 1981.

[17] P. Dewilde, A. Vieira and T. Kailath, On a generalized Szegö–Levinson realization algorithm for optimal linear predictors based on a network synthesis approach, IEEE Trans. Circuits & Syst. 25-9, (September 1978) 663–675.

[18] P. Dewilde and H. Dym, Schur recursions, error formulas, and convergence of rational estimators for stationary stochastic processes, IEEE Trans. Inf. Theory 27-4 (July 1981) 446–461.

[19] H. Lev-Ari and T. Kailath, Schur and Levinson algorithms for nonstationary processes, Proc. Int. Conf. Acoust. Speech & Signal Process., Atlanta, March 1981 pp. 860–864.

[20] H. Lev-Ari and T. Kailath, On generalized Schur and Levinson–Szegö algorithms for quasi-stationary processes. Proc. 20th Conf. on Decision and Control, San Diego, December 1981 pp. 1077–1080.

[21] H. Lev-Ari and T. Kailath, Parametrization and modelling of non-stationary processes, IEEE Trans. Inf. Theory 30-1 (January 1984) 2–16.

7. Concluding remarks

In this course we have shown that Toeplitz matrices appeared as a useful description in many problems relating to signal and systems analysis, and, generally, are associated with stationarity and linearity properties.

A structural index δ, introduced by Kailath and co-workers [1, 2], defined by comparing a given matrix R with its shifted version along the diagonal,

$$R - ZRZ^T = G\Sigma G^T, \tag{7.1}$$

where Z is the shift operator and Σ is a signature matrix with rank δ, was a convenient tool for evaluating the distance to Toeplitz of this matrix. More generally, this index δ measures the complexity of a matrix in terms of generic components defined as the algebraic sum of squares of triangular Toeplitz matrices. Many usual covariance matrix estimators happen to have a low displacement rank in this sense.

This index being invariant under Schur complement, the Choleski factorization of R may be deduced from Σ orthogonal transformations on the generators (columns of G). At the same time, the close relationship between the displacement ranks of R and R^{-1} give general inversion formulas and algorithms for computing the Choleski factors of R^{-1} and R as well.

We have shown that many useful and well-known algorithms of the Levinson type originally developed for inverting Toeplitz matrices, could be extended to close-to-Toeplitz non-stationary covariance matrices using displacement ranks. Moreover, we have emphasized the fact that working on some impulse-response estimates (Choleski factors of R) was even more beneficial than working on the usual linear prediction coefficients of an AR model (Choleski factors of R^{-1}). The computational complexity of these algorithms is of the order δp^2 (where $p + 1$ is the dimension of the matrix) and they are therefore classified in terms of their distance to Toeplitz matrices.

Some important developments in the field were, however, not covered in this course.

(i) *Lattice structures.* The common framework underlying the factorizations of R and R^{-1} is the computation of the so-called "reflection coefficients" as "Schur parameters". The corresponding FIR on IIR filters are consequently amenable to the same lattice structure. These lattice structures may be extended to δ-stationary covariances [3–7]. It is also useful to consider special linear phase lattices including a symmetrical and an anti-symmetrical filter instead of the usual forward and backward models. This brings up even faster versions in the algorithms of the Levinson class.

(ii) *Non-symmetric and matrix multichannel extensions.* Non-symmetric Toeplitz matrices frequently occur when dealing with cross-covariances, instrumental variables, or when computing the AR part of an ARMA by going to the tail of the correlation sequence [8]. Non-symmetrical lattices are also useful for representing general ARMA models [9]. In the standard regular case, most of the above algorithms may be extended to multichannel vector time series and block Toeplitz matrices, including normalized versions [10].

(iii) *Recursive fast algorithms.* The previous linear predictors, the impulse-response model lattice filters, may be updated sequentially as new samples become available. This amounts to updating the generators of the inverse covariance matrix R_t^{-1} (or of R_t itself giving rise to "fast Kalman filter" algorithms). The corresponding adaptive algorithms remain optimal in the least-square sense for a computation cost (from $5p$ to $8p$ per step) just slightly above the suboptimal gradient versions ($2p$ per step) [11–18].

The above discussion also leaves some room for further research. As defined, the displacement rank being a structural index is a little too algebraic in nature and leaves no freedom for approximation. A naive estimator of the covariance (and even the unbiased estimator, for example) will not feature a low displacement rank. Even if we take into account that an exponential window may be included in the displacement rank definition, the class of admissible (pre-, post-) windows is too narrow. This is the reason why some fundamental work still has to be done on this important concept of displacement ranks:

(a) *Lyapunov functions.* It has been noticed early that the above definition (7.1) was nothing but a special case of the Lyapunov equation of state space analysis of linear discrete-time systems,

$$R - ARA^T = BB^T, \tag{7.2}$$

where R is the state covariance, A the state dynamic and B the input matrix. If A has all its eigenvalues inside the unit circle, then eq. (7.2) leads to the infinite covariance representation,

$$R = \left(\sum_{i=0}^{\infty} A^i B \right) \left(\sum_{i=0}^{\infty} A^i B \right)^T, \tag{7.3}$$

where the controllability matrix of the system is easily recognized. Displacement ranks are thus defined by a simple shift dynamics, thus implying a finite representation ($A = Z$ is nihilpotent) [19].

(b) *Other representation operators.* Among other examples of dynamic operators of interest in eq. (7.2), is the k-circulant matrix Q such that

$$Q^{n+1} = kI,$$

where $(n + 1)$ is the dimension of Q (note that for $k = 0$, $Q = Z$).

Using the same arguments as for shift displacement ranks, we have for $\delta = 1$,

$$R - QRQ^T = \mathbf{gg}, \tag{7.4}$$

which implies the representation ($|k| \neq 1$)

$$(1 - k^2)R = \sum_{i=0}^{n} (Q^i g)(Q^i g)^{\mathrm{T}}.\tag{7.5}$$

This representation is well adapted to estimators of the covariance matrix of the type introduced by T.W. Anderson (see section 1.3.), where the entries in he data matrix are periodized with or without time-reversal.

(c) *Generalized ranks.* It has been noted that the notion of rank, as used in the displacement rank definition, is a generalized concept due to the signature matrix Σ. This leaves us with an unclear interpretation problem associated to negative covariances (complex signals, impedance versus reactance, ...). Moreover, the matrix in the right-hand side of (7.1) is no longer a grammian matrix of linear dependence. The usual tools of algebra must be generalized as to include Σ (e.g. Σ orthogonal householder matrices). With the corresponding inner product, the norm of a vector may be zero without the vector being zero. As a consequence, the question of minimality, of constance of displacement rank after perturbation, are somewhat intricate. This raises the question of necessary and sufficient conditions for a matrix to be Toeplitz ($\delta = 2$ is not sufficient), and of defining classes of matrices which is congruent to Toeplitz [20].

(d) *Other displacement equations.* Sticking with the shift operator Z as a representation operator for R, the definition equation (7.1) can also be replaced in order to exhibit some other special structural features from this matrix:

$$R = RZ - ZR,$$
$$R = R - \lambda ZRZ^{\mathrm{T}}.\tag{7.6}$$

and

$$R = -Z^{\mathrm{T}}RZ + R - ZRZ^{\mathrm{T}},$$
$$R = R + \alpha_1 ZRZ^{\mathrm{T}} + \alpha_2 Z^2 RZ^{2\mathrm{T}} + \cdots\tag{7.7}$$

The first order description in eq. (7.6) is used for extending the displacement rank concept to exponential weighting (forgetting factor λ) windows [7]. The generalized polynomial description in eq. (7.7) with adjustable coefficients α_i, was used to model covariances of constant systems driven by time-varying (auto-regressive) gains [21].

The question that finally arises at the end of this series of lectures, is the following: is there any way to recognize that a given matrix has a fast algorithm? Other patterned matrices than close-to-Toeplitz matrices have already been identified as having fast inversion algorithms [22–25], and there is no evidence of a necessary and sufficient condition. The question is then twofold:

–finding a convenient definition of a generalized displacement concept, hopefully containing some free adjustable continuous parameters, providing a low rank representation when applied to the matrix.

–in relation to the objective (e.g. inversion, Choleski factorization, eigenvector computation, ...), using the invariance of this definition under a class of transformations (e.g. inversion, Schur complement, ...).

This surely leaves open a large number of directions for future investigation.

References

[1] T. Kailath, S.Y. Kung and M. Morf, Displacement ranks of matrices and linear equations, J. Math. Anal. & Appl. 68 (1979) 395–407; Bull Am. Math. Soc. 1 (September 1979) 769–773.

[2] H. Lev-Ari and T. Kailath, Lattice filter parametrization and modeling of non-stationary processes, IEEE Trans. Inf. Theory 30-1 (January 1984) 2–16.

[3] J. Makhoul, Stable and efficient lattice methods for linear prediction, IEEE Trans. Acoust. Speech & Signal Proc. 25-5 (October 1977) 423–428.

[4] B. Friedlander, Lattice fitlers for adaptive processing, Proc. IEEE 70-8 (August 1982) 829–867.

[5] B. Friedlander, Lattice methods for spectral estimation, Proc. IEEE 70-9 (September 1982) 990–1017.

[6] T. Kailath, Time-variant and time-invariant lattice filters for non-stationary processes, Proc. Symp. sur les Algorithmes rapides pour l'Analyse des Signaux et Systèmes, Aussois 1981; Outils et modèles mathematiques pur d'analyse des signaux et systèmes, Recherch Coopérative du Programme 567, Vol. 2, pp. 417–467, Ed. CNRS, 1982.

[7] H. Lev-Ari, T. Kailath and J. Cioffi, Least-squares adaptive lattice and transversal filters: a unified geometric theory, IEEE Trans. Inf. Theory, special issue Adap. Filt., 30-2 (March 1984) 222–236.

[8] B. Friedlander, Instrumental variable methods for ARMA spectral estimation, IEEE Trans. Acoust. Speech & Signal Proc. 31-2 (April 1983) 404–415.

[9] J. Leroux, Non-symmetric lattice structure for pole-zero filters, Signal Process. 6-4 (August 1984) 331–335.

[10] M. Vax and T. Kailath, Efficient inversion of Toeplitz–block Toeplitz matrices, IEEE Trans. Acoust. Speech & Signal Proc. vol 31-5 (October 1983) 1218–1221.

[11] L. Ljung, M. Morf and D. Falconer, Fast calculation of gain matrices for recursive estimation schemes, Int. J. Control 27-1 (1978) 1–19.

[12] D. Lee, M. Morf and B. Friedlander, Recursive least-squares ladder estimation algorithms, IEEE Trans. Acoust. Speech & Signal Proc. 29-3 (June 1981) 627–641.

[13] C. Samson, A unified treatment of fast algorithms for identification, Int. J. Control 35-5 (May 1982) 909–934.

[14] G. Carayannis, D. Manolakis and N. Kaloupsidis, A fast sequential algorithm for least-square filtering and prediction, IEEE Trans. Acoust. Speech & Signal Proc. 31-6 (December 1983) 1395–1402.

[15] B. Porat, Pure time updates for least-squares lattice algorithms, IEEE Trans. Autom. Control 28-8 (August 1983) 865–866.

[16] D. Lin, On digital implementation of the fast Kalman algorithm, IEEE Trans. Acoust. Speech & Signal Proc. 32-5 (October 1984) 998–1005.

[17] J. Cioffi and T. Kailath, Fast recursive least-squares transversal filters for adaptive filtering, IEEE Trans. Acoust. Speech & Signal Proc. 32-2 (April 1984) 304–337.

[18] P. Fabre and C. Gueguen, Improvement of the fast recursive LS algorithms via normalization; a comparative study, IEEE Trans. Acoust. Speech & Signal Proc., accepted for publication.

[19] Y. Genin, P. Vandooren, T. Kailath, J.M. Delosme, M. Morf, On Σ lossless transfer functions and related questions, J. Linear Algebra Appl. 50 (April 1983) 251–275.

[20] P. Delsarte, Y. Genin, Y. Kamp, On the class of positive definite matrices equivalent to Toeplitz matrices, to appear.

[21] C. Gueguen, Sur une mesure du caractère non stationnaire, 9ième Colloque GRETSI, Vol. 1, Nice, Mai 1983 pp. 27–33.

[22] G. Carayannis, N. Kaloupsidis, D. Manolakis, Fast recursive algorithms for a class of linear equations, IEEE Trans. Acoust. Speech & Signal Proc. 30-2 (April 1982) 227–239.

[23] B. Picinbono, Fast algorithms for Brownian matrices, IEEE Trans. Acoust. Speech & Signal Proc. 31-3 (June 1983) 512–514.

[24] H. Krishna, S. Morega, Fast $0(n)$ complexity algorithms for diagonal innovation matrices, IEEE Trans. Acoust. Speech & Signal Proc. 32-6 (December 1984) 1189–1193.

[25] B. Picinbono, M. Benidir, On a class of signals and matrices giving very fast algorithms for linear problems, Signal Process. 9-4 (December 1985) 215–224.

COURSE 13

VLSI ARRAY PROCESSORS FOR DIGITAL SIGNAL PROCESSING*

SUN-YUAN KUNG

*University of Southern California, Signal & Image Processing Institute,
Department of Electrical Engineering
University Park / MC-0272, Los Angeles, CA 90089, USA*

*This research was supported in part by the National Science Foundation under Grant
ECS-82-13358, and by the Office of Naval Research under Contracts N00014-85-K-0469
and N00014-85-K-0599.

J.L. Lacoume, T.S. Durrani and R. Stora, eds.
Les Houches, Session XLV, 1985
Traitement du signal / Signal processing
© *Elsevier Science Publishers B.V., 1987*

Contents

1. Introduction

High-speed signal processing depends critically on parallel processor technology. In most applications, general-purpose parallel computers cannot offer satisfactory real-time processing speed due to severe system overhead. Therefore, to real-time digital signal processing (DSP) systems, special-purpose array processors have become the only appealing alternative. It is important to note that, in designing or using such array processors, most signal processing algorithms share the critical attributes of regularity, recursiveness, and local communication. These properties are effectively exploited in the innovative systolic and wavefront array processors. These arrays maximize the strength of VLSI (Very Large Scale Integration) in terms of intensive and pipelined computing and yet circumvent its main limitation on communication. The application domain of such array processors covers a very broad range, including digital filtering, spectrum estimation, adaptive array processing, image/vision processing, and seismic and tomographic signal processing. This article provides a general overview of VLSI array processors and a unified treatment from algorithm, architecture and application perspectives.

The basic discipline in a top-down design methodology—as depicted in fig. 1—depends on a fundamental understanding of algorithm, architecture, and application. Note that the boundary between software and hardware has become increasingly vague in the environment of VLSI system design. This enhances the already prevailing roles of the algorithm analyses and the mappings of algorithms to architectures. Therefore, a very broad spectrum of innovations will be required for obtaining highly-parallel array processing, e.g. new ideas on communication/computation trade-offs, parallelism extractions, array architectures, programming techniques, processor/structure primitives and numerical performances of DSP algorithms.

784

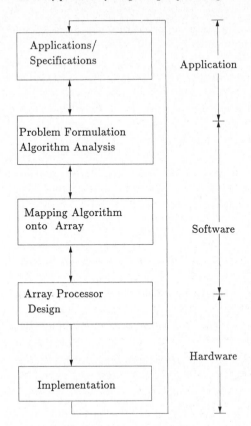

Vertically Integrated VLSI System Design

Fig. 1. Top-down design integration: In striving for a cohesive exploration of the overall implications of VLSI, cross-disciplinary discussion on application, algorithm and architecture is necessary. In fact, *integration* will be the keyword in VLSI. This means that innovations on a very broad spectrum of disciplinaries, including algorithm analyses, parallelism extractions, array architectures, programming techniques, functional primitives, structural primitives and numerical performance of DSP algorithms will be needed.

1.1. Impacts of VLSI device technology

VLSI architecture enjoys a major advantage of being very scalable technologically (Mead and Conway 1980). This means that the efforts of architecture redesign will be very minor when the device technology is scaled down to ultra-submicron level. However, as chip size is increased the interconnection problems will become very severe. Before long, chip cost, performance and speed will be determined primarily by interconnection delay and area. Therefore, VLSI device technology does not simply offer a promising future, but also creates some new design constraints. For example, modular building blocks and the alleviation of the burden of global interconnection are often essential in VLSI design.

Scaling effects.   In the scaling of geometry, we often assume that all the dimensions, as well as the voltages and currents on the chip, are scaled down by a factor α ($\alpha > 1$ implies that sizes or levels are shrinking). When scaling down the linear dimensions of a transistor by α, the number of transistors that can be placed on a chip of given size scales up by α^2. If the average interconnection length is not scaled down with the same factor α, the interconnection delay may actually increase. When the delay time of the circuit depends largely on the interconnection delay (instead of on the logic-gate delay), minimal and local interconnections will become an essential factor for an effective realization of the VLSI circuits.

1.2. VLSI architectural design principles

VLSI architectures should exploit the potential of the VLSI technology and also take into account: (i) the layout constraint and the resultant interconnection costs in terms of area and time; and (ii) the cost of VLSI processors as measured by silicon area and pin count. VLSI architecture design strategies stress modularity, regularity of data and control paths, local communication, and massive parallelism.

There are several key properties of VLSI (Randell and Treleaven 1982):
(1) Design complexity is critical.
(2) Wires occupy most space on a circuit.
(3) Communication problems degrade operation.
(4) Concurrency is required for performance.

(5) Input/output and computation must be balanced.
In general, the criteria for VLSI architectures are:
(a) Modularity and effective utilization of building blocks.
(b) Simple and regular data and control paths.
(c) Localized or reduced interconnections.
(d) Multiple use of each input data item to minimize I/O problem.
(e) Extensive concurrency to improve performance.
Some design principles are summarized below.

Homogeneity. In VLSI, there is an emphasis on keeping the overall architecture as *regular* and *modular* as possible, thus reducing the overall complexity. For example, memory and processing power will be relatively cheap as a result of high regularity and modularity. Even in the communication or wiring, a careful algorithmic study may help create some form of regularity. This depends on the special arrangements, realized in the course of topological mappings from algorithms to architectures.

Pipelining. In many DSP applications, throughput rate often represents the overriding factor dictating the system performance. In order to optimize throughput, a different design choice is often made than that of minimizing the total processing time (latency). Pipeline techniques fit naturally in our aim of improving throughput rate. Especially, for a majority of signal processing algorithms, suitable pipelining techniques are now well-established. A prominent example is the systolic/wavefront array discussed in section 2.

For signal processing arrays, pipelining at all levels should be pursued. It may bring about an extra order of magnitude in performance with very little additional hardware. Although most of the current array processors stress only word-level pipelining, the new trend is to exploit the potential of multiple-level pipelining (i.e. combined pipelining in all the bit-level, word-level and array-level granularities).

Locality. "Principle of locality is seen at every level of VLSI design" (Seitz 1984). In systolic arrays, both spatial locality and temporal locality are stressed (S.-Y. Kung 1984). The notion of locality can have two meanings in array processor designs: *localized data transactions* and *localized control flow*. In fact, most recursive signal processing algorithms permits both locality features; and they are fully exploited in the design of wavefront arrays as we shall elaborate in section 2.

Communication. In VLSI technology, computations per se are becoming very inexpensive and easily affordable. Therefore, the most critical factor in VLSI design is: *communication*. Architectures which balance communication and computation and circumvent communication bottlenecks with minimum hardware cost will eventually play a dominating role in VLSI systems.

CAD tools. For special-purpose array processors, the system specifications required by the applications may change significantly if the development cycle is too long; therefore, fast-turnaround implementation is critical. CAD tools for all levels of array processor design are essential. For VLSI implementation, the development of simplified design rules and structured design methodology have already allowed system designers to design their chips quickly. Silicon compiler technology is also becoming mature. It is therefore important that a proper high-level (array processor) structured design and description tools be developed that will be compatible with the existing back-end (low-level) CAD tools.

2. VLSI array processors

A very promising solution to the real-time requirement of signal processing is to use special-purpose array processors and to maximize the processing concurrency by either pipeline processing or parallel processing, or both. As long as communication in VLSI remains restrictive, locally-interconnected arrays will be of great importance. An increase of efficiency can be expected if the algorithm arranges for a balanced distribution of work load while observing the requirement of locality, i.e. short communication paths. These properties of load distribution and information flow serve as a guideline to the designer of VLSI algorithms and eventually lead to new designs of architecture and language.

The first such special-purpose VLSI architectures are systolic and wavefront arrays, which boast tremendously massive concurrency. *The concurrency in the systolic/wavefront arrays is derived from pipeline processing or parallel processing, or both.* These processing strategies are illustrated in fig. 2. The notion of combined pipeline and parallel processing will become more evident when we demonstrate below how parallel processing "computational wavefronts" are pipelined successively through the processor arrays.

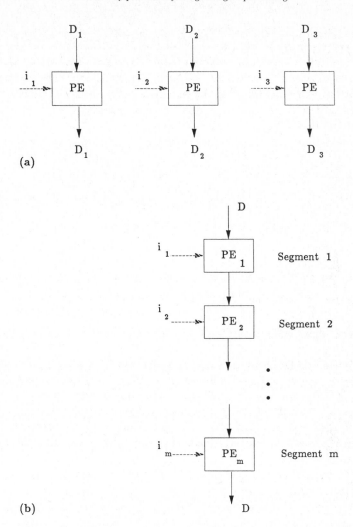

(a)

(b)

Fig. 2. Array processors derive a massive concurrency from both parallel and pipeline processing schemes (Hayes 1978). (a) Parallel processing means that all the processes defined in terms of the data (D_i) and the instructions (i_i) may directly access the m processors in parallel and keep all the processors busy. (b) Pipeline processing means that a process is decomposed into many subprocesses, which are pipelined through m processors, i.e. m segments aligned in a chain, and each subprocess will be processed one after another. For each subprocess coming out of the array, there will be a processor vacant and ready to receive and handle a new subprocess immediately. Therefore, all the m processors can again be kept busy all the time by the pipeline technique.

Array processors and algorithm expressions. A fundamental issue on mapping algorithms onto array is *how to express parallel algorithms in a notation which is easy to understand by humans and possible to compile into efficient VLSI array processors.* Thus, a powerful expression of array algorithms will be essential to the design of arrays. This course proposes primarily three ways of array algorithm expression: systolic, wavefront and signal-flow-graph (SFG) expressions.

To understand the roles of parallel and pipeline processing in array processors, we first examine a very popular systolic array processing example, as depicted in fig. 3.

The picture does not explicitly indicate the separate roles associated with either parallel or pipeline processing schemes. It is possible to clarify their individual roles by using a simpler SFG representation.

2.1. Signal flow graph (SFG)

The prime tool advocated here is that of the signal flow graph, consisting of *nodes* and *edges* (Mason 1953, S.-Y. Kung 1984), as illustrated in fig. 4. The SFG representations have been popularly used for signal processing flow diagrams, such as FFT, digital filters and many other domains of signal and system applications. For a reader from the signal processing community, it should be particularly enlightening to see how to apply SFG analysis to signal processing architecture designs.

The descriptions of array processing activities, in terms of the SFG representations, are often easy to comprehend. A typical example used for illustrating a two-dimensional array operation is matrix multiplication, as discussed in fig. 5 in section 3.

The abstraction provided by the SFG is very powerful to use, and yet the transformation of a SFG description to a wavefront or systolic array can be accomplished automatically, as discussed in section 3.1. In fact, *a systolic array can be considered as an SFG array in combination with pipelining and retiming.* To accomplish an ultimate abstraction, the SFG array often exhibits only the parallel processing part explicitly while leaving the pipelining part implicitly expressed. (The pipelining will then later be made explicit. Examples will be provided below.) In our discussion, we shall first map parallel algorithms onto SFG arrays and then convert them into systolic arrays.

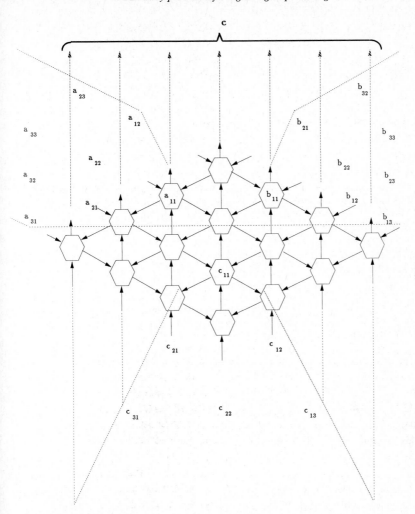

Fig. 3. A hexagonal systolic array for matrix multiplication. The space–time activities associated with the data movements are implied by the initial data distribution and the speed and direction of data movements. However, the problem is that the picture does not explicitly depict the parallel and pipeline processing schemes used.

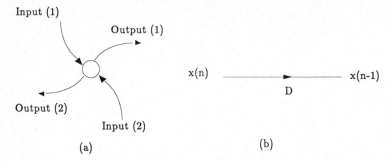

Input (1)

Output (1)

x(n) ⟶ x(n-1)

D

Output (2)

Input (2)

(a)   (b)

Fig. 4. Examples of signal flow graph notation: (a) An operation node with (two) inputs and (two) outputs; (b) an edge as a delay operator. *Notations:* In general, a *node* is often denoted by a circle representing an arithmetic or logic function *performed with zero delay*, such as multiply, add, etc. (see fig. 2a). An *edge*, on the other hand, denotes either a function or a delay. Unless otherwise specified, for a large class of signal processing SFG's, the following conventions are adopted for convenience. When an edge is labeled with a capital letter D (or D', $2D$, etc.), it represents a time-delay operator with delay-time D (or D', $2D$, etc.) (see fig. 2b). A node is considered to be delay-free, unless otherwise specified. In fact, the SFG representation derives its power from the assumption that the computations in the *node* are delay free, warranting simpler snapshot descriptions than the systolic counterpart. Consequently, the undertaking of tracing the detailed space–time activities associated with pipelining is simplified.

2.2. Systolic array

Systolic arrays are very amenable to VLSI implementation. It is especially suitable to a special class of computation-bound algorithms and takes advantage of their regular, localized data flow. "A systolic system is a network of processors which rhythmically compute and pass data through the system. Physiologists use the word 'systole' to refer to the rhythmically recurrent contraction of the heart and arteries which pulses blood through the body. In a systolic computing system, the function of a processor is analogous to that of the heart. Every processor regularly pumps data in and out, each time performing some short computation, so that a regular flow of data is kept up in the network" (Kung and Leiserson 1978).

For example, it is shown by H.T. Kung and Leiserson (1978) that some basic "inner product" processing elements (PE's) (each performing the operation $Y < -Y + A * B$) can be *locally* connected together to perform digital filtering, matrix multiplication and other related operations. In general, the data movements in a systolic array are pre-arranged and are described in terms of the "snapshots" of the activities. (For examples, see fig. 6 in section 3.)

A systolic array often represents a direct mapping of computations onto processor arrays. It will be used as an attached processor of a host computer, as will be discussed in a later section. The basic principle of systolic design is that all the data, while being "pumped" regularly and rhythmically across the array, can be effectively used in all the PE's. The systolic array features the important properties of modularity, regularity, local interconnection, a high degree of pipelining and highly synchronized multiprocessing. It is also scalable architecturally, i.e. the size of the array may be indefinitely extended, as long as the system synchronization can be maintained. There exists very extensive literature on the subject of systolic array processing, the reader is referred to Fisher and Kung (1985) and the references cited therein.

There are no formal or coherent definitions of the systolic array in literature. In order to have a formal treatment of the subject, however, we shall adopt the following definition.

Definition. A systolic array is a computing network possessing the following features:

(a) *Synchrony.* The data are rhythmically computed (timed by a global clock) and passed through the network.

(b) *Regularity (i.e. modularity and local interconnections).* The array should consist of modular processing units with regular and (spatially) local interconnections. Moreover, the computing network may be extended indefinitely.

(c) *Temporal locality.* There will be at least one unit-time delay allotted so that signal transactions from one node to the next can be completed.

(d) *Pipelinability (i.e. O(M) execution-time speed-up).* A good measure for the efficiency of the array is the following speed-up factor, T_s/T_a, where T_s is the processing time in a single processor, and T_a is the processing time in the array processor. A systolic array should exhibit a *linear-rate pipelinability*, i.e., it should achieve an $O(M)$ speed-up, in terms of processing rates, where M is the number of processor elements (PE's).

One problem, however, is that the data movements in a systolic array are controlled by global timing-reference "beats." In order to synchronize the activities in a systolic array, extra delays are often used to ensure correct timing. More critically, the burden of having to synchronize the entire computing network will eventually become intolerable for very-large-scale or ultra-large-scale arrays.

2.3. Wavefront array

A simple solution to the above-mentioned problems is to take advantage of the control-flow locality, in addition to the data-flow locality, inherently possessed by most algorithms of our interest. This permits a data-driven, self-timed approach to array processing. Conceptually, this approach substitutes the requirement on correct "timing" by correct "sequencing." This concept is extensively used in dataflow computers and wavefront arrays.

A dataflow multiprocessor (Dennis 1980) is an asynchronous, data-driven multiprocessor which runs programs expressed in data-flow graph form. Since the execution of its instructions is "data-driven," i.e., the triggering of instructions depends only upon the availability of operands and resources required, unrelated instructions can be executed concurrently without interference. The principal advantages of data-flow multiprocessors are simple representations of concurrent activity, relative independence of individual PE's, greater use of pipelining and reduced use of centralized control and global memory.

However, for a general-purpose data-flow multiprocessor, the interconnection and memory conflict problems remain very critical. Such problems can be greatly alleviated if *modularity* and *locality* are incorporated into data-flow multiprocessors. This motivates the concept of the wavefront array processors (WAP).

The derivation of a wavefront processing consists of three steps: (a) express the algorithms in terms of a sequence of recursions, (b) each of the recursions is mapped to a corresponding computational wavefront, and (c) successively pipeline the wavefronts through the processor array. (A simple matrix multiplication example is discussed in section 3.)

Definition. A wavefront array is a computing network possessing the following features:

(a) *Self-timed, data-driven computation*: No global clock is needed, since the network is self-timed.

(b) *Modularity and local interconnection*: Basically the same as in a systolic array. However, the wavefront array can be extended indefinitely without having to deal with the global synchronization problem.

(c) $O(M)$ *Speed-up and Pipelinability*: Similar to the systolic array.

Note that the major difference distinguishing a wavefront array from a systolic array is the data-driven property. Consequently, the temporal locality condition (see (c) in the definition of systolic array) is no longer needed, since there is no explicit timing-reference in the wavefront arrays. By relaxing the strict timing requirement, there are many advantages gained, such as speed and programming simplicity. A wavefront processing example for matrix multiplication is discussed in fig. 7 section 3.

As a justification for the name "wavefront array," we note that the computational wavefronts are similar to electromagnetic wavefronts, since each processor acts as a secondary source and is responsible for the propagation of the wavefront. The pipelining is feasible because the wavefronts of two successive recursions will never intersect (by Huygens' wavefront principle), thus avoiding any contention problems. It is even possible to have wavefronts propagating in several different fashions, e.g., in the extreme case of non-uniform clocking, the wavefronts are actually crooked. What is necessary and sufficient is that the order of task sequencing must be correctly followed. The correctness of the sequencing of the tasks is ensured by the wavefront principle [S.-Y. Kung et al. 1982].

The wavefront processing utilizes both the localities of data flow *and* control flow inherently existing in many signal processing algorithms. Since there is no need of synchronizing the entire array, a wavefront array is truly architecturally scalable. In fact, it may be stated that *a wavefront array is a systolic array in combination with the data flow principle.*

3. Array processors for matrix multiplications

In this section, a simple matrix multiplication example is used to illustrate the SFG, systolic, and wavefront array processing.

Let $A = \{a_{ij}\}$ and $B = \{b_{ij}\}$ and $C = A \times B = \{c_{ij}\}$ all be $N \times N$ matrices. Matrix A can be decomposed into columns A_i and matrix B into rows B_j, and therefore,

$$C = A_1 * B_1 + A_2 * B_2 + \cdots + A_N * B_N, \tag{3.1}$$

where the product $A_i * B_i$ is termed "outer product". The matrix multiplication can then be carried out in N recursions (each executing one outer product). It follows

$$C^{(k)} = C^{(k-1)} + A_k * B_k. \tag{3.2}$$

There will be N sets of wavefronts involved. More explicitly,

$$c_{i,j}^{(k)} = c_{i,j}^{(k-1)} + a_i^{(k)}b_j^{(k)}, \qquad a_i^{(k)} = a_{ik}, \qquad b_j^{(k)} = b_{kj}, \qquad (3.3)$$

for $k = 1, 2, \ldots, N$.

SFG array processing

The notion of SFG array processing allows an extensive use of broad-casting, since a node or a zero-delay edge is considered to be delay-free. As a result, a very straightforward SFG array for the matrix multipli-cation algorithm is derived in fig. 5.

Since systolic array contains both parallel and pipeline processing, a SFG array can be considered a systolic array without explicit pipelining. Therefore, the SFG will explicitly exhibit only the parallel processing activities. Note that there is no pipelining assumed in the above SFG example.

To make the system more realizable in a locally interconnected array, we note that

systolic array = SFG array + pipelining;

i.e. we need to assign proper delays on the edges. Moreover, the processing time required at each processing node will have to be reas-signed. (This is done as opposed to the zero-delay assumption idealized in SFG forms.) All these imply a (slow-down) readjustment of input data rate, a necessary price to pay in a local-interconnected and pipeline network. This is explained in the following.

Systolic array processing

For the matrix multiplication algorithm, a square systolic array, the corresponding data arrangement are proposed as shown in fig. 6. The question now is: How does such a scheme actually produce the desired multiplication results? A popular way to give a demonstration is to display the space–time activities in the first few consecutive "beats" like those displayed in fig. 6. Suppose the design is correct, the pre-arranged data will meet the designated partners, perform the appropriate oper-ations and yield desired "products." The complete activities as well as the general rule can then be derived by induction. Below in this section, we shall discuss a simple (and possibly automatic) conversion from an SFG array into a systolic array, which may alleviate the burden of verification.

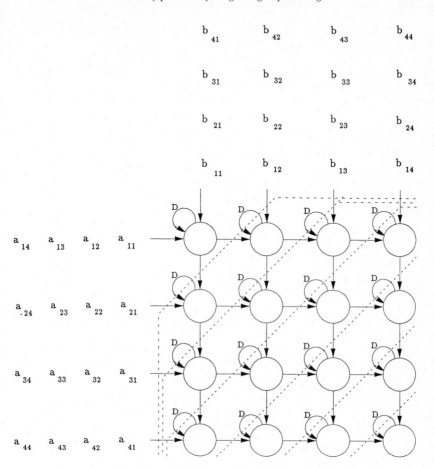

Fig. 5. An SFG array for matrix multiplication: A straightforward SFG array design is to broadcast the columns A_i and rows B_i (cf. eqs. 3.2 and 3.3) instantly along the square array as shown in the figure. Multiply the two data meeting at node (i, j) and add the produce to $c_{ij}^{(k)}$, the data value currently residing in a register in node (i, j). Finally, the new result will update the register via a loop with a delay D and get ready to interact with the new arriving operands. As all the column and row input data continue to arrive at the nodes, all the outer-products will be sequentially summed. Although this design is not directly suitable for a VLSI circuit design, due to the use of global communication, it may be converted to a systolic array as shown in fig. 4, or a wavefront array as shown in this figure. A simple conversion strategy will be provided in section 3.

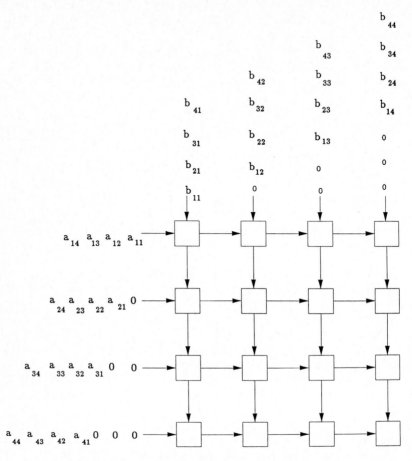

Fig. 6. A systolic array for matrix multiplication. For this example, a two-dimensional square array forms a natural topology for the matrix multiplication problem. The figure specifically shows a 4 × 4 array of processing elements (PE's). All the PE's (represented by the square boxes) uniformly consume and produce data in one single time-unit. In terms of one "snapshot" of the activities the input data (from matrices A and B appearing at the left and top parts of the figure) are pre-arranged in an orderly sequence. The C data stay temporarily within the PE's and will be pumped out from one side of the array. Due to the systolization rules as discussed in section 3, the inputs from different columns of B and rows of A will have to be adjusted by a certain number of delays before arriving at the array. This is why some extra zeros are introduced here.

Wavefront array processing

The notion of computational wavefront may be better illustrated by the example of the matrix multiplication algorithm. The topology of the matrix multiplication algorithm can be mapped naturally onto the square, orthogonal $N \times N$ matrix array of the wavefront array processor (WAP), as in fig. 7. To create a smooth data movement in a localized communication network, we propose a notion of the computational wavefronts. A wavefront in a processor array is corresponding to a mathematical recursion in the algorithm. Successive pipelining of the wavefronts through the computational array will accomplish the computation of all recursions.

More elaborately, the computational wavefront for the first recursion in matrix multiplication is now examined. Suppose that the registers of all the processing elements (PE's) are initially set to zero:

$$C_{ij}^{(0)} = 0, \quad \text{for all } (i, j).$$

The entries of A are stored in the memory modules to the left (in columns), and those of B in the memory modules on the top (in rows). The process starts with PE $(1, 1)$ and

$$C_{11}^{(1)} = C_{11}^{(0)} + a_{11} * b_{11}$$

is computed. The computational activity then propagates to the neighboring PE's $(1, 2)$ and $(2, 1)$, which will execute:

$$C_{12}^{(1)} = C_{12}^{(0)} + a_{11} * b_{12},$$

and

$$C_{21}^{(1)} = C_{21}^{(0)} + a_{21} * b_{11}.$$

The next front of activity will be at PE's $(3, 1)$, $(2, 2)$ and $(1, 3)$, thus creating a computation wavefront traveling down the processor array. Once the wavefront sweeps through all the cells, the first recursion is complete. As the first wave propagates, we can execute an *identical* second recursion in parallel by *pipelining* a second wavefront *immediately after* the first one. For example, the $(1, 1)$ processor will execute

$$C_{11}^{(2)} = C_{11}^{(1)} + a_{12} * b_{21}$$

$$\vdots$$

$$C_{ij}^{(k)} = a_{i1} * b_{1j} + a_{i2} * b_{2j} + \cdots + a_{ik} * b_{kj}$$

and so on.

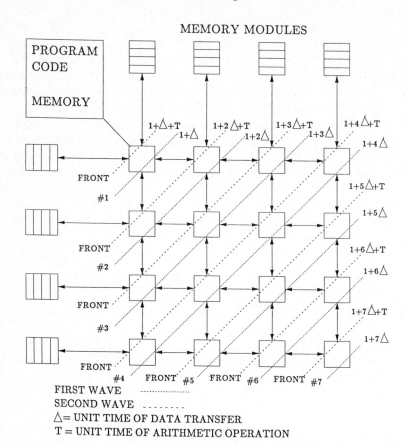

MEMORY MODULES

PROGRAM
CODE

MEMORY

FRONT #1

FRONT #2

FRONT #3

FRONT #4   FRONT #5   FRONT #6   FRONT #7

FIRST WAVE   ··················
SECOND WAVE   ─ ─ ─ ─ ─
△ = UNIT TIME OF DATA TRANSFER
T = UNIT TIME OF ARITHMETIC OPERATION

Fig. 7. Wavefront processing for matrix multiplication. In this example, the wavefront array consists of $N \times N$ processing elements with regular and local interconnections. The figure shows the first 4×4 processing elements of the array. The computing network serves as a (data) wave propagating medium. Hence, the hardware will have to support pipelining the computational wavefronts as fast as resource and data availability allow, which can often be accomplished simply by means of a handshaking protocol, such as that proposed in Kung et al. (1982). The (average) time interval (T) between two separate wavefronts is determined by the availability of the *operands* and *operators*. In this case, T is equal to the time needed for the arithmetic operations: multiply and add. The speed of wavefront propagation is determined by the time interval Δ, which in this case is equivalent to the data transfer time. Note that, in this wavefront processing example, *parallel processing is along the anti-diagonal PE's as each of the PE's may be processed independently. On the other hand, pipeline processing occurs along the diagonal direction.*

The notion of computational wavefront allow us to deal with the question of separating the roles of pipeline and parallel processing. Generally speaking, parallel processing activities always occur at the PE's on the same "front," while pipelining activities are often perpendicular to the "fronts." With reference to the wavefront processing example in fig. 7, PE's on the anti-diagonal lines perform parallel processing, since each of the PE's processes information independently. On the other hand, pipeline processing occurs along the diagonal direction, the computational wavefronts are piped along the diagonal direction. The data and sequence dependences are thus preserved.

4. Mapping algorithms onto arrays

4.1. Systolization of SFG computing networks

A cut-set systolization procedure (S.-Y. Kung 1984). A cut-set in an SFG is a minimal set of edges which partitions the SFG into two parts. The systolization procedure is based on two simple rules:

Rule (1) Time-scaling: All delays D may be scaled, i.e., $D \rightarrow \alpha D$, by a single positive integer α. Correspondingly, the input and output rates also have to be scaled down by a factor α. The time-scaling factor (or, equivalently, the slow-down factor) α is determined by the slowest (i.e. maximum) loop delay in the SFG array.

Rule (2) Delay-Transfer: Given any cut-set of the SFG, we can group the edges of the cut-set into *in-bound edges* and *out-bound edges*, depending upon the directions assigned to the edges. Rule (2) allows advancing k time-units on all the out-bound edges and delaying k time-units on the in-bound edges. It is clear that, for a (time-invariant) SFG, the general system behavior is not affected because the effects of lags and advances cancel each other in the overall timing. Note that the input-input and input-output timing relationships will also remain exactly the same only if they are located on the same side. Otherwise, they should be adjusted by a lag of $+k$ time-units or an advance of $-k$ time-units.

As an illustration, with reference to fig. 5, the dashed lines indicate a set of possible cuts. A simple procedure involving delay-transfer of one time-unit (D) yields the systolic array shown in fig. 6.

Example. Multiplication of a banded matrix and a full rectangular matrix.

Let us look at a slightly different, but commonly encountered, type of matrix multiplication problem. This involves a banded-matrix A, $N \times N$, with bandwidth P, and a rectangular matrix B, $N \times N$:

$$
AB =
\begin{bmatrix}
x & x & x & & & & \\
x & x & x & x & & & \\
x & x & x & x & x & & \\
& x & x & x & x & x & \\
& & x & x & x & x & x \\
& & & x & x & x & x & x \\
& & & & \cdots\cdots\cdots\cdots
\end{bmatrix}
\begin{bmatrix}
x & x & x & x & x & \cdots & x \\
x & x & x & x & x & \cdots & x \\
x & x & x & x & x & \cdots & x \\
x & x & x & x & x & \cdots & x \\
x & x & x & x & x & \cdots & x \\
x & x & x & x & x & \cdots & x \\
x & x & x & x & x & \cdots & x
\end{bmatrix}
$$

(A banded matrix A is one which has nonzero elements only on a finite "band" along the diagonal.)

This situation arises in many application domains, such as DFT and time-varying (multi-channel) linear filtering, etc. In most applications, $N \gg P$ and $N \gg Q$; therefore, *it is very uneconomical to use $N \times N$ arrays for computing $C = A \times B$.*

Fortunately, with a slight modification to the SFG in fig. 5, the same speed-up in performance can be achieved with only a $P \times Q$ rectangular array (as opposed to an $N \times Q$ array). This is shown in fig. 8a.

A systolization example. If local interconnection is preferred, the proposed procedure given above can then be used to systolize the SFG array in fig. 8a and yield the data array as depicted in fig. 8b. The systolization procedure is detailed below:

(a) Time-scaling: According to the systolization rule (1), the slow-down factor α is determined by the maximum loop delay in the SFG array. Refer to fig. 8a, any loop consists of one up-going and one down-going edges yield a (maximum) delay of two. This is why the final pipelined systolic array has to bear a slow-down factor $\alpha = 2$. The pipelining rate is 0.5 word per unit-time for each channel.

(b) Delay transfer: Apply rule (2) to the cut-sets shown in fig. 8a. The systolized SFG will have one delay assigned to each edge and thus represent a localized network. Also based on rule (2), the inputs from different columns of B and rows of A will have to be adjusted by a certain number of delays before arriving at the array. By counting the

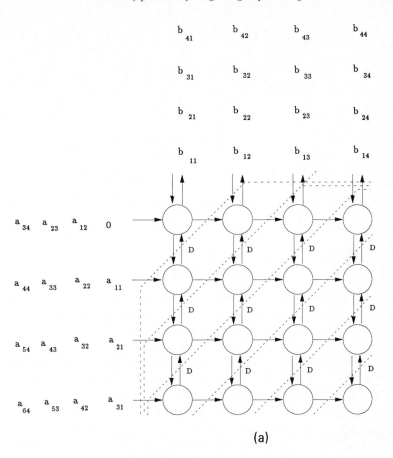

(a)

Fig. 8a. An SFG array for banded-matrix and matrix multiplication. Now, the left memory module will store the matrix A along the band-direction, and the upper-module will store B the same as before. Note that the major modification to the array is that, between the recursions of outer products, there should be an upward shift of the partial sums. This is because the input matrix A is loaded in a skewed fashion. The final result (C) will be output from the I/O ports of the top-row PEs.

cut-sets involved in fig. 8a, it is clear that the first column of B needs no extra delay, the second column needs one delay, the third needs two (i.e., attributing to the two cut-sets separating the third column input and the adjacent top-row processor), etc. Therefore, the B matrix will be skewed as shown in fig. 8b. A similar arrangement can be applied to the input matrix A.

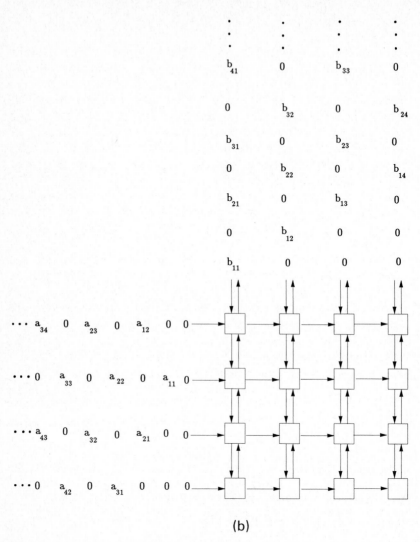

(b)

Fig. 8b. The systolic array design as a result of applying the systolization procedure to fig. 8a. Note that the pipelining rate is proportional to α^{-1}, and here $\alpha = 2$.

4.2. Converting an SFG array into a wavefront array

The wavefront propagation is very similar to the previous case. Since in wavefront processing the (average) time interval, T, between two separate wavefronts is determined by the availability of the *operands* and *operators*. (In this case, there is a feedback loop, constituted by one down-going and one up-going edges as shown in fig. 8.) For example, in the node$(1,1)$, the second front has to wait till the first front complete all the following steps: (a) propagate a data downwards (processing time: Δ), (b) perform the arithmetic operations at node$(2,1)$ (processing time: T_{MA}), and (c) return the result upwards to node$(1,1)$ (processing time: Δ). Once the result is returned to node$(1,1)$, the second front can be immediately activated. The activations of all the later fronts follow exactly the same procedure; therefore, the (average) time separating two consecutive fronts is $T = T_{MA} + 2\Delta$.

In fact, every regular and modular SFG array can be converted into a wavefront array. In a self-timed system, the exact timing reference is ignored; instead, the central issue is sequencing. Getting a data token in a self-timed system is equivalent to incrementing the clock by one time-unit in a synchronous system. Therefore, the conversion of an SFG into a data-driven system involves substituting the delay operators D by "handshaked delay" registers. (A "handshaked delay" register is a device which prevents any incoming data from directly passing through *until* the handshaking flag signals a "pass.") For example, applying this conversion process to the SFG array in fig. 5 (also, equally applicable to that in fig. 8a) for matrix multiplication yields the wavefront array shown in fig. 7 (S.-Y. Kung 1984).

4.3. How to improve the pipelining rate

In the above case, the maximum loop delay is two. Thus the slow-down factor $\alpha = 2$ and the pipelining rate is reduced by one half. Now the question is: what is the best pipelining rate achievable for the algorithm? To improve the rate we would like to reduce the slow-down factor down to $\alpha = 1$. This is possible only when all the loops are eliminated. For the banded-matrix and matrix multiplication problem, one has to resort to a major modification on the SFG array. The trick is to load the input B matrix from the bottom and travel upwards, as illustrated in fig. 9a, instead of from the top and travel downwards as in fig. 8a. Here it is a valid change, because as long as broadcasting is assumed in the SFG

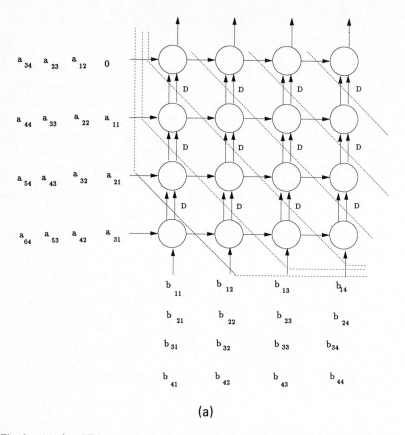

$$a_{34} \quad a_{23} \quad a_{12} \quad 0$$

$$a_{44} \quad a_{33} \quad a_{22} \quad a_{11}$$

$$a_{54} \quad a_{43} \quad a_{32} \quad a_{21}$$

$$a_{64} \quad a_{53} \quad a_{42} \quad a_{31}$$

$$b_{11} \qquad b_{12} \qquad b_{13} \qquad b_{14}$$

$$b_{21} \qquad b_{22} \qquad b_{23} \qquad b_{24}$$

$$b_{31} \qquad b_{32} \qquad b_{33} \qquad b_{34}$$

$$b_{41} \qquad b_{42} \qquad b_{43} \qquad b_{44}$$

(a)

Fig. 9a. Another SFG array for banded-matrix and matrix multiplication. Note that the additional modification to fig. 8a is that the input B matrix is now loaded from the bottom traveling upward, instead of from the top traveling downward.

model, it does not matter whether the (same) data are loaded top-down or bottom-up. But the effects on α and thus the achievable pipelining rates are totally different. Note that in the modified SFG array there exist no loops. This implies that $\alpha = 1$. Consequently, the cuts as shown in fig. 9a do not call for any time rescaling, it needs only a delay transfer of D. After the cut-set delay transfer procedure, the resultant systolic array is depicted in fig. 9b. Note that this new version offers a through-put-rate of one word per unit-time for each channel, which is *twice as fast* as the previous systolic design.

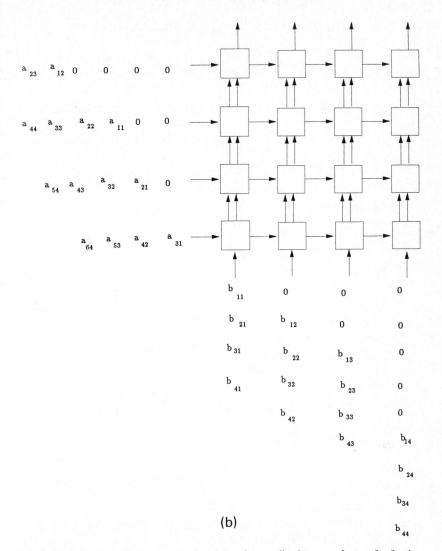

(b)

Fig. 9b. The systolic array as a result of applying the systolization procedure to fig. 9a. As the pipelining rate is proportional to α^{-1}, and here α is reduced to 1 (as opposed to $\alpha = 2$ in fig. 8b), this systolic array can run twice as fast as the previous design.

4.4. Two versions of mapping recursive algorithms onto arrays

Regarding the match between the indexing of the algorithm and that of the array, there are two types of mapping from recursive algorithms onto arrays:

Version I: Match PE space (PE row number) to data elements (rows of matrix).

Version II: Match PE space (PE row number) to recursion number.

Two examples of SFG networks for LU decomposition. In the LU decomposition, a given matrix C is decomposed into

$$C = A \times B, \tag{4.1}$$

where A is a lower-triangular matrix and B is an upper-triangular matrix. The recursions involved are

$$b_j^{(k)} = C_{kj}^{(k-1)}, \tag{4.2a}$$

$$a_i^{(k)} = \frac{1}{C_{(kk)}^{(k)}} C_{ik}^{(k-1)}, \tag{4.2b}$$

$$C_{ij}^{(k)} = C_{ij}^{(k-1)} - a_i^{(k)} b_j^{(k)}, \tag{4.2c}$$

for $k = 1, 2, \ldots, N$, $\quad k \leqslant i \leqslant N$,
$\qquad\qquad\qquad\quad k \leqslant j \leqslant N$.

Verifying the procedure by tracing back eq. (4.2a), we note that

$$C = C_{ij}^{(0)} = \sum_{k=1}^{N} a_i^{(k)} b_j^{(k)}, \tag{4.3}$$

where $A = \{a_{mn}\} = \{a_m^{(n)}\}$, and $B = \{b_{mn}\} = \{b_n^{(m)}\}$ are the outputs of the array processing.

Comparing with eq. (4.3), eqs. (4.2) are basically a reversal of the matrix multiplication recursions. Two versions of SFG array representation for the above iterations are discussed in fig. 10.

Let us compare the two versions of LU-decomposition array algorithms. Version I enjoys the following advantages: (a) there is no need of processor re-programming (no change of processor functions), and (b) it is easily adaptable to the banded-matrix LU decomposition problem. On the other hand, Version II also offers very attractive advantages: (a) there is no need of diagonal connections, and (b) there is a better utilization efficiency as it yields the same throughput rate with only 50%

of PE's. Moreover, as we shall discuss below, the systolized array of Version II enjoys a better pipelining factor, α, equal to 1, compared to a value of 3 for Version I, although they both result in the same latency.

4.4.1. Systolization of arrays for LU decomposition

Apply the systolization procedure for fig. 10a, the cut-sets as shown in fig. 11a will call for a triple scaling of $D \rightarrow 3D$. (This is because each north-west-bound delay edge is "cut" twice.) The procedure leads to an array configuration depicted in fig. 11a, which is topologically equivalent to the two-dimensional hexagonal array proposed by H.T. Kung and Leiserson (1978) and by H.T. Kung (1982).

Applying the cut-set procedure to fig. 10b, a systolic array for LU-decomposition of Version II can be obtained. The resultant configuration of the array is depicted in fig. 11b.

4.5. Other systolic arrays

There are many systolic arrays and algorithms which are more complicated than the ones for the matrix multiplication example discussed above. These include systolic algorithms for convolution, correlation, FIR, IIR, and lattice filtering for one- or two-dimensional signals. Also worth mentioning are a number of important matrix operations such as linear system solution, least-square solution, Toeplitz system solution, and eigenvalue and singular value decompositions.

5. Comparisons between systolic and wavefront arrays

The main difference between the two array processing schemes is that the wavefront arrays operate asynchronously (at the price of handshaking hardware), while the systolic arrays pulse globally in synchronization with a global clock.

The systolic array features the very important advantages of modularity, regularity, local interconnection, highly pipelined multiprocessing, and continuous flow of data between the PE's. It has a good number of DSP applications. However, the wavefront design offers some additional useful advantages:

Maximum pipelining. A major thrust of the wavefront array derives from its maximizing the pipelinability by exploring the data-driven nature inherent in many parallel algorithms. This becomes specially useful in the case of uncertain processing times used in individual PE's.

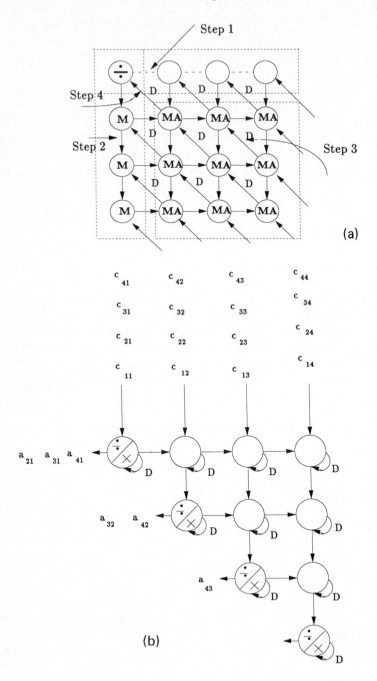

(a)

(b)

As reported in [McWhirter 85], wavefront pipelining may yield a significant speed-up, compared with pure systolic pipelining. [In the special simulation example used by Broomhead et al. (1985), the improvement is a factor of almost two.]

Architectural extendibility. The wavefront array also highlights the extendibility of the array size, since it can get around the global synchronization requirements. Whereas the asynchronous model in the wavefront arrays incurs a fixed time delay overhead due to the handshak-

---

Fig. 10. Two versions of SFG arrays for LU decomposition. *Version I*: Match PE space (PE row number) to data elements (rows of matrix). *Version II*: Match PE space (PE row number) to recursion number.

(a) Version I:

Step I: $b_j^{(k)}$ are to be propagated downward.

Step II: The data $a_i^{(k)}$ are to be propagated rightward.

Step III: Multiply the above data meet at node (i, j) and subtract the product from $C_{ij}^{(k)}$, the current data sitting in (i, j), and update the (i, j) register.

Step IV: By moving the updated $(N - 1) \times (N - 1)$ submatrix to the north-eastern corner after a delay D, the second recursion is now readied. Note that the second recursion will now initiate at node $(1, 1)$, (rather than at node $(2, 2)$). (There is a hardware advantage to have all recursions initiated in node $(1, 1)$, since the divider and other special boundary nodes may be re-used again and the need to change the node functions is thus avoided.)

(b) Version II: The input is a full matrix C of size $n \times n$. The final result on the B matrix stays in the nodes, and the A matrix is the output from the diagonal nodes.

The first row of the array processes the first recursion ($k = 1$), the second row processes the second recursion ($k = 2$), and so on. The activities in the first row PE's are described below.

When the first row of C is input into the first row of PE's, PE$(1, 1)$ computes c_{11}^{-1}, and all other PE's do not operate. Then, all data in the first row of PE's are delayed and fed back to themselves for the next computation. When the 2nd row data arrive at the 1st row of PE's, PE$(1, 1)$ switches its function to multiplication and computes $-c_{11}^{-1}c_{21}$ and broadcasts it to the whole row, and PE$(1, j)$ computes $c_{1j} - c_{21}c_{2j}/c_{11}$ and sends the result to the 2nd row of PE's, ready for the second recursion (i.e. $k = 2$). The same process applies to other rows of data when they arrive at the first row.

For $k = 2, 3, \ldots$, since all recursions are basically the same in operation, the same process applies to other rows of data when they arrive at their respective rows, i.e. 2nd, 3rd rows of PE's. Note that during the processes of triangularization, the size of the matrix is shrinking. So, the first recursion has size n, the second has size $(n - 1)$, and so on. This is why it leads to triangular array.

Note also that the function change from division to multiplication for diagonal PE's can be very easily done, if we use CORDIC arithmetic unit (Ahmed 1985). All that is needed is changing one control bit.

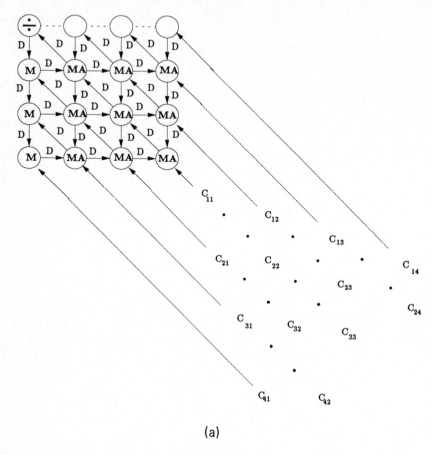

(a)

Fig. 11. Systolic arrays for LU decomposition: (a) version I, and (b) version II.

ing processes, the synchronization time delay in the systolic arrays is primarily due to the clock skew which changes dramatically with the size of the array. This latter phenomenon will be a potential barrier in the design of ultra-large-scale synchronous computing systems.

Programming simplicity. The notion of computational wavefront also facilitates the programmability of array processors. By tracing the wavefronts, the description of the space-time activities in the array may be significantly simplified. The parallel processing language, Occam, is very suitable for programming wavefront arrays.

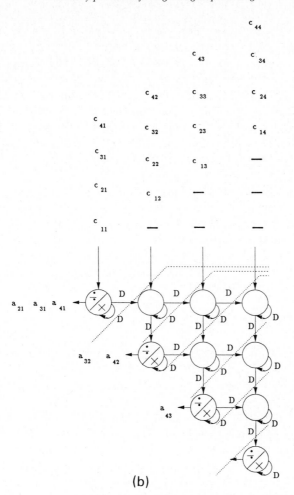

(b)

Fig. 11. continued

Fault-tolerance of array processors. To enhance reliability of computing systems, real-time signal processing architectures demand a special attention in run-time fault tolerance. However, 2-D systolic arrays are in general not feasible for run-time fault tolerance design since it requires a global stoppage of PE's when any failure occurs. It is known that certain fault tolerance issues (roll-back, suspension of execution, etc.) are simpler to handle in data flow architecture than in other multi-processors

(Dennis 1980). Since wavefront arrays incorporate the data-driven feature into the arrays, they pose similar advantages in dealing with time uncertainties in the fault tolerance environment. For example, once a fault is detected, further propagation of the wavefront will be automatically suspended, according to the wavefront principle. More specifically, due to its asynchronous nature, one only needs to stop the faulty PE, and all subsequent PE's would automatically stop as a ripple. Systolic arrays, in comparison, would probably require a global "error-halt" signal to be broadcasted, and the corresponding roll-back problem would be far more cumbersome.

In summary, a systolic array is useful when the PE's are simple primitive modules, since the handshaking hardware in a wavefront array would represent a non-negligible overhead for such applications. On the other hand, a wavefront array is more favored when the PE's involve more complex modules (such as multiply-and-add and lattice or rotation operations), or when a robust and reliable computing environment (such as fault-tolerance) is essential.

6. Algorithm design criteria

An effective design of algorithms for array processing hinges upon a full understanding of the problem specification, mathematical analyses, parallelism analysis and the practicality of mapping the algorithms onto real machines.

Parallel array algorithm design is a new area of research study, which has profited from the theory of signals and systems and has been influenced by linear algebraic numerical methods. In a conventional algorithm analysis, the complexity of an algorithm depends on the computation and storage required. The modern concurrent computation criterion should include one more key factor: *communication*. In the array algorithms design, the major factors comprise computation, communication, and memory.

The key aspects of parallel algorithms under VLSI architectural constraints are presented below:

(1) Maximum concurrency. The algorithm should be structured to achieve maximum concurrency and/or maximum throughput. (Two algorithms with equivalent performance in a sequential computer may fare very differently in parallel processing environments.) An algorithm will be favored if it expresses a higher parallelism which is exploitable by the computing arrays.

Example. A very good example is the problem of solving Toeplitz systems, for which the major algorithms proposed in the literature are the Schur algorithm and the Levinson algorithm (S.-Y. Kung and Hu 1983). The latter is by far more popular in many spectrum estimation applications such as the maximum entropy method (MEM) (Burg 1975) or the maximum likelihood method (MLM) (Capon 1969). In terms of sequential processing, both the algorithms require the same number of operations. However, in terms of the achievable concurrency when executed in a linear array processor, the Schur algorithm displays a clear-cut advantage over the Levinson algorithm. More precisely, using a linear array of N PE's the Schur algorithm will need only $O(N)$ computation time, compared with $O(N \log N)$ required for the Levinson algorithm. Here N represents the dimension of the Toeplitz matrix involved. For a detailed discussion, see the work by S.-Y. Kung and Hu (1983).

(2) Maximum pipelinability and balancing computation and I / O. Most signal processing algorithms demand very high throughput and are computation intensive compared with the input/output (I/O) requirement. Pipelining is essential to the throughput of array processors. The exploitation of the pipeline technique is often very natural in regular and locally-connected networks; therefore, a major part of concurrency in array processing will be derived from pipelining. In general, the pipelining rate is determined by the "maximum" loop delay in the SFG array. To maximize the rate, one must select the best among all possible SFG arrays for any algorithm. The pipeline techniques are especially suitable for balancing computation and I/O because the data tend to engage as many processors as possible before they leave the array. This helps in reducing the I/O bandwidth for outside communication.

Example. Note that, for the banded-matrix and matrix multiplication algorithm, the systolic array shown in fig. 9b offers throughput-rate twice as fast as the design in fig. 8b.

(3) Trade-off between communication and computation costs. To make the interconnection network practical, efficient and affordable, regular communication should be encouraged. Key issues affecting the regularity include local versus global, static versus dynamic, and data-independent versus data-dependent interconnection modules. The criterion should maximize the tradeoff between interconnection cost and throughput. For example, to conform with the communication constraints imposed by

VLSI, a lot of emphasis has recently been placed on a special class of local and recursive algorithms.

Example. Trade-off between computation and communication costs.

When the communication in VLSI systems is emphasized, the trade-off between computation and communication becomes a central issue (cf. fig. 12). The preference on regularity and locality will have a major impact in designing parallel and/or pipelined algorithms. Comparing DFT versus FFT, the computations are $O(N^2)$ versus $O(N \log N)$ in favor of FFT. However, the DFT enjoys a simple and local communication, while the FFT involves a global interconnection, i.e. the PE's retrieve their data from far away elements. In the trade-off of computation versus communication costs, the choice is no longer obvious.

(4) Numerical performance, quantization effects and data dependency. Numerical behavior depends on many factors such as the word-length of the computer, and the algorithms used. Unpredictable data dependency may severely jeopardize the processing efficiency of a highly regular and structured array algorithm. Effective VLSI arrays are inherently highly pipelined and hence require well structured algorithms with predictable data movements. Iterative methods with dynamic branching, dependent on data produced in the middle of the process, are less suited for pipelined architecture.

Example. Comparison of linear system solvers.

It is well known that there are three major numerical algorithms for solving a linear system of equations; namely, the LU decomposition, the Householder QR (HQR) and the Givens QR (GQR) decomposition algorithms (Stewart 1973). From a numerical performance point of view, a HQR or a GQR decomposition is often preferred over an LU decomposition for solving linear systems. As to the *maximum concurrency achievable by array processing*, the GQR algorithm achieves the same 2-D concurrency as the LU decomposition with the same complexity in a modular, streamlined fashion. They both hold a decisive advantage over the HQR method in terms of maximum concurrency achievable. Therefore, the Givens QR method is superior to the LU decomposition and the Householder QR method when both numerical performance and massive parallelism are considered. Note that the price, however, is that the GQR method is computationally more costly than the other two methods (cf. the third column of table 1.)

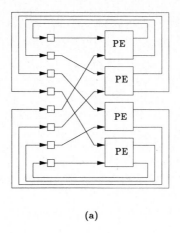

(a)

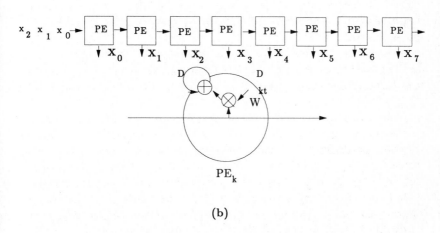

(b)

Fig. 12. Array processor architectures for (a) FFT algorithm; and (b) DFT algorithm. Note that, in terms of total processing times, the ratio is log N versus N in favor of FFT. On the other hand, the FFT array requires a *global* (Perfect–Shuffle) communication network. In contrast, the DFT array can be easily systolized and implemented in a modular processor array with *local* communication. Note that the weighting factors W^{kt} are time-varying; and the Fourier transform output $\{X_k\}$ stays in the PE and will eventually be pumped out.

Table 1
Comparison of linear system solvers. They key issues are data dependency, numerical
performance, maximum concurrency and the total number of computations.

Algorithm	Numerical performance	Maximum concurrency achievable	Number of operations
LU decomposition	Bad*	2-D array	$(N^3/3)$
HQR decomposition	Good	1-D array	$(2N^3/3)$
GQR decomposition	Good	2-D array	$(4N^3/3)$

*The numerical performance of LU decomposition may be improved by using a
pivoting scheme (Stewart 1973). However, this necessitates control on the magnitude of
a pivot, jeopardizing the otherwise smooth data flow in the array.

7. Implementation considerations of array processor systems

An array processor is in general used as an attached processor, enhancing the computing capability of the host machine. The major components of an array processor system consist of:
(1) host computer;
(2) interface system, including buffer memory and control unit;
(3) connection networks (for PE-to-PE and PE-to-memory connections);
(4) processor array, comprising a number of processor elements with
    local memory.

A possible overall system configuration is depicted in fig. 13, where the
design considerations for the four other major components are further
elaborated. In general, in an overall array processing system, one seeks to
maximize the following performance indicators: computing power, using
multiple devices; communication support, to enhance the performance;
flexibility, to cope with partitioning problems; reliability, to cope with
fault-tolerance problems; and practicality and cost-effectiveness.

The four items mentioned above will now be discussed.

Host computer. The host computer should provide batch data storage,
management, and formatting; it should determine the schedule program
which control the interface system and connection network; and it
should generate and load object codes to the PE's. A very challenging
task to the system designer is to identify a suitable host machine for
interfacing with high-speed array processor units.

Interface system. The interface system, connected to the host via host
bus, has the functions of down-loading and up-loading data. Based on

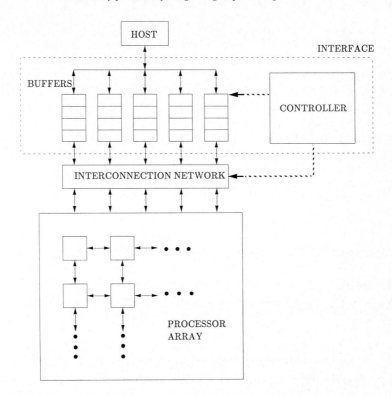

Fig. 13. An example of array processing system configuration. An array processor is often used as an attached processor linked with a host through an interface system.

Host computer: The host computer supports data storage and formatting, and schedules program management.

Interface system: The interface system, connected to the host via host bus, has the functions of down-loading and up-loading data. Based on the schedule program, the controller monitors the interface system and the array processor. The interface system should also furnish an adequate hardware support for many common data management operations.

Connection network: Connection networks provide a set of mappings between processors and memory modules to accommodate certain common global communication needs. Incorporating certain structured interconnections may significantly enhance the speed performance of the processor arrays.

Processor arrays: The overall array system may consist of one or more locally-interconnected processor arrays.

the schedule program, the controller monitors the interface system and array processor. The interface system should also furnish an adequate hardware support for many common data management operations. A very challenging task for the system designer is the management of blocks of data. Another task is to make sure the memory (buffer) unit is able to balance the low bandwidth of system I/O and the high bandwidth of array processors.

Connection network. Connection networks provide a set of mappings between processors and memory modules to accommodate certain common global communication needs. Incorporating certain structured interconnections may significantly enhance the speed performance of the processor arrays.

In array processing, data are often fetched and stored in parallel memory modules. However, the fetched data must be realigned appropriately before they can be sent to individual PEs for processing. When some global interconnection algorithms are to be executed effectively, or certain array reconfigurations or partitioning are to be supported, then such alignments also become necessary. They are implemented by the routing functions of the interconnection network, such as wrap-around connection (e.g. spiral or torus) or perfect shuffle, or hyper-cube network, etc. It is shown by Wu and Feng (1980) that the shuffle-exchange network can realize arbitrary permutations of N data in $3 \log N$ passes. However, using such network requires a major effort for generating the somewhat complicated control codes. Cross-bar connection network, when affordable, represent another convenient choice, which offers maximum flexibility, at the expense of extremely high hardware costs.

In short, the performance for VLSI arrays depends critically on the communication cost for data transactions. The degree of achievable parallel processing also depends heavily on the effectiveness of matching communication need as called for by the parallel algorithm with the interconnectivity as offered by the real array architecture. It is therefore desirable to have a formal way of characterizing the communication requirement and evaluating its cost.

Processor arrays. For simplicity, only one processor array is physically depicted in fig. 13. However, the concept of networking several processor arrays has now attracted a good deal of attention. For example, when a problem is decomposible into several subproblems, to be executed one after the other, it will be useful to have each subproblem executed in its own processor array, while utilizing the network to facilitate the data pipelining between the arrays. This suggests a pipelining scheme in the array-level, which may increase the processing speed-up by one more order of magnitude.

8. DSP-oriented array processor chips

The implementation of VLSI chips and the structure of array computing systems depends largely on the established switching technologies. Another important design consideration is to determine the appropriate level of granularity of the processor elements (PE's) composing the array, cf. fig. 14.

For some low-precision digital and image processing applications, it is advisable to consider very simple processing primitives. A good example of a commercial VLSI chip is NCR's geometric arithmetic parallel processor, or GAPP, which is composed of a 6 by 12 arrangement of single-bit processor cells. Each of the 72 processor cells in the NCR45CG72-device contains an ALU, 128 bits of RAM and bi-directional communication lines connected to its four nearest neighbors: one to the north, east, south and west (Davis and Thomas 1984). Each instruction is broadcasted to all the PE's, making the array to perform like a SIMD (single-instruction-multiple-data) machine.

Many DSP applications require the PE's to include more complex primitives, such as multiply-and-add modules. An example of a commercial chip with a larger granularity is INMOS's Transputer (Wilson 1984). Transputer is an OCCAM-language based design, which provides hardware support for both concurrency and communication—the heart of array computing. It has a 32 bit processor capable of 10 MIPS, 4 Kbytes of 50 ns static RAM, and a variety of communications interfaces. It adopts the now popular RISC (reduced-instruction-set-computer) architecture design. The INMOS links, with built-in handshaking circuits, are the hardware representation of the channels for communications. Furthermore, its programming language, OCCAM, is very suitable for programming wavefront-type array processing. Therefore, the transputer can be readily adopted for the construction of (asynchronous) wavefront arrays.

Other examples of commercially available VLSI chips, worthy of consideration for array processor implementations, are NEC's data flow chip μpd7281 (Chase 1984), TI's programmable DSP chip TMS320, and

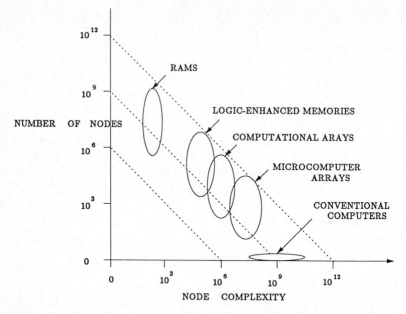

Fig. 14. Different levels of granularity of PE in array system, adapted from Seitz (1984). An example of smaller PE granularity is NCR's GAPP. GAPP array lies in the intersection domain between the logic-enhanced-memory group and the computational array group as shown in the figure. Such kind of simple processor primitives are often preferred in many low-precision image processing applications. An example of a larger PE granularity is INMOS' Transputer. Many DSP applications require fast multiply/accumulate, high-speed RAM, fast coefficient table addressing; a possible choice appears to be an enhanced version of transputer. Referring to the figure, a transputer array belongs largely to the micro-computer array domain. Because of the built-in asynchronous communication hardware, transputers are very suitable for the implementation of wavefront arrays.

recent 32-bit processors such as AMD 29325, Motorola 68020, and Weitek's 32-bit (or the new 64-bit) floating-point chips (Ware et al. 1984). Many DSP applications require very special features such as fast multiply/accumulate, high-speed RAM, fast coefficient table addressing, etc. Therefore, the development of a series of customized special-purpose chips for DSP array processors should be given a high priority by the VLSI and DSP research community.

9. Building block primitives for DSP array processors

Large design and layout costs suggest using repetitive modular structures, i.e. a few different types of simple (and often standard) cells, in a VLSI design. Thus, we must identify the primitives that can be implemented efficiently and that would optimally realize the potential of the VLSI device technologies.

From computer components perspective, a typical set would include primitives for arithmetic and logic, accessing data items, controlling I/O. From DSP applications perspective, a major portion of algorithms can be reduced to a basic set of matrix operations and other related algorithms. These facts can be exploited in order to simplify the hardware design. We propose the functional classification of building blocks for designing signal processing array processors:

(1) Arithmetic units; e.g. adders, parallel multiplier, multiplier–accumulator, cordic processor, residue-number arithmetic units and Boolean algebra.

(2) Memory/storage units; e.g. RAM, ROM, table look-up, stacks (LIFO, FIFO), buffer memory and cache.

(3) Communication/interconnection modules; e.g. handshaking, I/O port, DMA (direct memory access), packet-switching (communication control), cross-bar, perfect shuffle, hyper-cube, spiral/torus links and multi-broadcasting.

(4) DSP functions; e.g. inner product, sorters, interpolation, (1-D and 2-D) FIR, IIR filters, (1-D and 2-D) median filters, (1-D and 2-D) correlator, (1-D and 2-D) convolution and (1-D and 2-D) DFT/FFT.

(5) Matrix operations; e.g. matrix–vector multiplication, matrix–matrix multiplication, LU decomposition, QR decomposition, eigenvalue/singular value decompositions, linear system solver, least-square solver, null-space solver and Toeplitz system solver.

(6) Others; e.g. clock generator, A/D, D/A converters and self-testing units.

10. Application domain of systolic / wavefront arrays

The power and flexibility of the systolic/wavefront type arrays and MDFL programming are best demonstrated by the broad range of the signal processing and scientific computation algorithms suitable for the systolic/wavefront arrays (S.Y. Kung 1984). Such algorithms can be roughly classified into three groups:

(1) Basic matrix operations: (a) matrix multiplication, (b) banded-matrix multiplication, (c) matrix–vector multiplication, (d) LU decomposition, (e) LU decomposition with localized pivoting, (f) Givens algorithm, (g) back substitution, (h) null-space solution, (i) matrix inversion, (j) eigenvalue decomposition, and (k) singular value decomposition.

(2) Special signal processing Algorithms: (a) Toeplitz system solver, (b) 1-D and 2-D linear convolution, (c) circular convolution, (d) ARMA and AR recursive filtering, (e) linear-phase filtering, (f) lattice filtering, (g) DFT, and (h) 2-D correlation (image matching).

(3) Other algorithms: PDE (partial difference equation) solution, sorting, transitive closure and shortest-path problems.

These signal processing algorithms share the critical attributes of regularity, recursiveness and local communication, which are effectively exploited in the systolic and wavefront array processors. These arrays maximize the strength of VLSI in terms of intensive and pipelined computing and yet circumvent its main limitation on communication. The application domain of such array processors covers a very broad range, including digital filtering, spectrum estimation, adaptive array processing, image/vision processing, and seismic and tomographic signal processing.

Furthermore, if the overall array system consists of a flexible *interconnection network*, then the communication constraint can be significantly relaxed. The array processors will become applicable to many global-type algorithms, such as the FFT algorithm, sorting, the Kalman-filter network, etc.

11. An applicational example for adaptive noise cancellation

From the applicational system perspective, one of our objectives is to upgrade modern passive sonar systems by introducing high-speed array processors for front-end processing and spectrum estimation computations. The new system should then possess real-time and high-performance processing capabilities. We shall use an example for adaptive noise cancellation. More specifically, we discuss the McWhirter's algorithm based on least square minimization using QR decomposition (McWhirter 1983).

Given any $N \times p$ matrix X with $N > p$ and an N-element vector y, find the p-element vector of weights w which minimizes $\|e\|$, where

$$e = Xw + y$$

and $\| \cdot \|$ denotes the usual Euclidean norm. The problem may be solved by the method of orthogonal triangularization (QR decomposition) which is numerically well-conditioned and may be described as follows. An orthogonal matrix Q is generated such that

$$QX = \begin{bmatrix} R \\ 0 \end{bmatrix}, \quad \text{and} \quad Qy = \begin{bmatrix} u \\ v \end{bmatrix},$$

where R is a $p \times p$ upper triangular matrix and u is a p-element vector and 0 is the zero matrix. It follows that the least-squares weight vector w must satisfy the equation

$$Rw + u = 0,$$

which may readily be solved by back-substitution.

The process of recursive least-squares minimization may be carried out using a triangular wavefront processor, based on pipelining a sequence of Givens rotations, viz. an elementary orthogonal transformation of the form:

$$\begin{bmatrix} c & s \\ -s & c \end{bmatrix} \begin{bmatrix} 0 & \cdots & 0, r_i & \cdots & r_k & \cdots \\ 0 & \cdots & 0, x_i & \cdots & x_k & \cdots \end{bmatrix}$$
$$= \begin{bmatrix} 0 & \cdots & 0, r_i & \cdots & r_k' & \cdots \\ 0 & \cdots & 0, 0 & \cdots & x_k' & \cdots \end{bmatrix} \tag{11.1}$$

The elements c and s may be regarded as the cosine and sine, respectively, of a rotation angle which is chosen to eliminate the leading element of the lower vector. It is illustrated in fig. 15 for the case $p = 4$ (cf. Broomhead et al. 1985).

The Givens rotation method is recursive in the sense that the data from the matrix X are introduced row by row, and as soon as each row x_n^{T} has been absorbed into the computation, the resulting triangular matrix $R(n)$ represents an exact QR decomposition for all data processed up to that stage.

The boundary cell in each row (indicated by a large circle in fig. 15) computes the rotation parameters c and s appropriate to the internally stored components and the vertically propagating data vector. These rotation parameters are then passed horizontally to the right.

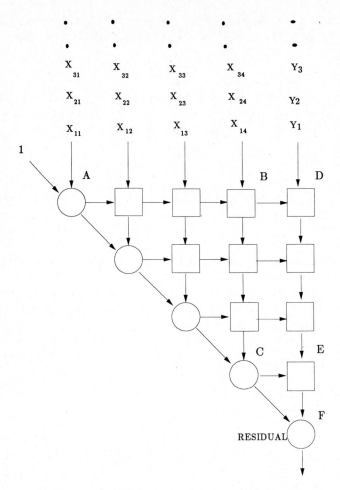

Fig. 15. Wavefront array for recursive least-squares minimization, adapted from [Broomhead et al. 85]. As a least-square solver array, each of the non-diagonal PE's performs a basic Givens rotation, shown in eq. (11.1). Note that the diagonal (circle) PE's are responsible for generating the rotation angle parameters c and s. The parameters are then propagated rightwards to all the PE's along the same row to be used for the rotation operations.

The internal cells (indicated by squares) are subsequently used to apply the same transformation to all other elements of the received data vector. Note that when the triangular array illustrated in fig. 15 is operated in the wavefront processor mode, the operation of PE is no longer controlled by a globally distributed clock signal. Instead, each PE receives its input data from the specified directions, performs the specified functions and delivers the appropriate output values to neighboring PE's. The operation of each cell is controlled locally and depends on the necessary input being available and on its previous outputs having been accepted by the appropriate nearest neighbors. As a result it is not necessary to impose a temporal skew on the input data matrix. For example, let us consider the top processor in the right-hand column. In the wavefront case, this will not operate on its first input data sample y, until the required rotation parameters c and s are available from the neighboring processor on its left.

As each row x_n^T of the matrix X moves down through the array, it interacts with the previously stored triangular matrix $R(n-1)$ and is eliminated by the sequence of Givens rotations. As each element y_n moves down through the right-hand column of processors, it undergoes the same sequence of Givens rotations, interacting with the previously stored vector $u(n-1)$ and generating the updated vector $u(n)$ in the process. It follows that the exact least-squares solution could be derived at every stage of the process by solving the corresponding triangular linear system for $w(n)$ (Gentleman and Kung 1981).

In many least-squares applications the primary objective is to compute the sequence of residuals

$$e_n = x_n^T w(n) + y_n,$$

while the associated weight vectors $w(n)$ are not of direct interest. It has recently been shown (McWhirter 1983) how the residual e_n at each stage of the computation may be generated quite simply using the array in fig. 15 without any need to solve the associated triangular linear system for $w(n)$. The parameter x_{out} produced by the bottom processor in the right-hand column of cells is simply multiplied by the parameter y_{out} which emerges from the final boundary processor and this produces the residual directly (Broomhead et al. 1985).

12. Summary

We have witnessed the rapid growth of signal processing and computing technology that followed the invention of the transistor in the 1940's and integrated circuits in the late 1950's. The emergence of new VLSI technology, along with modern engineering workstations, CAD tools, and other hardware and software advances in computer technology, virtually assure a revolutionary information processing era in the near future. The signal processing community will very soon face a prevailing impact of the modern VLSI technologies via future integration of computers, communications, control, command, intelligence and information.

VLSI technology, starting as a device research area, provides opportunities and constraints which will open up new areas of research in computer architecture. From a scientific research perspective, a close interaction between VLSI and array architecture research areas will be essential. This paper has identified several novel architectures which maximize strengths of VLSI in terms of intensive and pipelined computing and yet circumvent its main limitations on reliability and communication. In the author's opinion, the research and development in the array processors do not only benefit from the revolutionary VLSI technology; it will also play a central role in shaping the course of algorithmic, architectural and applicational trends of future supercomputer technology.

References

Ahmed, H.M., 1985, Alternative arithmetic unit architectures for VLSI digital signal processors, in: VLSI and Modern Signal, eds. S.Y. Kung, H.J. Whitehouse and T.K. Kailath (Prentice Hall, Englewood Cliffs, NJ).

Broomhead, D.S., J.G. Harp, J. McWhirter, K.S. Palmer and J.G.B. Roberts, 1985, A practical comparison of the systolic and wavefront array processing architectures, Proc. IEEE Int. Conf. on Acoustics, Speech and Signal Processing, Tampa, FL (November, 1985) pp. 296–299.
also in: Proc. IEEE, Workshop on VLSI Signal Processing, Los Angeles (November 1984).

Burg, J.P., 1975, Maximum Entropy Spectral Analysis, Ph.D. Thesis (Stanford University, Stanford, CA).

Capon, J., 1969, High-resolution frequency–wavenumber spectrum analysis. Proc. IEEE **57**, 1408–1418.

Chase, M., 1984, A pipelined data flow architecture for digital signal processing the NEC uPD7281, in: IEEE Workshop on VLSI Signal Processing, Los Angeles (November 1984).

Davis, R.H., and D. Thomas, 1984, Systolic array chip matches the pace of high-speed processing, Electron. Des. **October**, 207–218.

Dennis, J.B., 1980, Data flow supercomputers, IEEE Computer Magazine **November**, 48–56.

Fisher, A.L., and H.T. Kung, 1985, Special-purpose VLSI architectures: general discussions and a case study, in: VLSI and Modern Signal Processing, eds. S.Y. Kung, H.J. Whitehouse and T.K. Kailath (Prentice Hall, Englewood Cliffs, NJ).

Gentleman, W.M., and H.T. Kung 1981, Matrix triangularization by systolic array, SPIE Proc. Real-Time Signal Processing IV, Bellingham, WA (1981) p. 298.

Hayes, J.P., 1978, Computer Architecture and Organization (McGraw-Hill, New York).

Kung, H.T., 1982, Why systolic architectures?, IEEE Computer Magazine 15(1), 37–47.

Kung, H.T., and C.E. Leiserson, 1978, Systolic arrays (for VLSI), in Sparse Matrix Symposium, 1978 (SIAM, Philadelphia, PA) pp. 256–282.

Kung, S.Y., 1984, On supercomputing with systolic/wavefront array processors, Invited paper, Proc.. IEEE 72(7), 867–884.

Kung, S.Y., and Y.H. Hu, 1983, A highly concurrent algorithm and pipelined architecture for solving Toeplitz systems, IEEE Trans. Acoust. Speech & Signal Process. 31(1), 66–76.

Kung, S.Y., K.S. Arun, R.J. Gal-Ezer and D.V. Bhaskar Rao, 1982. Wavefront array processor: language, architecture and applications, IEEE Trans. Comput., special issue on Parallel and Distributed Computers C-31(11), 1054–1066.

Mason, S.J., 1953, Feedback theory—some properties of signal flow graphs. Proc. IREE 41, 920–926.

McWhirter, J.G., 1983, Recursive least-squares minimization using a systolic array, in: SPIE Proc. Real-Time Signal Processing VI, San Diego (1983) p. 105.

Mead, C., and L. Conway, 1980, Introduction to VLSI Systems (Addison-Wesley, New York).

Randell, B. and P.C. Treleaven, 1982, VLSI Architecture (Prentice Hall, New York).

Stewart, G.W., 1973, Introduction to Matrix Computations (Academic Press, New York).

Seitz, C., 1984, Concurrent VLSI architectures, Invited paper, IEEE Trans. Comput. C-33(12), 1247–1265.

Ware, F., L. Lin, R. Wong, B. Woo and C. Hansen, 1984, Fast 64-bit chip set gangs up for double-precision floating-point work, Electronics July, 99–103.

Wilson, P., 1984, Thirty-two bit micro supports multiprocessing, Comput. Des. June, 143–150, 1984.

Wu, C.L., and T.Y. Feng, 1980, On a class of multistage interconnection networks, IEEE Trans. Comput. C-29(8), 694–702.

COURSE 14

SIGNAL PROCESSING ON A PROCESSOR ARRAY

Stewart F. REDDAWAY

ICL, Defense Technology Centre
Eskdale Road, Winnersh, Wokingham, Berk, RG11 5TT, UK

J.L. Lacoume, T.S. Durrani and R. Stora, eds.
Les Houches, Session XLV, 1985
Traitement du signal / Signal processing
© *Elsevier Science Publishers B.V., 1987*

Contents

1. Introduction

This work is based on experience of using, and planning to use, three generations of Distributed Array Processor (DAP) (see refs. [1-4]) from ICL. These are SIMD machines; other machines with some similarity are:

CLIP 4 (University College, London),
MPP (NASA and Goodyear),
GRID (GEC),
SCAPE (Brunel University).

There are also some similarities with systolic and wavefront arrays.

Section 2 is an overview of DAP, while section 3 is an examination of some application case studies.

2. Overview of Distributed Array Processors (DAP)

2.1. Digital Signal Processing (DSP) and programmable array processors

A wide range of digital signal processing can be implemented on a standardised architecture. Compared with custom hardware this allows:
(a) lower risk,
(b) more functional agility,
(c) functional evolution during in-service lifetime,
(d) keeping the same software through technology upgrades.

From another viewpoint, algorithms, languages and architecture all influence each other, and architecture is influenced by Very Large Scale Integration (VLSI).

2.2. DAP background

There is a considerable history of processor array machines, and this section gives a brief background on the machines the author has been

834

involved with. The connection with signal processing is more recent.

The pilot DAP system was completed in 1976 and had 1024 Processing Elements (PEs) arranged 32 × 32. This led to a similar first-generation DAP product with 4096 PEs organised 64 × 64 made from MSI technology, and connected as a backend to an ICL 2900 mainframe. These large DAPs (330 PCBs) were marketed with a number crunching emphasis and 6 of them were delivered during 1980–1983. One of these machines, supporting about 200 users, is at Queen Mary College, London, where there is a DAP Support Unit to help (mainly academic) users. These machines have lived up to their technical expectations with high performance on a wide range of mainly scientific applications.

The first second-generation, or MiniDAP machines were delivered at the end of 1985. The main focus of this course is a MiniDAP LSI machine known as MilDAP, which occupies 16 PCBs and has a 32 × 32 array of PEs. Its initial host machine is a PERQ, but it is less tightly coupled to it than with first generation, and the MilDAP will go into equipment without a host system. An important new feature is the (programmable) Fast I/O hardware which opens the way to high data rate applications such as graphics, signal and image processing.

Work has started on a military R & D contract, known as VDAP, to investigate machines, with several different array sizes, that exploit VLSI by having much more powerful PEs*.

2.3. DAP architecture

The MilDAP has a 32 × 32 array of one-bit processing elements (PEs) (see ref. [4]) which each have access to 8 or 16 Kbits of RAM. Communication can be two-dimensional (or with a little software help one-dimensional), or, as we shall see, global. The two-dimensional N–S and E–W features are nearest-neighbour connections and highways, both with bit-serial PE connection. The high data throughput and computing capacity is achieved by the parallel operation of the array: whether the operation is a data shift between PEs or an internal instruction, every PE participates, subject only to a veto imposed by an "activity" bit which can locally inhibit PEs. DAP processing is controlled by a Master Control Unit (MCU). This can form a matrix by broadcasting a 32-bit

---

*See note on p. 857.

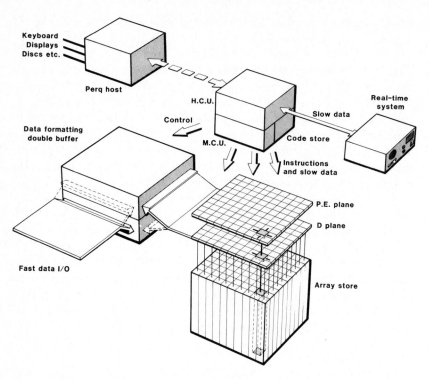

Fig. 1. Schematic representation of MilDAP showing the relationships between the Host Connection Unit (HCU), Master Control Unit (MCU) and the Processing Element (PE) plane; and the D input/output plane and the fast data I/O (FIO) unit.

vector along the highways to every row or every column of PEs, and the AND of a matrix can be performed in row or column directions to provide a 32-bit vector to the MCU. If the MCU then branches on that vector being all TRUE we have a global test in just 2 instructions. A scalar can be globally broadcast. The DAP can validly be regarded as a partitioned memory with processing power distributed across it, avoiding limitations to available memory bandwidth.

Figure 1 shows a simplified schematic of the MilDAP developed jointly by ICL and RSRE; it has a cycle time of 150 ns using gate array and standard RAM technology and occupies about 1 cu. ft. The Fast I/O unit includes two buffers, each of $16 \text{ K} \times 32$ bits. The Fast I/O unit can be "programmed" to achieve a wide range of data reorganisations. At the corner-turning level, this is important since processing is fastest for data stored "vertically" below each processing element, whereas external data is usually in a word parallel or "horizontal" format. Higher-level reorganisations are also very powerful in matching convenient I/O and data structures for efficient processing. During data input, words are written into one of the two buffers through a 32-bit wide bidirectional interface, and then read out (reorganised) into the D bit-plane (see fig. 1), along 32 "row" channels by fast data shift operations independent of other array operations. Once the D-plane is full, an array cycle is stolen to write the data plane of 1024 bits to a main memory store plane. Even at the maximum data transfer rate of 40 MBytes/s, less than 5% of array processing cycles are stolen. The procedure reverses for data output.

The bit-serial but highly word-parallel nature of MilDAP processing means that the machine is not committed to a particular data representation. This allows accuracy-speed tradeoffs right down to one-bit working for logical comparisons which can be made at speeds of several Gbits/s.

The program-development system is based on a separate PERQ host computer which provides the compiler, assembler, library of routines and other utilities including a DAP simulator*. Programs are down-loaded from PERQ during development and from a convenient storage medium such as a disc or ROM in stand-alone use.

2.4. Programming MilDAP

MilDAP has a very efficient assembly language (APAL) [5] and a high-level language DAP-Fortran [6] which is a parallel extension of Fortran, the main difference being the ability to process complete arrays as single items. As an example, consider the Fortran code:

```
DO 9 M = 1,32
DO 9 N = 1,32
   IF (A(M,N).GE.O) GOTO 9
   A(M,N) = A(M,N) + 5
9 CONTINUE
```

This is very inefficient on a serial machine partly due to the IF. In the corresponding DAP-Fortran statement,

```
A(A.LT.O) = A + 5
```

---

*The program development system is now also available on VAX and SUN workstations.

the Boolean matrix "A.LT.0" is used as a mask so that only those values of A corresponding to a "TRUE" value are changed. The contrast with sequential machines on conditional operations is important: the sequential machine performs a conditional jump, whereas the DAP will typically use activity control to perform a masked assignment. In addition to the basic functions which have been extended to take vector and matrix arguments, a large number of other functions are provided as standard. These include (a) functions to create arrays from scalars, (b) row or column broadcasting, (c) maximum element, and (d) row, column and matrix sums and many others. DAP-Fortran inspired many of the parallel processing facilities planned for Fortran 8X.

Considerable improvements in speed can often be achieved by coding critical sections in APAL. APAL, for example, allows full advantage of variable wordlengths. An example of APAL code for the addition of two 5-bit integer matrices is

```
AT

CF

DO 5 TIMES

QS 4 (M3 - )

:SICPCQS 4 (M4 - )
```

where the MCU register M3 points to the sign bit of the first matrix, and M4 to that of the second matrix. An explanation of the above code is as follows:

(1) Set activity control register A to "true".
(2) Clear Carry register C.
(3) Loop over number of bits.
(4) Fetch first matrix, starting at low-order end.
(5) Do the add, overwriting the second matrix under activity control and preserving carry in C for the next pass of the loop. The effect of the negative sign is to automatically decrement the effective plane address.

Program development is performed on an ICL PERQ II workstation which is connected to the MilDAP via the Host Connection Unit (HCU). All the supporting software, DAP-Fortran compiler, APAL assembler and consolidator is run on the PERQ which accepts and displays diagnostic and trace information from the MilDAP. The PNX (UNIX) operating system provides file handling and screen editing

Table 1
MilDAP/MiniDAP and DAP-3 performance on matrix operations

Precision	Operation	MilDAP/MiniDAP (150 ns cycle)	DAP-3 (100 ns)
logical	logical	> 1 GOP	> 1 GOP
8-bit integer	add	280 MOP	420 MOP
	Mult.(M–M)	42 MOP	60 MOP
	(M–constant)	100–200 MOP	140–280 MOP
16-bit integer	add	140 MOP	210 MOP
	Mult.(M–M)	10 MOP	14 MOP
	(M–constant)	30–100 MOP	40–140 MOP
32-bit real	add	8.6 MFLOPs	12 MFLOPs
	mult.(M–M)	5.1 MFLOPs	7 MFLOPs
	square	10.9 MFLOPs	16 MFLOPs
	squareroot	7.4 MFLOPs	11 MFLOPs
	divide	4.1 MFLOPs	6 MFLOPs
	log (base e)	4.2 MFLOPs	6 MFLOPs
	max. of 1024 numbers	41 MFLOPs	60 MFLOPs
data	scan for match	6×10^9 bits/s	9×10^9 bits/s

facilities. In addition, a MilDAP simulator, written in "C" allows program development (including timing estimates) on workstations without a DAP. A comprehensive library of standard functions such as FFTs, matrix operations, random number generators, etc. is also available. The MilDAP can be used in field situations in a stand-alone mode without the PERQ.

2.5. Performance

The best serial algorithms are often not the best parallel algorithms. An example of an inefficient serial algorithm which is particularly suited to parallel computation is Batcher's "bitonic sort" [6], which has been applied to DAP in ref. [7]. To sort N items, this method requires $(\log N)/4 \times$ more comparisons than an efficient serial algorithm, but this is a small price to pay for a regular algorithm with $N/2$ parallelism.

The bit-serial MilDAP has some unusual relative speeds for functions; see table 1 for some matrix operations. Other performance figures for MilDAP and DAP-3 are given in table 2.

Table 2
MilDAP/MiniDAP and DAP-3 performance

Operation	MilDAP/MiniDAP (150 ns)	DAP-3 (100 ns)
Floating point–Peak MFLOPs		
24 bit	13	19
32 bit	9	13
64 bit	4	6
Fixed point–Peak MOPs		
logical	2000	3000
8 bit	260	400
16 bit	140	210
32 bit	70	110
FFT examples (APAL coding)		
(1) 1024 64-point FFTs		
complex 10-bit input		
time (ms)	8.5	5.6
overall (MOPs)	160	240
(2) 512 × 512 2D-FFTs		
real 8-bit input		
time (ms)	100	70
overall MOPs	80	120

Basic techniques of array processing on DAP are covered elsewhere; for example, ref. [3] describes maximum element, scalar-matrix multiplication, square root and "recursive doubling".

2.6. Applications overview

Most, but not all, applications where the emphasis is on performance and where the processing load is non-trivial have good DAP solutions. The following is a semi-random list of areas:

matrix arithmetic,
field problems,
sorting,
text processing (e.g. compilation),
speech,
image processing,
graphics,

pattern recognition,
CAD,
searching,
picture searching,
encryption,
AI/Prolog/Lisp?

Applications need significant work expressible as array operations. This can be not only arithmetic, but also many other forms, such as:

data movement,
table look-up,
searching,
sorting,
conditional operations.

Applications are best approached top down, with the whole application (or a large part of it) considered as to how it best maps on to a processor array. If this high-level mapping is done well, the details often fall nicely into place. If too small a part of an application is considered, it is often more difficult to get a good mapping.

2.6.1. Matrix algebra examples

Before coming on to more specifically signal processing we look at some matrix algebra.

A DAP-Fortran code for matrix multiplication when the matrices match the DAP size is:

```
C = 0
DO 10 K = 1, 32
10 C = C + MATC(A(, K))*MATR (B(K,))
```

Figure 2 shows the data movements involved with, for example, a column of matrix **A** being expanded to a temporary matrix. The time for such a code on MilDAP is about 15 ms. Larger matrices involve handling indexed sets of 32×32 sub-matrices, with various possibilities for the mapping into sub-matrices. (See section 3.4 on image processing for a discussion on sheet and crinkled mapping.)

Matrix inversion is discussed in refs. [2] and [8]. For both inversion and multiplication, data movement overheads can be reduced for larger matrices.

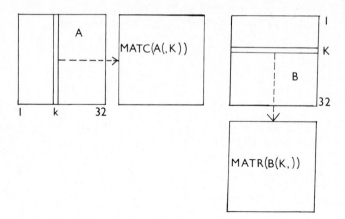

Fig. 2. Data movements for matrix multiplication.

2.6.2. Overview of a radar application

This application has been implemented on MilDAP and is considered in more detail below. Here we use it to illustrate some points.

The implementation is an air-to-air search mode of a multi-mode airborne radar; the same DAP hardware can be programmed to process many different modes. This mode has a medium pulse repetition frequency, and the characteristics are shown in table 3. The performance on the stages is shown in table 4. Note that no stage dominates the time, and so a special-purpose piece of hardware suitable for just one type of processing (for example FFT) could not have a dramatic impact, quite

Table 3
Characteristics of MPRF air-to-air search mode

Algorithm	Types of processing
Spectral decomposition	
Window and FFT	unconditional $+, -, *$
Log modulus	special
CFAR	unconditional $+, -$
	conditional $+, -$
	logical
	:
Ambiguity resolution	table look-up
	data movement
	:

Table 4
MPRF radar performance; 60 000 radar complex
(2 × 10 bits) points at 0.5 μs sampling

Stage	MilDAP	
	Time (ms)	MOPs
"Input"	1.4	
Window		
64-Point FFT	9.0	150
(10–14 bit) fixed point		
Log modulus	3.8	
CFAR	8.2	200
Ambiguity resolution	< 2.6	
(depends on # targets)		
Total	< 25.0	
Time available	35	

apart from the cost of interfacing. The wide range of different types of processing required is typical of the increasing sophistication of signal processing applications; this means that special hardware solutions have to be increasingly complex, in contrast to solutions with programmable hardware.

It should also be noted that the implementation batched up 60 000 points for processing. This achieves greater throughput by allowing the exploitation of high-level parallelism and at the same time using efficient serial algorithmic techniques at the detailed level; this also minimises communication between PEs, and in this example each 64-point FFT is done serially entirely in a single PE, with 1000 FFTs being done in parallel across the array of PEs.

Signal processing typically involves several very different stages, some with data-dependent (conditional) processing and often with some Boolean processing. Data precision is generally fairly low. With specialised hardware a sequence of special boxes or chips are needed, and much emphasis must be given to transfers between these; performance is governed by the speed of the slowest unit, and flexibility is lacking. With a programmable SIMD processor array such as DAP we have a "one box" solution which offers good flexibility and overall performance.

2.7. Some principles of application mapping

It is important to minimise communication and I/O. The following (related) techniques can be useful.

Go for high-level parallelism. Low-level algorithms are often best done serially, with many of them being done in parallel. This may involve (a) using more than the minimum memory, or (b) increasing the latency (delay between input and output); but throughput is increased.

Localise the processing. For example, in applying neighbourhood operations over whole images "crinkling" the image so that one PE deals with a local area is superior to mapping a "sheet" of the image across the array of PEs. The latter involves work to join the sheets, more data movement, and often extra arithmetic.

If data to be combined is held in a sequentially accessed memory, then "communication" is achieved in the sequence of addresses given to the memory. For example, the patterns of "movements" needed for FFT or for Batcher-bitonic sorting are "free" when built into the addresses used in the code, provided the combining data is all in the same PE, and (from high-level parallelism) similar operations are replicated across the array.

"Oversize" problems are advantageous. For example dealing with 256×256 matrices on a 32×32 DAP. This allows the above techniques to be used. Algorithms are a combination of serial and parallel.

In moving from one processing stage to another it is sometimes worthwhile reorganising the data in the array. (A uniform spread of data across the PEs is retained.) An example is if one stage is one-dimensional processing on a multi-dimensional array, and another stage is two or more dimensional neighbourhood processing. The former stage should have data from one line in the dimension being processed in as few PEs as possible, while the latter should have a local section along all directions being processed in as few PEs as possible so as to minimise routing. The MPRF radar is an illustration.

I/O can be minimised by doing as much of an application as possible in one array, with no intermediate I/O. Having memory inside the array is very important, both for data and constants. In MilDAP, the use of processing cycles for data movement can be minimised by using the Fast I/O to re-order the data on input or output.

2.8. Some comparisons with systolic and wavefront arrays

There are some similarities between the processor arrays described in this course and systolic and wavefront arrays, but there are important differences.

With DAP a large memory mixed with processing power helps the use of the techniques in section 2.7 so as to minimise communication and maximise throughput. Most systolic/wavefront arrays do not have large memories distributed in the array.

Systolic/wavefront arrays tend to be tailored to a particular algorithm or algorithm class, and hence the emphasis is on hardware. This contrasts with the fixed design of the DAP with the emphasis on programming the applications. Hence DAP tends to tackle complete applications and systolic/wavefront arrays tackle parts of applications.

Systolic/wavefront arrays deal with individual or local parts of algorithms very fast, with much of the work being communication. With the use of high-level parallelism on DAP the individual or local parts of algorithms may be serial and slow; but they are efficient, with minimum communication and high overall throughput.

3. Application case studies

System requirements have been programmed for MilDAP to explore the best ways of mapping algorithms onto the machine and to quantify its performance on total problems. The performance rating of any machine is rarely achieved in practice, and with DAP it is particularly necessary to overcome the suspicions: (a) that algorithms with branching instructions cannot exercise all the PEs effectively, and (b) that the fixed size of the array of PEs limits us to problems with special array sizes. These suspicions are shown to be unfounded by the wide diversity of algorithms described in this course.

3.1. MPRF radar

Modern airborne radars require complex algorithms applied to data of MHz bandwidths. Moreover radars are now multi-functional with variable parameters such as Pulse Repetition Frequency (PRF) and pulse width, and have distinct operational modes like ground attack, air interception, terrain following, synthetic aperture mapping, etc. This necessitates a programmable processor.

One of the most demanding tasks is the "medium PRF" aircraft interception mode. This comprises several algorithms which exercise the processor by: (a) input and output of data; (b) arithmetically intensive processing on multiple FFTs; (c) data dependent, arithmetic and logical, local area operations for Constant False Alarm Rate (CFAR) processing in strongly varying clutter environments; (d) data reorganisation and logical comparisons to resolve ambiguities in both range and velocity

(Doppler effect). This application was programmed for MilDAP by Reddaway et al. [9] in order to expose any deficiencies, and to demonstrate versatility combined with high throughput.

The radar input to the fast I/O buffer is 10 + 10 bit complex samples at maintained rates of 2 MHz (5 MBytes/s). In addition to this, housekeeping instrumentation data at a much slower rate will be input via a separate channel. The radar PRF varies from pulse burst to pulse burst through a cycle of 5 different values, and blocks of data are combined together in the fast I/O hardware. This organises the data into the most efficient distribution in the DAP store, and in this case inputs half a bit-plane of 32 × 32 bits to the array memory for each "stolen" clock cycle. Data is processed in batches of 10 pulse bursts; while processing is proceeding, the data for the next batch is being input to the DAP array memory.

Bursts of 64 pulses sampled in up to 128 range gates are windowed using fixed 10-bit real weight vectors, Fourier-transformed to 14 + 14 bit complex spectra and converted to 10-bit log-modulus power estimates (70 dB dynamic range) in 64 Doppler cells. The total number of range gates over the five bursts totals rather less than 512, including an allowance for duplication of range gates to cater for cyclic wrap-around in range. Thus 2 sets of 5 bursts are laid out across the 1024 PEs with all 64 points for each range gate in the same PE; the fast I/O subsystem has also introduced index reversed ordering for the 64 points so that after the FFT the ordering is natural. The FFTs are entirely contained within single PEs, and are done with full serial optimisation for radix 8; parallelism is obtained from doing 1000 independent FFTs in parallel. Extensive use is made of efficient scalar–matrix multiplications in the FFT computations, because each PE is using the same multiplier coefficients concurrently; a code generator automatically minimises the add/subtract work by analysing the bit-patterns of the constant multipliers. During the FFT, there is precision growth at the MS end of the fixed-point numbers, and noisy bits are dropped at the LS end. The log-modulus function is specially coded; it has some similarity to conversion from fixed to floating point. The spectrum analysis phase is done at a rate equivalent to 220 ns per sample, or 45% of the corresponding data collection interval. Performance could be improved somewhat by doing 60 or 72 point FFTs by Winograd or prime factor techniques.

CFAR detection includes (i) smoothing in the two-dimensional range–Doppler domain using a simple two-dimensional filter with cyclic boundaries, (ii) time averaging where clutter discontinuities are found, (iii) comparing the signal with the local clutter level, (iv) removing detections where the known angle of look and aircraft velocity result in high-clutter levels, and (v) eliminating returns which behave in range or velocity in a way inconsistent with expected target behaviour. The three-dimensional filters are continually updated on the basis of recent history. The data is rearranged between the spectrum analysis and CFAR processing, so that a single PE holds 16 Doppler points from 4 range gates instead of 64 Doppler points from 1 range gate; this reduces both communication and arithmetic in the two-dimensional neighbour-hood operations. (A square 8 × 8 section of points would be ideal, but the cost of the extra data rearrangement means the optimum is 16 × 4.) The DAP activity control is efficient in dealing with the data dependent processing. The CFAR computation takes 140 ns per sample (28%).

The medium PRF mode is doubly ambiguous. The apparent detections are first collapsed in Doppler, keeping the Doppler location of the detection with the peak power value, and are then "unfolded" in range to form a binary vector showing the possible target ranges corresponding to the observed detections out to 1280 range cells. Five of these range patterns, each from a different PRF pulse burst, are examined, and a 3 out of 5 detection criterion applied to ranges where a toleranced coinci-dence is found. Finally, these potential target detections are checked for consistency in Doppler by referring back to their original frequency values (before collapsing) and generating by table look-up the possible real frequency bin. These table entries are logical maps of the real velocity (frequency) span divided into 1024 bins, and stored as DAP logical matrices with the possible bins marked (up to 50 per ambiguous frequency bin, when all allowance for tolerance is included). These Doppler frequencies are then checked for coincidences, again with a 3/5 criterion, to finally confirm the targets. This comparing of (Boolean) detections is particularly efficient with 1-bit PEs. Resolving the radar ambiguities takes less than 40 ns per sample (8%).

The above items total 400 ns per sample, whilst the sample separation is 500 ns. Extra dead time exists between pulse bursts so there is 30% rather than 20% spare processing capacity. This shows that MilDAP provides a flyable processor adequate for this extremely demanding role.

3.1.1. Some MPRF radar performance derivations
More details of some of the algorithms and performance derivations follow here.

The operation counts for a 64-point FFT on complex data are given in table 5. The general rotations use 3 multiplies and 3 adds. The multiplies use the scalar-matrix code generator. The average precision is about 12

Table 5
64-Point FFTs–Operation counts and performance for 1024 off

Operation	Number of complex operations	Number of simple arithmetic operations	
		multiply	add
Complex adds	384	–	768
45° Rotations	36	72	72
General rotations	44	132	132
Total		204	972
Approximate average cycles/ operation		100	35
Totals		20400	34000
			54400 cycles total

bits. (At each stage the precision grows 1 bit at the MS end, and loses 0.5 bit at the LS end). The cycle count is for 1024 independent FFTs; overheads are about 10% and the time is 9 ms with a 150 ns cycle time.

The log modulus was specially coded and takes 390 cycles per result matrix. For 64 result matrices and 150 ns cycle time, this gives 3.74 ms.

The data rearrangement, using a methodology known as "musical bits" [10] takes place in two stages, one involving a route distance of 1 PE and one a distance of 2 PEs. The costs are 4 and 5 cycles/BIP (BIt-Plane). At this point there are 640 BIPs, so the time is 0.87 ms.

In the CFAR, 5×5 cell averaging is done with an algorithm that for one dimension adds the points in pairs resulting in half as many results (half an add per point), and then adds the results so as to get all sums of 4 starting at 2-point intervals (half an add per point). Each of these sums is used to form 2 sums of five by adding the appropriate original point (1 add per point). Thus, there are 2 adds per point for one dimension; applying the same procedure in the other dimension gives all 25-point sums at a cost of 4 adds/point. The central point is subtracted and the resulting 24-point sum is divided by 3 (effectively = 24), equivalent in cost to 2 adds.

The total arithmetic is thus about 7 adds per point. The precision growth during the adds is limited by dropping at intervals some LS bits; the average precision is about 11 bits, with a matrix add costing about 32 cycles. The arithmetic time is thus about 2.2 ms. However, the routing and end effects due to each PE holding only a 16×4 section increase this to about 4 ms. This represents about half of the total CFAR cost.

Finding the maximum Doppler cell is a local operation and therefore is not like the DAP-Fortran maximum functions that apply to one DAP matrix. The heart of the work is within one PE where each point is compared with the maximum so far, and (conditionally) copied in if it is bigger. The comparison costs 2 cycles/BIP and the copy costs 2.5 cycles/BIP. In total there is about 20% extra work because one range gate is spread over 4 PEs. This work thus takes about 0.52 ms, to which must be added a small amount for finding the position of the maximum Doppler cell in each range gate.

3.2. *Large transforms*

Doing one 64-point FFT in every PE was dealt with in the MPRF radar. Here we consider some big transforms.

First, consider the following transforms with 64 K complex data points:

$64 \times 32 \times 32$	3D
64×1024	2D
32×2048	2D
64 K	1D

MilDAP would map these across its 32×32 array and down the memory of each PE to a depth of 64 points. The best approach is to perform a 64-point FFT within the PEs, as in the radar case; this is followed by using "musical bits" to transpose the cuboid of points so that lines of 32 points in one of the array directions end up "vertically" in single PEs. (The data can be considered as 2 cubes $32 \times 32 \times 32$ stacked above each other. Each cube is transposed or rotated about 1 axis.) Using the data now in each PE, two 32-point FFTs are performed (in every PE). The procedure is now repeated by transposing about the other array direction and doing two more 32-point FFTs in every PE.

As described, a $64 \times 32 \times 32$ FFT has been performed, with the data left in index reversed order. (If necessary this can be unscrambled with more "musical bits".) The other FFTs can be performed by inserting appropriate twiddle multiplications.

The cost for a careful APAL coding, in the style of the MPRF radar, and for an average precision of 16 bits (11 bits growing to 20 bits) is as follows:

– 64-Point FFTs with 16-bits average precision costs about 80 000 cycles.

- Transposition costs about 30 cycles/BIP, so with 2 K BIPs this is 60 K cycles.
- Two 32-point FFTs with 16-bit average precision costs about 64 000 cycles.

The cost of a twiddle multiplication is quite high, as these multiplies must be matrix–matrix; at 16-bit precision the cost for a complete twiddle multiply is about 80 K cycles. Thus the cost of the three-dimensional transform is about 328 000 cycles (49 ms), for the two-dimensional transforms about 408 000 cycles (61 ms) and for the one-dimensional transform about 488 000 cycles (73 ms).

Other mappings are possible. For example the two- and three-dimensional transforms can save some multiplications by having some of the multiplies within the 32- or 64-point transforms combined across the dimensions of the FFT.

With a 512×512 FFT that is mapped with the two-dimensional structure matching the DAP two-dimensional array, one line of 512 points maps to a row of 32 PEs 16 points deep in memory; the two dimensions give a total of 256 points per PE. There are two equally good ways of proceeding.

Both methods start by performing 16×16 two-dimensional FFTs on the data in each PE; advantage can be taken of combining the multiplies across the dimensions. The first method then rotates in new 16×16 sections into each PE and repeats the 16×16 FFTs, and then completes the transform by rotating in the final 2×2 sections. The second method rotates lines of 32 points into PEs and performs 32 point FFTs, and then repeats this in the other array direction. In image processing the data will be real, and adapting the algorithms is left as an exercise for the reader; the time for 16-bit average precision is about 100 ms.

Another image-processing example is the generation of cosine transforms on all 8×8 segments of a 512×512 image (for the purpose of compression). If the data is 8-bit there are 2048 BIPs, and if the mapping is "crinkled" no data routing is needed. On either MilDAP or a third-generation 8×8 DAP the time should be about 10 ms.

Multiple independent transforms where it is not possible to have one transform per PE should have transforms confined to as few PEs as possible, usually as square a section of PEs as possible. For example, 128 independent 512-point transforms should have a 512-point transform in a 4×2 section of PEs. (Sometimes earlier or later processing may dictate a linear section, 8×1 in this case.)

Prime factor (or Winograd) can give a modest performance improvement. Individual transforms confined within a PE are no problem. In other cases it is even more advantageous than for power of 2 FFTs to have as many points per PE as possible. For factors that cannot initially be mapped in the memory of one PE, the points are "rotated" into position to do that prime factor transform.

The first prime factor example is for 64 independent 1680-point ($3 \times 5 \times 7 \times 16$) transforms. Each transform is put in a 4×4 section of PEs, with 105 points per PE. When the 16-point transforms are needed, "musical bits" techniques are used to rotate the 16 points into "vertical" lines. This is repeated 7 times until there are 112 matrices of points; some PEs in the top 16 matrices are empty. The "inefficiency" in the processing is confined to the 16-point transforms and is only $112 - 105 = 7$ out of 112; overall this is a negligible 2% in this case.

The second prime factor example is a single two-dimensional 16×5040-point transform mapped onto a third-generation 16×16 DAP, giving 315 points deep in each PE. To do the 16-point transforms, they are rotated into position; this is done 20 times to generate a total of 320 points deep, an "inefficiency" of 5 in 320.

3.3. Convolutions

The use of transforms to perform (large) convolutions is well known, as it reduces computational complexity from order N^2 to order $N \log N$. The FFTs described above can be used for this purpose.

Number theory can be used in two ways. One is in the prime-factor and Winograd transforms mentioned above. The other is in the number representations used; these Number Theoretic Transforms (NTT) do arithmetic modulo special numbers, such as Fermat numbers ($2^{2^n} + 1$). The advantages are that the multiplications in the transforms can be reduced to special forms of add/subtract (efficient to implement on DAP) and the results are exact (no rounding error). The main disadvantage is that higher precision must be used, both to have unambiguous results and to cater for long transforms without having to use multi-dimensional methods.

One application implemented a few years ago was the checking of large Mersenne numbers (of form $2^p - 1$) to see if they are prime. The standard test involves repeated squaring of numbers of p bits precision; with p about 100 000 this is a major task. The number can be divided into appropriate precision "digits" and the square done by convolution methods. With Fermat Number Transforms (FNT) good results can be achieved, and it is just worth going to a recursive, or multi-dimensional, approach similar to the Schonhage–Strassen algorithm for very high

Table 6
Checking Mersenne numbers for primes, with p around 86 000

Computer	Time (min.)
CRAY-1 direct	140
CRAY-1 divide + conquer	70
CYBER 205 transform	60
DAP(4096PE) transform	38

precision multiply. A one-dimensional approach was implemented on DAP, and in 1983 the present author compiled table 6 showing the speed of checking Mersenne numbers for different implementations on different machines. The first-generation (64×64) DAP checked 16 numbers at a time.

Both FFTs and FNTs have been studied for an area correlation application in image processing (see section 3.4.3); the performance is similar.

3.4. Image processing

Since most images have many more pixels than the DAP has PEs, we have a choice in the way we section the image and pack it into the array store. Two extreme mappings are "crinkled" and "sheet" [11]. With the crinkled mapping (for 32×32 MilDAP) the image is divided into 32×32 local areas (each of 8×8 pixels for a 256×256 image), with each area put into a single PE. A DAP matrix is a regular sample of the whole image. With the sheet mapping, the image is divided into 32×32 sheets of pixels which are then mapped onto DAP matrices. For local operations on whole images the "crinkle" mapping is normally best in that it minimises the number of data shifts required and avoids sub-image boundaries. For processing where it may be desirable to do some work exclusively on a limited area, then the sheet mapping allows this; for example, some approaches to blob extraction favour a sheet mapping. Sheet and crinkled mappings can be advantageously combined. For example, an image can be divided into 64×64 sheets which are then crinkled into the 32×32 array; this limited crinkling avoids the worst performance effects of the sheet mapping on whole image local operations. The fast I/O can input most mappings "free" to the processing.

Table 7
Processing 256 × 256 images of 8 bits

Operation	Time (ms)	
	MilDAP/MiniDAP	DAP-3
3 × 3 Sobel edge detection	0.8	0.54
3 × 3 Median filtering	6	4
Histogram	16	11
Grey-scale normalisation	2	1.4
Summing or differencing	0.22	0.15
Shrink or expand (1-bit data)	0.6 per stage	0.4
2D FFT (8-bit growing to 17-bit)	22	15

Some times for APAL coding on 256 × 256 8-bit images are given in table 7. These benchmarks for MilDAP show that the machine is able to achieve on-line processing for requirements relating to images in TV format. Furthermore, the capacity for inputting data at 40 MBytes/s is more than adequate.

3.4.1. Sobel edge detector
The Sobel operator is a very special form of a 3 × 3 convolution operator:

$$\begin{array}{rrr} 1 & 2 & 1 \\ 0 & 0 & 0 \\ -1 & -2 & -1 \end{array}$$

The number of adds per point with careful use of intermediate results is 3. For a 512 × 512 image and a fully crinkled mapping the neighbour routing is only about a 10% overhead. Thus for 8-bit pixels and 150 ns cycle time, the total time for the application of the Sobel operator in one direction is $16 \times 16 \times 3 \times 8 \times 3 \times 150 \times 10^{-9} = 2.8$ ms. [There are 16 × 16 pixels/PE, 3 adds/pixel, 8 bits/pixel and (2.5 or) 3 cycles per BIP for the adds.]

For a combined mapping with 64 × 64 sheets, the time nearly doubles to about 5.5 ms due to work stitching the sheets together and extra routing. For a fully sheet mapping, the time nearly doubles again to about 10 ms.

3.4.2. Histograms
It is not obvious how to get a good histogram algorithm on DAP, and three very different algorithms are considered in ref. [11]. The pixel-

decoding method is best for short pixels and surprisingly good results can be achieved. It works by decoding each pixel matrix into each of its logical matrix grey levels, and then summing the logicals for each grey level.

It is important that the summation is done within the PEs first, as this produces a large reduction in the amount of data, before summing across the PEs. For the latter, the parallelism can be maintained in the later stages of summation by performing in parallel the summation of an increasing number of different grey levels. For large images the total cost can be kept down to around 6 or 7 cycles per grey level logical matrix. For 8-bit pixels the time is about 60 ms for a 512 × 512 image and 16 ms for a 256 × 256 image. For 6-bit pixels, these times are, of course, 4 times faster, and techniques can often be used to limit the number of grey-scale values needed.

VDAP is planned to have some local index capability which will greatly speed up histograms.

3.4.3. Image understanding

Area correlation finds the best fit of a two-dimensional image against two-dimensional reference data. If no rotation or scale change is involved, this is a two-dimensional convolution which can be done by transforms. For example, fitting a 300 × 150 image to a 512 × 256 reference (to give a 213 × 107 result) could be done in 100 ms; finding the maximum, and hence the best-fit position, takes only another 0.3 ms. Similar methods can also be used for Difference Of Gaussians (DOG) convolution operators.

Another image-understanding algorithm being studied with encouraging results is a statistical classifier based on features extracted from blobs, which in turn were formed by multi-channel DOG analysis and area filling (see section 3.6). More syntactical methods are a matter for future research.

3.5. Speech recognition

The widely-used dynamic programming method known as Dynamic Time Warping (DTW) has been implemented in DAP Fortran plus a little APAL [12]. Performance using 16-bit data and a Euclidean distance measure (sum of the squares of the differences at each frequency) is over 5 words per second with a vocabulary of 102 words, assuming 19

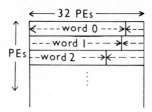

Fig. 3. Mapping of word templates for speech recognition.

frequency channels and an average of 30 spectra per word. Data as short as 4 bits is often used, and this offers nearly an order of magnitude performance improvement on DAP, because time is dominated by the square operation. With 8-bit data and APAL coding about 15 words per second is achieved (1.5 words per second with a 1000-word vocabulary). The DAP mapping is illustrated in fig. 3. With an average of 30 spectra per word, 34 words are evaluated simultaneously. The computation of the minimum-cost function is illustrated in fig. 4. The cost is accumulated by moving from corner to corner using a local decision function.

Recently an advanced technique [4] has been studied that further improves speed. The template word data is stored in a way that uses extra storage and lends itself to a table look-up technique for the

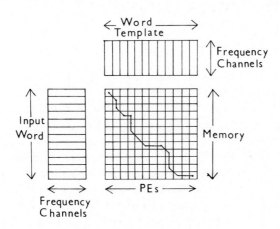

Fig. 4. Dynamic time warping.

Table 8
Some basic graphics operations

Operation	Speed for typical cases
Line drawing	8 Mpixels/s
Picture composition (Raster-Op)	500 Mbits/s (up to 3000 M)
Area filling	50 Mbits/s
3D Coordinate rotation (16 bit)	1 Mpoints/s (10% load for 2000 points @ 50 Hz)

differencing and squaring operations; a single input value is used as a global index to perform a matrix of table interpolations. For 6-bit data, this pushes performance up to about 50 words/s with a 100-word vocabulary (5 words/s for a 1000-word vocabulary, although the current MilDAP will be short of space).

3.6. Graphics

As an illustration of the potential of a programmable array, brief consideration is given to graphics, which can sometimes be an important function to combine with signal processing. MilDAP output can be via the fast I/O, which can be interfaced to a raster display, either via a frame store or else more directly. Performance varies with the parameters of particular cases, but (theoretical) figures for some basic graphics operations on MilDAP are given in table 8.

3.7. Some other applications

Some other applications are best dealt with mainly by way of references. Sorting has been mentioned already.

Very high-speed high-quality random number generation is used extensively in Monte Carlo calculations and is also sometimes useful in signal processing. Generation at speeds up to 800 million per second on a 64×64 DAP is described in ref. [13]; this speed is an order of magnitude faster than any other machine. The use of these random numbers in a very high-speed Monte Carlo simulation (the Ising model) is described in ref. [14], where speeds on a 64×64 DAP an order of magnitude faster than the best achieved on a CYBER 205 are reported. The use of DAP in Electronic Support Measure (ESM) is reported in refs. [1], [3] and [15]. A promising application area is radar plot extraction, track formation and track combination; for some of this work refer to ref. [16].

Note added in proof

A new company, AMT, is producing a product, DAP-3, hosted on a microVAX or SUN workstation. DAP-3 has more memory and a faster cycle time (100 ns) than MilDAP/MiniDAP, and performance figures for this machine are included in some tables.

References

[1] S.F. Reddaway, DAP–a Distributed Array Processor, in: Proc. 1st Ann. Symp. on Computer Architecture, eds. G.J. Lipovski and S.A. Szygerda, IEEE Catalog No. 73CH0824-3C, Comput. Architect. News 2-4 (1973) 61–65.

[2] P.M. Flanders, D.J. Hunt, S.F. Reddaway and D. Parkinson, Efficient high speed computing with the distributed array processor, in: High Speed Computer and Algorithm Organisation, eds. D.J. Kuck, D.H. Lawrie and A.H. Sameh (Academic Press, New York, 1977) pp. 113–128.

[3] P. Simpson, J.B.G. Roberts and B.C. Merrifield, MilDAP: Its architecture and role as a real time airborne digital signal processor, in: Proc. AGARD Conf. on The Impact of VHPIC on Radar, Guidance and Avionics Systems, Lisbon, May 1985, pp. 23-1–23-17.

[4] S.F. Reddaway, J.B.G. Roberts, P. Simpson and B.C. Merrifield, Distributed array processors for military applications, in: MILCOMP85, London 1985 (Microwave Exhibitions and Publishers Ltd, Tunbridge Wells, 1985) pp. 74–82.

[5] MiniDAP: APAL Language, ICL Techn. Publ. R30014/02 (Literature and Software Operations, ICL, Reading, UK, 1986).

[6] MiniDAP: DAP FORTRAN Language, ICL Techn. Publ. R30011/02 (Literature and Software Operations, ICL, Reading, UK, 1986).

[7] P.M. Flanders and S.F. Reddaway, Sorting on DAP, in: Parallel Computing 83, eds. M. Feilmeir, G. Joubert and U. Schendel (North-Holland, Amsterdam, 1984) pp. 247–252.

[8] S.F. Reddaway, Distributed array processor, architecture and performance, in: Nato ASI Series, Vol. F7, High-Speed Computation, ed. J.S. Kowalik (Springer, Berlin, 1984) pp. 89–98.

[9] S.F. Reddaway, A.L.G. Flanagan, D.J. Hunt, and J. Morris et al., unpublished work (1983).

[10] P.M. Flanders, A Unified Approach to a Class of Data Movements on an Array Processor, IEEE Trans. Comput. C 31-9 (1982) 809–819.

[11] S.F. Reddaway, DAP and its application to image processing tasks, paper 6.3 at the Image Processing Symposium RSRE Malvern, July 1983.

[12] P. Simpson and J.B.G. Roberts, Speech recognition on a distributed array processor, Electron. Lett. 19 (1983) 1018–1020.

[13] K.A. Smith, S.F. Reddaway and D.M. Scott, A very high performance pseudo-random number generation on DAP, Comput. Phys. Commun. 37 (1985) 239–244.

[14] S.F. Reddaway, D.M. Scott and K.A. Smith, A very high speed Monte Carlo simulation on DAP, VAPP II Conference, Oxford, August 1984, Comput. Phys. Commun. 37 (1985) 351–356.

[15] J.B.G. Roberts, P. Simpson, B.C. Merrifield and J.F. Cross, Signal processing applications of a distributed array processor, IEE. Proc. F 131 (1984) 603–609.

[16] R.W. Gostick and K.S. MacQueen, Radar tracking on the DAP, to be published.